Environmental Sustainability and Energy Economy

Environmental Sustainability and Energy Economy

Guest Editors

Susana Silva
Erika Laranjeira

MDPI

Basel • Beijing • Wuhan • Barcelona • Belgrade • Novi Sad • Cluj • Manchester

Guest Editors
Susana Silva
Faculty of Economics
University of Porto
Porto
Portugal

Erika Laranjeira
Faculty of Economics and
Bussiness Sciences
Lusíada University - North
Porto
Portugal

Editorial Office
MDPI AG
Grosspeteranlage 5
4052 Basel, Switzerland

This is a reprint of the Special Issue, published open access by the journal *Energies* (ISSN 1996-1073), freely accessible at: https://www.mdpi.com/journal/energies/special_issues/FL9BB4SZ54.

For citation purposes, cite each article independently as indicated on the article page online and as indicated below:

Lastname, A.A.; Lastname, B.B. Article Title. *Journal Name* **Year**, *Volume Number*, Page Range.

ISBN 978-3-7258-8199-4 (Hbk)
ISBN 978-3-7258-8200-7 (PDF)
https://doi.org/10.3390/books978-3-7258-8200-7

Contents

About the Editors

Susana Silva

Susana Silva holds a graduation (2007) and a PhD (2012) in Economics from the Faculty of Economics of the University of Porto. There, she is also an Associated Professor and she teaches several courses such as Environmental Economics, Energy Markets and Public Economics. Her research interests include the relationship between the economy, the environment, and energy, and, in particular, environmental sustainability in the energy sector. She has participated in several international conferences and published articles in international, peer-reviewed journals, conference proceedings and book chapters.

Erika Laranjeira

Erika Laranjeira is an Assistant Professor of Economics at Lusíada University, where she teaches Microeconomics and Macroeconomics. She holds a degree and PhD in Economics from the Faculty of Economics of the University of Porto, where she developed her academic and research background in economics. Her research interests include environmental and natural resource economics, health economics, and sustainable fashion. Her academic work focuses on interdisciplinary approaches to contemporary economic challenges, particularly the relationship between environmental issues, public health, and sustainable development.

energies

MDPI

Article

The Quantile Effects of Climate and Global Economic Policy Uncertainties on Renewable Energy Demands in the BRICS Nations: The Role of Green Ecological Policies

Dima AlAyoubi [1,*], Tarik Atan [1] and Majdi Awad [2,3]

1 Department of Business Administration, Cyprus International University, 99010 Nicosia, Cyprus
2 Faculty of Business, Lebanese University, Tripoli 1300, Lebanon
3 Faculty of Business, East International University, Tripoli 1300, Lebanon
* Correspondence: dima.ayoubi@hotmail.com

Abstract: Renewable energy has an important role in attaining environmental quality. Numerous studies have determined the factors driving the diminishment or promotion of renewable energy consumption (REC) globally. Despite the findings of prior studies, the specific effects of climate policy uncertainty (CPU) and global economic policy uncertainty (GEPU) on REC have not been probed in the case of BRICS economies. Furthermore, fewer considerations have been taken into account to explore whether green environmental policies (EPs) have a moderating role. Therefore, the present work contributes to filling the gap by providing an empirical model and assessing the panel quantile data and fixed-effect methods used between 2010 and 2020. The findings reveal that the expansion of the financial market, inflowing FDI, rising international remittances, developing environmental innovation, and increasing CPU are significant factors promoting REC while unfavorably increasing CO_2, economic activity, and GEPU impact. Notably, the results underscore that EPs have a significant positive effect, and stricter ecological strategies lead to a rise in REC. In addition, the findings highlight that by implementing stricter EPs, the effect of CPU on REC becomes more positive, and the negative effect of GEPU on REC is more controlled. In particular, the significant interaction effects highlight that the beneficial role of EPs is more pronounced in nations with strong EPs relative to low EPs. Overall, EPs have a substantial beneficial role in promoting REC when CPU and GEPU arise. These findings are reliable after using numerous measurements and methods, and several policy recommendations are suggested to governments, policymakers, and regulatory bodies.

Keywords: renewable energy demand; green policies; climate policy uncertainty; global economic policy uncertainty; BRICS

1. Introduction

Over the last decade, authorities and policymakers have put into place various plans and strategic activities to promote renewable energy (RE) globally. These actions include tax benefits, cheaper credits, and incentive plans aiming to encourage households to switch to RE, motivate firms to raise R&D in green energy and encourage investors to invest in RE plans to decrease climate change, reduce CO_2 emissions, and also attain ecological sustainability [1–5]. Although these actions are useful for promoting RE [6], there is a significant USD 1.1 trillion in funds still required by 2030 to control greenhouse emissions and achieve sustainable development goals (SDGs) [7].

Because of rising worldwide climate change and diminishing air quality, the subject of green energy and RE has become a trending subject at the global level. Consequently, shifting economies from conventional fossil fuels to RE sources is the most important decision in their agendas to reduce climate change and accelerate achieving SDGs [8]. Unlike traditional energy, RE is replenishable and also has ecologically friendly advantages.

Therefore, a strand of the literature has been created essentially to investigate the factors that help promote or diminish renewable energy consumption (REC) in various countries using a demand modeling method. For instance, the findings of prior studies have revealed that capital development (e.g., [9–11]) promotes REC by providing easier external financing at lower costs for investors and households. Likewise, refs. [12,13] revealed that environmental innovations promote REC by increasing energy efficiency and accelerating the shift of the economy to a green country. Moreover, refs. [14,15] highlight that economic openness promotes REC by providing more availability to external financing.

Furthermore, prior works have revealed that economic activity (e.g., [16,17]) and international remittances (e.g., [18,19]) promote REC by increasing the income level of investors and households. Additionally, past works have documented that CO_2 [15,17] diminishes REC. Moreover, earlier studies (e.g., [11,20,21]) have shown that green ecological strategies (EPs) promote REC by enhancing RE capacity growth. However, some works (e.g., [22,23]) have underscored how global economic policy uncertainty (GEPU) diminishes REC by decreasing investor and household income (demand side) [24] and raising investment costs in the private sector [25].

What about the determinants of REC in the BRICS countries, and which factors diminish or promote REC? Understanding this question is essential since it helps policymakers to increase the proportion of REC and help accelerate the energy transition from fossil fuels to green energy, which will eventually lead to attaining ecological sustainability. Brazil, Russia, India, China, and South Africa, which are considered BRICS, represent five massive and emerging economies in the world. Over the last decades, these economies have had outstanding economic development and endeavor to attain SDGs. For example, China applied numerous environmental strategies to encourage green energy in 2006 and 2009 and also used subsidy agendas in 2003 and 2010. This led to, for instance, China increasing its regional share of hydroelectricity production from 2.9% in 1973 to 30.1% in 2019. Also, it helped China to increase its regional share of wind electricity from 1.9% in 2005 to 28.4% in 2019 (https://www.iea.org/reports/key-world-energy-statistics-2021/supply, accessed on 1 June 2024).

Although some efforts have been made in some emerging economies to attain ecological sustainability, there are some significant challenges and problems that reduce the acceleration of the transition from traditional to green energy. For example, RE sources are not fully utilized in some nations, such as India and South Africa, and their share of fossil energy is relatively high [15]. Likewise, energy demands and carbon emissions are still soaring in developing nations, like BRICS, compared to advanced nations due to the rapid growth of energy-intensive businesses and a shortage of energy productivity (e.g., [26]). For example, China accounted for 4.7% of crude oil production regionally in 2020, which was 1.9% in 1973. In addition, the investment in RE schemes decreased to USD 15.4 B in India in 2018 and USD 91.5 B in China in 2019, leading to many RE schemes remaining uncompleted [27,28]. Therefore, BRICS is an interesting case study, and it is important to identify which determinants lead to the promotion or diminishment of REC in BRICS.

This work endeavors to answer the subsequent questions in BRICS settings: (i) What are the determinants of REC? (ii) How specifically do climate policy uncertainty (CPU) and GEPU influence REC? (iii) Do EPs have a moderating effect on CPU, GEPU, and REC?

To attain these objectives, the current research uses novel data for BRICS nations between 2010 and 2020. Likewise, this research also contributes by performing both fixed effects (FEs) and quantile panel data (PQ) techniques to find reliable results. Furthermore, unlike past work (e.g., [11,17]), the current research contributes by probing the elements of the REC in BRICS environments and by focusing mainly on the roles of CPU, GEPU, and EPS, which have not been studied deeply. We employ a novel BRICS EPS index, which uses data collected from the OECD Stat website and which is broadly used for strategy examination aimed at climate change and air pollution-lessening policies. This unique research on BRICS economies can provide findings that are valuable to the work of authorities, policymakers, and executives to upsurge the share of REC and ultimately attain SDGs.

The current research yields some important highlights. The research uncovered that the expansion of the capital market, inflowing FDI, rising international remittances, boosting environmental innovation, and rising CPU are significant drivers promoting REC. This indicates that policymakers and regulatory bodies should make external financing more accessible for investors and households. This would allow them to borrow funds at lower costs, thereby expediting the completion of renewable energy schemes. Likewise, the results denote that representatives should pay more attention to R&D development and enhance country-level competitiveness factors (e.g., private and public institution qualities, economic stability, political stability, infrastructure, and education) to expedite shifting from traditional energies to green energies and attract cross-border capital inflows. Furthermore, the findings highlight that rising CO_2, economic activity, and GEPU have detrimental impacts on the REC of BRICS environments. Moreover, the findings underline that stricter EPS leads to a rise in the REC, and by implementing a stricter EPS, the impact of CPU on REC becomes more positive, and the negative impact of GEPU on REC is more controlled. The findings recommend policymakers design and implement stricter environmental policies to moderate the impact of GEPU on the REC. Policymakers, by developing effective EPS, could accelerate climate change alleviation strategies by promoting REC, particularly when CPU and GEPU arise.

The structure of this study is organized as follows: Section 2 reviews previous works. Section 3 discusses the data and methodology used. Sections 4 and 5 detail the univariate and multivariate analyses, followed by a robustness test. Finally, Section 6 presents the conclusions.

2. Literature Review

A portion of energy research in the literature has focused on identifying the key factors that influence renewable energy consumption (REC) across various economies using demand modeling methods. Previous research has primarily indicated that several significant drivers impacting REC globally include capital market development, financial openness, international remittances, ecological innovation, carbon dioxide (CO_2) emissions, and overall economic activity. Additionally, earlier studies have shown that global Carbon Pricing Units (CPUs) and Global Environmental Policy Units (GEPU) also play an important role in affecting REC. In the following paragraphs, we will discuss in detail how these factors are connected to REC.

2.1. Country-Level Factors

According to previous studies [11,29], the development of financial markets can facilitate the availability of external funds at lower financing costs for both investors and households, thereby promoting renewable energy commercialization (REC). A country with a well-developed capital market can expedite the implementation of renewable energy

(RE) projects and enhance access to a variety of financial products and services. This, in turn, can inspire firms to increase research and development (R&D) activities in clean energy, improve ecological knowledge, and expand RE initiatives.

Additionally, earlier research [30,31] demonstrated that financial openness, characterized by the inflow of foreign capital such as foreign direct investment (FDI), provides cheaper external funds for investors and households, which can also promote REC. A nation with policies supporting financial openness can facilitate the dissemination of better management practices and greener technologies, ultimately leading to an increase in both the supply and demand for renewable energy [32].

Numerous studies (e.g., [11,19]) have shown that international remittances can increase household income levels, which, in turn, may promote renewable energy consumption (REC). As remittances flow into households, they encourage families to invest in renewable energy plans. This financial support allows them to enhance their knowledge and skills related to renewable energy and implement green systems in their homes [33].

Similarly, research by [12,13] indicates that ecological innovations, which improve energy productivity, can also promote REC. Countries with high levels of technological innovation are likely to reduce the investment costs associated with new renewable energy initiatives. This shift can help transition a nation from a traditional economy to a greener one [34,35]. As a result, such changes lead to decreased CO_2 emissions and assist in achieving the Sustainable Development Goals (SDGs) [13].

Numerous studies have shown that CO_2 emissions and economic activity negatively impact renewable energy consumption (REC). This can be attributed to some governments promoting fossil fuel energy through financial incentives, which lowers the price of carbon emissions and ultimately reduces REC. However, other research indicates that economic growth can promote REC, with more developed nations requiring increased energy production from both non-renewable and renewable energy (RE) sources.

Additionally, certain studies [36,37] suggest that environmental policies, such as taxes, can encourage investors and households to support REC. Research [11] has found that implementing green ecological methods can lead to an increase in REC. Countries with stricter environmental policies are likely to see enhanced growth and consumption of renewable energy capacity [21,36]. Importantly, the works by [38–40] indicate that the effectiveness of ecological strategies is influenced by the type of renewable energy sources utilized.

2.2. Global-Level Factors

Several studies have examined the impact of global-level factors on renewable energy consumption (REC). For example, the earlier research by [11,23] indicated that global economic policy uncertainty (GEPU) negatively affects renewable energy from both the supply and demand perspectives. An increase in GEPU tends to lower the income levels of investors and households, which, in turn, slows down the transition to a green economy in the long run [24]. Additionally, GEPU raises investment costs for the private sector, leading to a decrease in the supply of renewable energy [25].

Conversely, other studies have emphasized the significant role of global climate policy uncertainty (CPU) in stimulating both renewable energy demand and supply. For instance, research by [41,42] found that an increase in CPU encourages firms to invest more in green energy initiatives (supply side) and boosts the green energy sector supported by rising CPU. Moreover, reference [8] documented that CPU promotes REC (the demand side) in the long term. Reference [43] revealed that during extreme climate events, the causal connection between CPU and renewable energy consumption strengthens considerably. Additionally, reference [44] showed that CPU has a favorable impact on REC in the United

States. A summary of the literature is presented in Table 1. Several studies have examined the impact of global-level factors on renewable energy consumption (REC). For example, the earlier research by [11,23] indicated that global economic policy uncertainty (GEPU) negatively affects renewable energy from both the supply and demand perspectives. An increase in GEPU tends to lower the income levels of investors and households, which, in turn, slows down the transition to a green economy in the long run [24]. Additionally, GEPU raises investment costs for the private sector, leading to a decrease in the supply of renewable energy [25].

Table 1. Summary of the literature.

Authors	Case-Study	Factors	Methods
[8]	USA	Economic growth, crude oil prices, CPU	Autoregressive distributed lag (ARDL)
[10]	European Union countries	Energy prices, income, financial development, FDI	FE
[11]	OECD countries	Financial expansion, FDI, remittances, environmental innovation, CO_2, EPS, economic activity, GEPU	FE, PQ
[12]	OECD countries	Economic expansion, trade openness, technological innovation	ARDL, pooled mean group, mean group, FE
[14]	Asian countries	Equity market expansion, FDI, trade openness, economic growth	PQ, FE, GMM
[15]	BRICS	Financial expansion, technological innovations, FDI, economic growth, financial stability	FE, PQ
[17]	OECD countries	Geopolitical risks, CO_2, economic globalization, income	GMM
[22]	USA	Economic policy uncertainty (EPU), population, real R&D expenditure, crude oil price, economic expansion	Nonparametric tests, nonparametric regressions
[23]	China	EPU, FDI, economic size, financial development	Nonlinear ARDL
[24]	Emerging markets	Financial development, geopolitical risk, FDI, energy prices, income	GMM

Several studies have identified various country-specific and global factors that influence the rise or decline of renewable energy consumption (REC) worldwide. However, the specific impacts of Consumer Price Uncertainty (CPU) and global economic policy uncertainty (GEPU) on REC, particularly within the context of BRICS nations, have not been thoroughly examined. Furthermore, there has been limited exploration into whether economic policy stability (EPS) serves as a moderator between CPU, GEPU, and REC. This study aims to address these gaps by utilizing novel panel data from emerging BRICS countries and applying the Pooled Ordinary Least Squares (POLS) and fixed effects (FE) methods for the period from 2010 to 2020.

3. Methodology

3.1. Data Description

This research examines the factors influencing renewable energy consumption (REC) in BRICS nations over the period from 2010 to 2022. However, due to data accessibility issues and inconsistencies among various databases, the final analysis is limited to the years 2010 to 2020.

Table A1 in Appendix A lists the BRICS nations and presents the average values of the key determinants for each country. According to the data, Brazil has the highest average for LnREC (3.831) and LnFDI (1.270). South Africa, on the other hand, leads in LnSMC (5.490), indicating that it has the most developed capital market. Additionally, China has the highest averages for LnTINV (12.265) and $LnCO_2$ (16.110). It is noteworthy that Brazil has the lowest average for LnREM (−1.951), while Russia shows the highest average for LnGDPC (8.941). This research examines the factors influencing renewable energy consumption (REC) in BRICS nations over the period from 2010 to 2022. However, due to data accessibility issues and inconsistencies among various databases, the final analysis was limited to the years 2010 to 2020.

Following earlier studies, we identified the main factors that influence renewable energy consumption (REC). This study collected data on various factors, including renewable energy consumption (REC), capital market development (SMC), trade openness (FDI), remittances (REMs), technological innovation (TINV), carbon dioxide emissions (CO_2), and economic output (GDP) from the World Bank. Specifically, we obtained Consumer Policy Uncertainty (CPU) and Global Economic Policy Uncertainty (GEPU) data from the policy uncertainty website.

The CPU index is constructed based on reporting terms such as "uncertainty", "carbon dioxide", "global warming", and "regulation", as mentioned in articles from eight leading U.S. newspapers. The GEPU index, on the other hand, is the GDP-weighted average of national Economic Policy Uncertainty (EPU) indices for 21 nations. Each national EPU index reflects the relative frequency of articles in local newspapers that discuss three terms: economy, policy, and uncertainty.

Furthermore, for further analysis, we selected the environmental policy stringency (EPS) index from the OECD Stat website. Additionally, we gathered data for the World Uncertainty Index (WUI) and Geopolitical Risk Index (GPR) from the policy uncertainty website. The WUI is calculated based on indices of economic uncertainty for 143 nations, employing frequency counts of the term "uncertainty" in the quarterly Economist Intelligence Unit (EIU) country reports. Similarly, the GPR is derived by counting the number of articles related to adverse geopolitical events in newspapers each month.

The explanations and sources for these elements are detailed in Table 2.

Table 2. Factors explanations.

Factors	Symbols	Definitions	Sources
Dependent variable:			
Renewable energy consumption	REC	REC (% of total final energy consumption)	World Bank (https://databank.worldbank.org/)
Independent variables:			
Financial development	SMC	Stock market capitalization (% of GDP)	
Trade openness	FDI	FDI, net inflows (% of GDP)	
International remittances	REM	Personal remittances received (% of GDP)	
Technological innovation	TINV	Total number of patent applications	
Carbon dioxide emissions	CO_2	CO_2 emissions (kt)	
Economic development	GDPC	GDP per capita (current US$)	
Climate policy uncertainty	CPU	CPU index	www.policyuncertainty.com
Worldwide policy uncertainty	GEPU	GEPU index	

Figure 1 presents a scatterplot matrix illustrating the relationships among the determinants studied. Additionally, Figure 2 displays the time series data for REC, CPU, and GEPU within the BRICS countries from 2010 to 2020. As shown in Figure 2, there is a positive correlation between REC, CPU, and GEPU, suggesting that increases in CPU and GEPU lead to a rise in REC. Furthermore, Figure 2 indicates that REC showed a continuous increase throughout the study period, while CPU experienced a decline during the periods of 2010–2013 and 2017–2018.

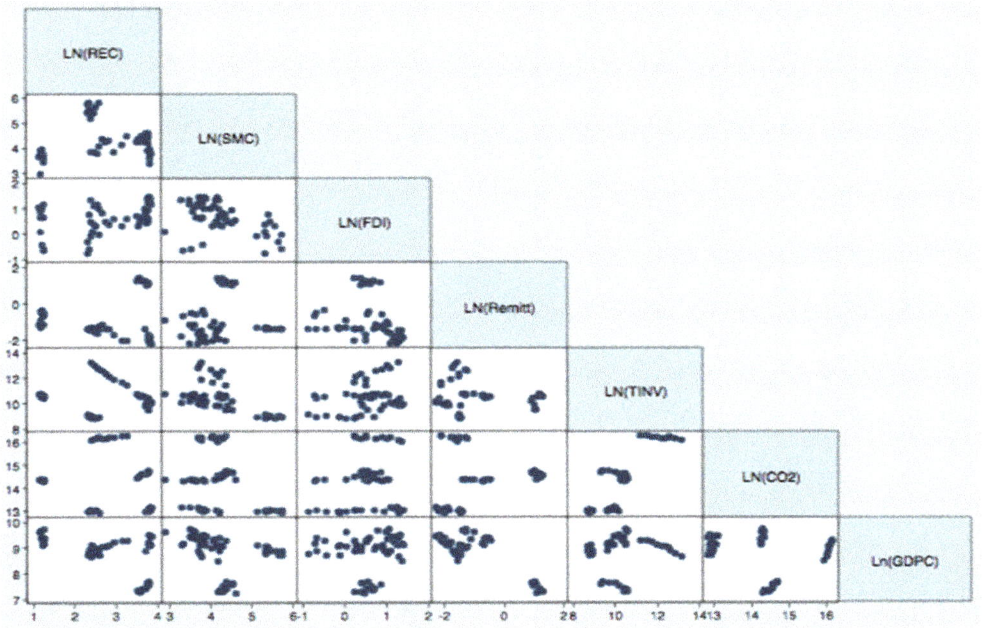

Figure 1. Scatterplot matrix.

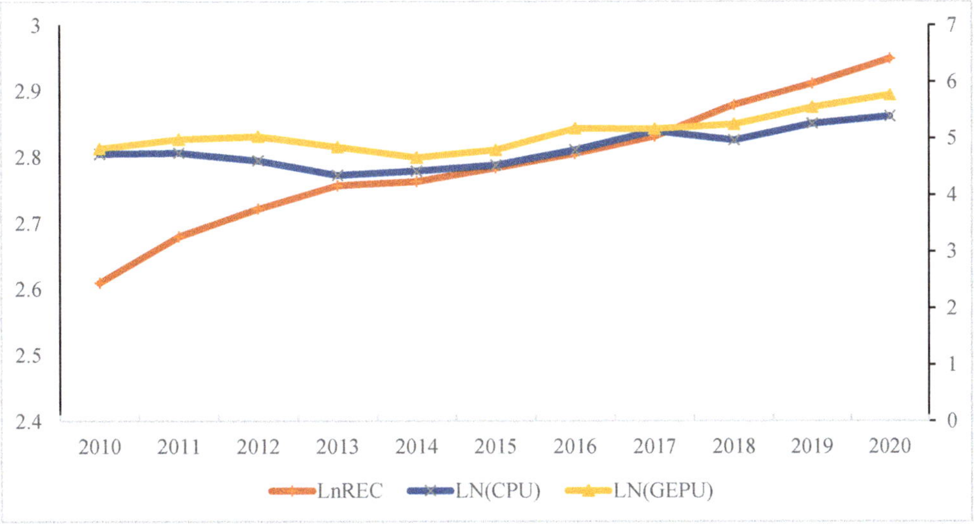

Figure 2. Time series plot of REC, CPU, and GEPU.

Figure 3 highlights the potential impact of various factors on renewable energy consumption (REC) in BRICS countries. As illustrated, certain factors such as Structural Market Conditions (SMCs), foreign direct investment (FDI), Consumer Price Utility (CPU), remittances (REMs), and total investment (TINV) are anticipated to have a positive effect on REC. Conversely, other factors like Global Energy Policy Uncertainty (GEPU), gross domestic product growth (GDPC), and carbon dioxide emissions (CO_2) may impact REC both positively and negatively. The analysis presented in Figure 3 was achieved through a series of pre-estimation tests, estimation analyses, and post-estimation tests to validate the results.

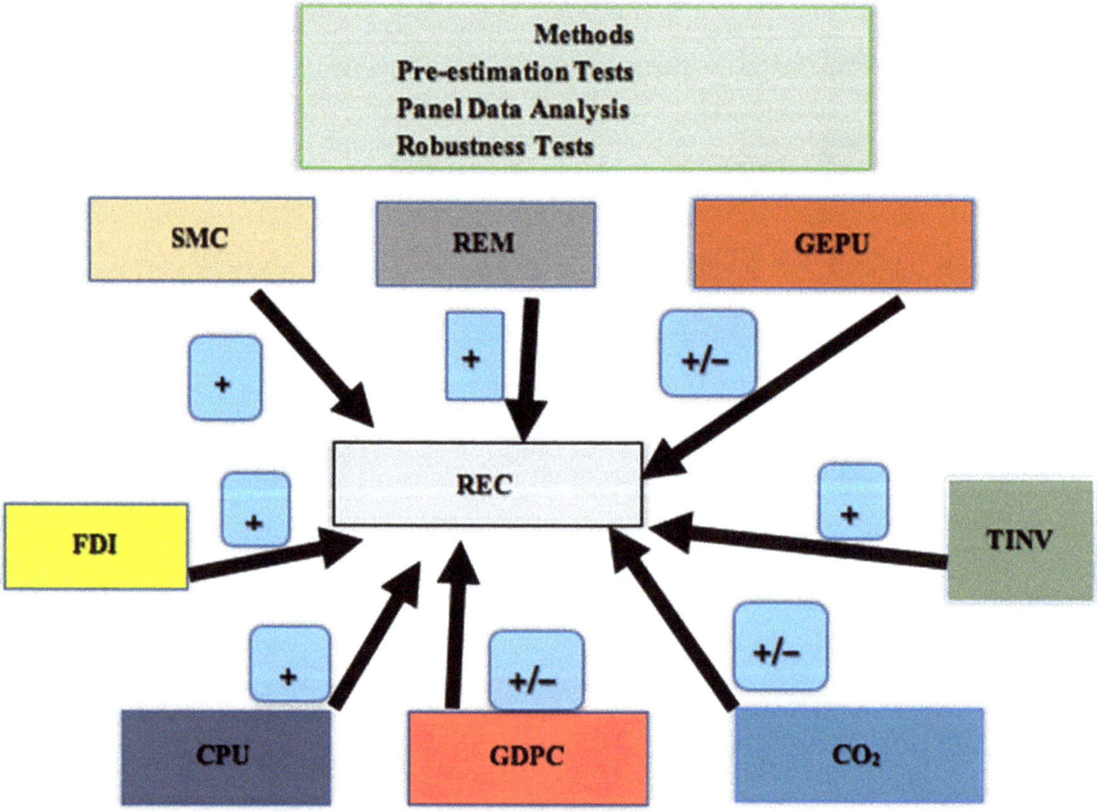

Figure 3. Workflow of the study.

Table 3 presents the results of the correlation analysis. It indicates that there is no evidence of multicollinearity among the factors, allowing for the inclusion of all potential determinants in the model estimates. Additionally, Table 3 shows that some correlations are statistically significant at the 1% level, with both positive and negative signs. Furthermore, the Variance Inflation Factor (VIF) suggests that the estimation models do not suffer from multicollinearity, even when incorporating all possible factors.

Table 3. Correlation analysis.

	SMC	FDI	REM	TINV	CO_2	GDPC	CPU	GEPU	VIF
SMC	1.000								1.12
FDI	−0.261 *	1.000							1.06
REM	0.062	−0.213 *	1.000						1.16
TINV	−0.277 *	0.294 *	−0.063	1.000					1.03
CO_2	0.332 *	0.084 *	0.194	0.358 *	1.000				1.18
GDPC	−0.292	0.104	−0.346 *	0.091	−0.163	1.000			1.04
CPU	0.181	−0.161	0.023	−0.192	0.007	−0.005	1.000		1.09
GEPU	0.153	−0.132	0.018	−0.195	0.011	0.002	0.434 *	1.000	1.15

Note: * is statistically significant at 1%.

3.2. Model and Methodology

Equation (1) is employed to perform empirical analyses.

$$REC_{it} = \alpha_0 + \alpha_1 SMC_{it} + \alpha_2 FDI_{it} + \alpha_3 REM_{it} + \alpha_4 TINV_{it} + \alpha_5 CO2_{it} + \alpha_6 GDPC_{it} + \alpha_7 CPU_t + \alpha_8 GEUP_t + \varepsilon_{it} \quad (1)$$

where country and time are represented correspondingly.

To mitigate the effect of outliers, the data elements are winsorized at the 1% level. Furthermore, each factor is normalized by applying the natural logarithm. In the estimation of Equation (1), we utilized the PQ method (see, e.g., [11,15,45]). This approach helps identify connections across multiple points in the distribution of the dependent variable (LnREC) for the estimation of Equation (1). Quantiles act as thresholds that divide a probability distribution into smaller, continuous intervals that each contain equal probabilities. The PQ technique allows us to understand the relationships among variables at different quantiles.

Additionally, we employed the fixed effects (FEs) approach while clustering standard errors at the country level to ensure reliable results. Studies conducted by [46,47] indicated that using the panel data methodology enhances estimation efficiency and addresses issues related to multicollinearity and heterogeneity. To mitigate the effect of outliers, the data elements are winsorized at the 1% level. Furthermore, each factor is normalized by applying the natural logarithm. For the estimation of Equation (1), we utilized the PQ method (see, e.g., [11,15,45]). This approach helped identify connections across multiple points in the distribution of the dependent variable (LnREC) for the estimation of Equation (1). Quantiles act as thresholds that divide a probability distribution into smaller, continuous intervals that each contain equal probabilities. The PQ technique allows us to understand the relationships among variables at different quantiles.

4. Univariate and Multivariate Analyses

Table 4 presents the descriptive statistics of determinants from 2010 to 2020. The average of $LnCO_2$ is 14.190, which is the highest, with values ranging from 12.882 to 16.208. Conversely, the average of LnREM is −0.931, and the lowest average ranges from −2.227 to 1.327. Following that, LnFDI has an average of 0.597, with a range between −0.824 and 1.420.

Additionally, Table 4 indicates that LnREM has the lowest minimum value at −2.227, while $LnCO_2$ has the highest maximum value at 16.208. Moreover, $LnCO_2$ exhibits the highest variation with a standard deviation of 1.172, followed closely by LnTINV at 1.170. In contrast, LnGEPU and LnCPU display the lowest variations, with standard deviations of 0.322 and 0.327, respectively.

Table 4. Descriptive statistics (2010–2020).

Elements	Mean	Median	St.Dev	Min	Max
LnREC	2.808	2.748	0.964	1.172	3.890
LnSMC	4.290	4.192	0.699	2.931	5.777
LnFDI	0.597	0.636	0.596	−0.824	1.420
LnREM	−0.931	−1.402	1.135	−2.227	1.327
LnTINV	10.347	10.294	1.170	8.736	13.174
LnCO_2	14.190	14.297	1.172	12.882	16.208
LnGDPC	8.761	8.999	0.731	7.208	9.679
LnCPU	4.810	4.741	0.327	4.351	5.387
LnGEPU	5.095	5.035	0.322	4.663	5.768

4.1. Multivariate Results

Before estimating the models, this study conducted cross-sectional dependence (CD) and panel unit root tests. Table 5 shows that the test values for all factors are statistically significant, indicating that any variation in one country will definitely impact the others in the sample. In the robustness section, we employed the CD post-estimation test to examine whether the estimated model encounters any dependency issues.

Table 5. Cross-sectional dependence.

	REC	SMC	FDI	REM	TINV	CO_2	GDPC	CPU	GEPU
Pesaran's test	13.802 **	24.725 *	38.551 **	44.275 *	19.459 *	33.239 *	24.337 **	19.117 *	42.722 *

Note: * and ** reveal statistical significance at the 1% and 5%, correspondingly.

We also conducted the panel unit root test as recommended by sources [48,49] to examine the stationarity of the factors. Specifically, the suggested methods allowed us to assess the stationarity of the factors while considering both trend and cross-dependence alternatives. As shown in Panels (A) and (B) of Table 6, the variables under investigation were stationary (I(1)), indicating that the panels did not contain unit roots.

Table 6. Panel unit root tests.

Factors	Panel (A): [48]		Panel (B): [49]	
	With Trend	With CD	With Trend	With CD
LnREC	−5.316 *	−4.325 *	−5.418 *	−6.343 *
LnSMC	−4.432 *	−6.437 *	−4.522 *	−4.532 *
LnFDI	−5.247 *	−4.662 *	−5.433 *	−2.217 **
LnREM	−4.566 *	−5.343 *	−4.684 *	−4.623 *
LnTINV	−6.323 *	−2.162 **	−6.341 *	−5.315 *
LnCO_2	−4.242 *	−4.527 *	−4.737 *	−3.224 *
LnGDPC	−5.538 *	−4.679 *	−6.354 *	−5.743 *
LnCPU	−5.477 *	−6.292 *	−2.213 **	−4.435 *
LnGEPU	−2.225 **	−2.212 **	−3.714 *	−2.146 **

Note: * and ** reveal statistical significance at the 1% and 5%, correspondingly.

We conducted a Granger causality test to examine the possibility of reverse causality among the factors. As shown in Table 7, the F-statistics were significant at the 1%, 5%, and 10% significance levels, indicating that causality flows from LnSMC, LnFDI, LnREM, LnTINV, LnCO_2, LnGDPC, LnCPU, and LnGEPU to LnREC for the BRICS economies. This suggests that reverse causality is likely to occur, and the tested model is not affected by potential endogeneity.

Table 7. Granger causality test.

H0			F-Statistics	[Prob]	Result
LnSMC	→	LnREC	6.421 *	[0.000]	Yes
LnFDI	→	LnREC	5.668 *	[0.000]	Yes
LnREM	→	LnREC	2.103 **	[0.024]	Yes
LnTINV	→	LnREC	5.451 *	[0.001]	Yes
$LnCO_2$	→	LnREC	4.873 *	[0.000]	Yes
LnGDPC	→	LnREC	5.932 *	[0.002]	Yes
LnCPU	→	LnREC	2.202 ***	[0.068]	Yes
LnGEPU	→	LnREC	2.224 **	[0.019]	Yes

Note: *, **, and *** denote 1%, 5%, and 10% statistical significance levels, respectively.

4.1.1. Estimation Results

Table 8 presents the determinants of renewable energy consumption (REC). The results indicate that the natural logarithm of Social Market Capitalization (LnSMC) has a positive effect on LnREC. However, the coefficient is statistically significant only in the 75th percentile (Q.75) at a significance level of 5% ($\alpha = 0.056$) and in the 95th percentile (Q.95) at a level of 10% ($\alpha = 0.053$). These findings are consistent with previous studies (e.g., [9–11,15]), suggesting that the development of financial markets plays a crucial role in promoting REC in BRICS economies. A well-developed financial market leads to greater availability of external financing and lower borrowing costs. As a result, investors are more likely to engage in renewable energy projects, which ultimately boosts REC and enhances ecological quality (e.g., [26,29]). Furthermore, a developed financial market encourages firms to invest more in research and development (R&D) for renewable energy initiatives, clean energy, and increasing environmental awareness, thereby contributing to higher REC in the long term.

Table 8. The determinants of REC: the role of CPU and GEPU.

Independent Factors	PQ				FE
	Q.25	Q.50	Q.75	Q.95	Coefficients
LnSMC	0.034	0.038	0.056 **	0.053 ***	0.022
	(1.44)	(0.78)	(2.14)	(1.68)	(1.18)
LnFDI	0.025	0.043 **	0.059 *	0.048 **	0.067 *
	(0.74)	(2.04)	(5.15)	(2.11)	(4.86)
LnREM	0.101 ***	0.122 **	0.138 **	0.144	0.011
	(1.73)	(2.15)	(2.08)	(1.33)	(1.54)
LnTINV	0.034 **	0.051 *	0.048 ***	0.042 *	0.028 **
	(2.18)	(3.87)	(1.71)	(4.66)	(2.19)
$LnCO_2$	−0.216	−0.354 **	−0.361 ***	−0.411 *	−0.244 **
	(−1.43)	(−2.14)	(−1.73)	(−5.18)	(−2.05)
LnGDPC	−0.041 **	−0.035 **	−0.026 *	−0.061 ***	−0.022
	(−2.14)	(−2.19)	(−4.18)	(−1.73)	(−1.43)
LnCPU	0.122 **	0.252 *	0.318 ***	0.311 ***	0.405 *
	(2.02)	(5.22)	(1.68)	(1.71)	(4.18)
LnGEPU	−0.232 **	−0.348 *	−0.316 **	−0.269 ***	−0.288 **
	(−2.12)	(−5.16)	(−2.02)	(−1.73)	(−2.19)
CD-test	—	—	—	—	(0.393)
Dummies	✓	✓	✓	✓	✓

Note: *, **, and *** denote 1%, 5%, and 10% statistical significance levels, respectively.

The results further highlight that foreign direct investment (LnFDI), remittances (LnREM), and total investment (LnTINV) have a positive impact on renewable energy consumption (LnREC), with statistically significant coefficients across various quantiles.

Based on the estimates from the fixed effects (FEs) method, a 1% increase in FDI and TINV results in an increase of 0.06% and 0.02% in renewable energy consumption, respectively. In agreement with previous studies, these findings indicate that cross-border capital inflows play a crucial role in enhancing renewable energy consumption in BRICS economies. The openness of these economies allows for foreign capital inflows, enabling investors to access additional external funds for renewable energy projects. Furthermore, this also facilitates the transfer of improved management practices and greener skills, which contribute to the growth of renewable energy both in the host country and globally [50]. The results indicate that international remittances play a significant role in promoting renewable energy consumption (REC) in BRICS economies, which is consistent with previous studies (e.g., [11,15,19,51,52]). Increased remittances raise household income levels, enabling families to invest in renewable energy technologies and fund renewable energy initiatives more easily. This shift gradually reduces the reliance on fossil fuels while increasing the use of renewable energy sources.

Additionally, the findings suggest that technological innovation is a crucial driver of REC in BRICS economies, aligning with earlier research ([12,53]). Enhanced technological innovation improves energy efficiency and accelerates the transition from conventional energy sources to clean and green alternatives. Consequently, this transition contributes to a decline in CO_2 emissions and supports the achievement of Sustainable Development Goals (SDGs) [13,54].

In contrast, Table 8 indicates that $LnCO_2$ has a negative impact on LnREC. According to the estimations from the fixed effects (FEs) method, a 1% increase in CO_2 results in a 0.2% decrease in renewable energy consumption (REC). These findings align with previous studies (e.g., [17]) suggesting that CO_2 is a significant factor adversely affecting REC in BRICS economies. This situation can be attributed to many governments promoting fossil fuel energy. For instance, OECD countries allocated nearly USD 108 billion to fossil fuel extraction in 2019, providing financial incentives for investments in this sector. This investment leads to a reduction in carbon emissions prices, undermining the effectiveness of ecological initiatives and obstructing the transition to a clean economy [55].

Additionally, consistent with the findings of Ref [14], gross domestic product per capita (GDPC) is also a significant factor negatively impacting REC in BRICS economies. However, some studies (e.g., [16,34,56]) have documented that REC can increase with a rise in real per capita income. Similarly, the research in Ref. [17] emphasizes that GDP per capita has a positive effect on REC in OECD nations.

Table 8 shows that LnCPU has a positive effect on LnREC at various quantiles in the BRICS nations: Q.25 ($\alpha = 0.122$), Q.50 ($\alpha = 0.252$), Q.75 ($\alpha = 0.318$), and Q.95 ($\alpha = 0.311$). According to the estimates from the fixed effects method, a 1% increase in CPU leads to a 0.4% increase in REC. This aligns with previous studies (e.g., [8,41]), indicating that CPU is a significant driver of REC in BRICS economies. An increase in CPU encourages firms to boost their investments in clean energy initiatives, which contributes to a higher REC in the long term. Furthermore, the research by [42] suggested that the green energy sector was enhanced by increasing CPU and the Paris Agreement (2015).

The results indicate that LnGEPU has a negative impact on LnREC at various quantiles: Q.25 ($\alpha = -0.232$), Q.50 ($\alpha = -0.348$), Q.75 ($\alpha = -0.316$), and Q.95 ($\alpha = -0.269$). Specifically, a 1% increase in GEPU is associated with a 0.28% reduction in REC within the BRICS nations. This aligns with previous findings from sources [11,22,23], suggesting that GEPU adversely affects REC. The rise in economic policy uncertainty (EPU) on an international scale negatively influences the income levels of investors and households, making them less motivated to replace fossil fuels with renewable energy sources (REs). Notably, the higher levels of GEPU result in a slower transition from traditional energy sources to REs.

It is important to mention that all estimated models incorporate time and country dummies, although their coefficients are not reported to save space. Additionally, the cross-sectional dependence (CD) post-estimation test produced a p-value of 0.393, which is not statistically significant, indicating that the estimation results are robust.

4.1.2. Additional Analysis: The EPS Interaction Effects with CPU and GEPU

This sub-section seeks to determine whether the environmental policy stringency (EPS) acts as a moderator between the relationships of Clean Production Utilization (CPU) and Green Energy Production Utilization (GEPU) in the context of Resource Efficiency and Conservation (REC). Understanding this issue carries significant policy implications for governments, as it can help improve REC by implementing effective green policies when both CPU and GEPU are present. To achieve this objective, we gathered EPS data from the OECD.Stat website. According to a report from the OECD Nations Department, the EPS index is calculated based on three sub-indices: market-based policies, non-market-based policies, and policies that support technology.

Table 9 examines the interaction effects of environmental policy strategies (EPS) with Carbon Pricing Units (CPUs) and Green Energy Policy Units (GEPUs) on renewable energy consumption (REC) in BRICS nations, taking into account various control variables. The findings indicate that EPS has a positive and statistically significant effect on REC. According to the existing literature, implementing green environmental strategies contributes to an increase in REC (e.g., [11,20,21]). Studies by [36,38] demonstrate that EPS effectively enhances REC, while the success of ecological strategies is influenced by the type of renewable energy sources utilized. Furthermore, reference [57] revealed that countries with lower levels of REC experience positive impacts from EPS.

Table 9. The nexus between EPS, CPU, GEPU, and REC.

Independent Factors	EPS		BRICS Economies
	Low EPS	High EPS	
LnCPU	0.118	0.233	0.338 **
	(2.02)	(1.17)	(2.11)
LnGEPU	−0.143 **	−0.231	−0.122 **
	(−2.14)	(−1.11)	(−2.18)
LnEPS	0.246	0.327 **	0.188 *
	(0.71)	(2.14)	(4.19)
LnCPU*LnEPS	0.028 **	0.033 *	0.017 ***
	(2.11)	(4.16)	(1.73)
LnGEPU*LnEPS	0.016 **	0.044 ***	0.022 *
	(2.19)	(1.73)	(2.02)
Control factors	✓	✓	✓
Dummies	✓	✓	✓

Note: *, **, and *** denote 1%, 5%, and 10% statistical significance levels, respectively.

Furthermore, Table 9 indicates that the coefficients for the interaction effects of (LnEPS*LnCPU) ($\alpha = 0.017$) and (LnEPS*LnGEPU) ($\alpha = 0.022$) are both positive and statistically significant in BRICS nations. This suggests that implementing stricter Environmental Policy Standards (EPSs) enhances the positive effect of CPU on renewable energy consumption (REC) and mitigates the negative impact of GEPU on REC. Therefore, EPS plays a crucial role in promoting REC when CPU and GEPU are present.

The analysis reveals that when data are divided based on the average Earnings Per Share (EPS), the interaction effects are significantly different between economies with low and high EPS. Specifically, the magnitude of the interaction effects for LnEPS*LnCPU ($\alpha = 0.028$) and LnEPS*LnGEPU ($\alpha = 0.016$) is relatively smaller in low EPS economies. By contrast, for high EPS economies, the coefficients are larger: LnEPS*LnCPU (α = 0.033) and LnEPS*LnGEPU (α = 0.044).

These findings suggest that the efficiency of EPS is crucial. The results indicate that CPU has a more positive effect on renewable energy consumption (REC), while GEPU has a less negative impact on environments with stricter green policies. Overall, the evidence demonstrates that strengthening EPS can promote REC by enhancing the positive effects of CPU and mitigating the negative effects of GEPU. Therefore, we recommend that policymakers design and implement stricter EPS to achieve higher levels of REC.

5. Robustness Check

First, we utilized alternative measurements, specifically the "stock market turnover ratio" (LnSMT) and the "world uncertainty index" (LnWUI), to assess financial market development and geopolitical economic policy uncertainty (GEPU), respectively. The data for these measurements were gathered from the World Bank and the policy uncertainty website.

Second, following the findings from the work referenced in [17], we estimated Equation (1) by introducing a new variable: the "geopolitical risk index" (LnGPR), with data collected from the same policy uncertainty website.

Third, we estimated Equation (1) using both pooled quantile (PQ) and fixed effects (FEs) regression methods to check for consistency in our results.

Finally, we conducted the CD post-estimation test [58] to determine whether the estimated model exhibited any dependency issues.

The findings presented in Table 10 align with previous studies, indicating that LnSMT, LnFDI, LnREM, LnTINV, and LnCPU have positive and statistically significant effects on renewable energy consumption (REC). Our results provide substantial evidence that improving access to external funds by developing financial markets and attracting foreign direct investment (FDI), increasing household income through international remittances, enhancing environmental innovation, and raising consumer purchasing power (CPU) contribute to promoting REC in BRICS countries.

Table 10. Robustness test.

Independent Factors	PQ				FE
	Q.25	**Q.50**	**Q.75**	**Q.95**	**Coefficients**
LnSMT	0.026	0.043 *	0.058 **	0.062 **	0.038 ***
	(0.78)	(5.66)	(2.08)	(2.13)	(1.73)
LnFDI	0.018	0.036 **	0.047	0.056 ***	0.034
	(1.15)	(2.14)	(0.88)	(1.68)	(1.24)
LnREM	0.121 ***	0.119 **	0.094	0.138 **	0.127 ***
	(1.71)	(2.14)	(1.42)	(2.13)	(1.73)
LnTINV	0.029 *	0.036 **	0.021 **	0.047 ***	0.012
	(4.16)	(2.19)	(2.05)	(1.69)	(0.44)
$LnCO_2$	−0.106	−0.155	−0.211	−0.388 **	−0.125
	(−1.12)	(−1.38)	(−1.58)	(−2.14)	(−1.41)
LnGDPC	−0.017	−0.024 **	−0.028 **	−0.043 ***	−0.035 *
	(−1.23)	(−2.17)	(−2.02)	(−1.71)	(−4.66)
LnCPU	0.117 ***	0.225 **	0.269 *	0.306 ***	0.241 **
	(1.71)	(2.12)	(4.76)	(1.69)	(2.11)
LnWUI	−0.224 **	−0.265 **	−0.327 ***	−0.237 **	−0.351 *
	(−2.13)	(−2.04)	(−1.69)	(−2.17)	(−4.88)
LnGPR	−0.121	−0.147	−0.172 **	−0.168 ***	−0.148 **
	(−0.77)	(−1.36)	(−2.12)	(−1.69)	(−2.14)
CD-test	—	—	—	—	(0.318)
Dummies	✓	✓	✓	✓	✓

Note: *, **, and *** denote 1%, 5%, and 10% statistical significance levels, respectively.

Conversely, the data reveal that $LnCO_2$, LnGDPC, LnWUI, and LnGPR negatively impact REC. Specifically, our findings demonstrate that increases in CO_2 levels, economic output, global uncertainty, and geopolitical risk contribute to a decline in REC within BRICS nations.

6. Conclusions

Several researchers (e.g., [11,15]) have examined the elements of renewable energy consumption (REC) in various countries. However, there has been less focus on this topic in the context of BRICS nations. To the best of the authors' knowledge, this study may be the first to investigate the moderating role of environmental policies (EPs) between Clean Production Units (CPUs), Green Energy Policy Uncertainty (GEPU), and REC in BRICS economies. This research aims to address these gaps by offering an empirical model that explores the determinants of REC in BRICS countries, utilizing traditional control factors. To achieve these objectives, this study employed the pooled quantile (PQ) and fixed effects (FEs) methods to analyze annual data from 2010 to 2020.

The findings indicate that the expansion of the capital market, an influx in foreign direct investment (FDI), increasing international remittances, advancements in environmental innovation, and rising capacity utilization (CPU) are significant factors that promote renewable energy consumption (REC). However, rising carbon dioxide (CO_2) levels, economic growth, and an increase in global economic policy uncertainty (GEPU) negatively affect REC in BRICS countries.

Additionally, the results emphasize that environmental policy stringency (EPS) has a positive and significant impact on REC, suggesting that stricter ecological policies enhance REC. Notably, EPS plays a significant moderating role between CPU, GEPU, and REC. This means that by implementing stricter EPS, the positive effect of CPU on REC can be amplified, while the negative impact of GEPU on REC can be mitigated.

Consequently, EPS has a crucial role in promoting REC when CPU and GEPU are present. Specifically, the interaction effects highlight that the positive influence of EPS is more pronounced in countries with high EPS compared to those with low EPS. The findings are robust, having been validated through multiple measures and various methodologies.

6.1. Policy Recommendations

Here, we provide several policy recommendations. First, the statistically significant findings regarding SMC, FDI, and REM suggest that policymakers should improve access to external funds for investors and households to promote renewable energy consumption (REC). This can be achieved by developing financial markets, fostering open economies, and enhancing the inflow of international remittances. By doing so, investors and households can access and borrow funds at lower costs, which encourages greater investment in renewable energy schemes.

Furthermore, the findings indicate that policymakers aiming to facilitate capital inflows into a host country should focus on enhancing competitiveness at the country level. This includes improving the quality of private and public institutions, ensuring financial and economic stability, maintaining political stability, and developing infrastructure (e.g., [59]).

Secondly, the significant positive impact of technological innovation (TINV) indicates to policymakers the importance of focusing on research and development (R&D) and financing renewable energy (RE) technologies. This focus can help expedite the transition from conventional energy sources to clean energy and contribute to achieving the Sustainable Development Goals (SDGs).

Moreover, the statistically significant findings regarding Environmental Policy Structures (EPSs) suggest that policymakers should design and implement stricter environmental policies. These policies can be market-based, non-market-based, or involve technological support, all aimed at promoting renewable energy consumption (REC). In countries with stricter environmental regulations, growth in RE capacity and consumption could be enhanced. This environment would also encourage both investors and households to actively promote REC.

Additionally, consistent with the study in ref [11], stricter environmental policies can help moderate the influence of Global Energy Price Uncertainty (GEPU) on the demand for RE. Therefore, by developing effective Environmental Policy Structures, policymakers can accelerate climate change mitigation strategies, particularly when confronted with Consumer Price Uncertainty (CPU) and GEPU.

6.2. Limitations of Study

Further studies could explore the findings of this research across both developing and developed economies. Additionally, it would be valuable to test the proposed model with various renewable energy (RE) sources. Incorporating the impact of economic globalization and sovereign Environmental, Social, and Governance (ESG) factors [60,61] on renewable energy consumption (REC) and different RE sources would also be beneficial. Moreover, future research should examine the short- and long-term relationships between these variables and REC.

Author Contributions: Conceptualization, D.A., T.A. and M.A.; methodology, D.A. and M.A.; software, D.A.; validation, D.A. and M.A.; formal analysis D.A.; investigation, D.A. and M.A.; resources, D.A. data curation, D.A.; writing—original draft preparation, D.A., T.A. and M.A.; writing—review and editing, D.A.; visualization, D.A. and M.A.; supervision, D.A.; project administration, D.A. and M.A. All authors have read and agreed to the published version of the manuscript.

Funding: This research received no external funding.

Data Availability Statement: The original contributions presented in this study are included in the article. Further inquiries can be directed to the corresponding author.

Conflicts of Interest: The authors declare no conflicts of interest.

Appendix A

Table A1. BRICS nations and the average of factors (2010–2020).

BRICS Countries	LnREC	LnSMC	LnFDI	LnREM	LnTINV	$LnCO_2$	LnGDPC
Brazil	3.831	3.853	1.270	−1.951	9.931	13.012	9.222
Russia	1.238	3.637	0.402	−0.841	10.518	14.306	9.375
India	3.717	4.416	0.556	1.156	10.160	14.554	7.413
China	2.801	4.055	0.787	−1.591	12.265	16.110	8.941
South Africa	2.455	5.490	−0.029	−1.427	8.862	12.967	8.852

References

1. Gozgor, G.; Lau, C.K.M.; Lu, Z. Energy consumption and economic growth: New evidence from the OECD countries. *Energy* **2018**, *153*, 27–34. [CrossRef]
2. Adams, S.; Nsiah, C. Reducing carbon dioxide emissions; Does renewable energy matter? *Sci. Total Environ.* **2019**, *693*, 133288. [CrossRef] [PubMed]
3. Zhou, A.; Li, J. Heterogeneous role of renewable energy consumption in economic growth and emissions reduction: Evidence from a panel quantile regression. *Environ. Sci. Pollut. Res.* **2019**, *26*, 22575–22595. [CrossRef]

4. Godil, D.I.; Yu, Z.; Sharif, A.; Usman, R.; Khan, S.A.R. Investigate the role of technology innovation and renewable energy in reducing transport sector CO_2 emission in China: A path toward sustainable development. *Sustain. Dev.* **2021**, *29*, 694–707. [CrossRef]
5. Addai, K.; Al Geitany, S.H.; Athari, S.A.; Farmanesh, P.; Kirikkaleli, D.; Saliba, C. Do Environmental Tax and Energy Matter for Environmental Degradation in the UK? Evidence from Novel Fourier-Based Estimators. *Energies* **2024**, *17*, 5732. [CrossRef]
6. Bloomberg New Energy Finance. *Global Trends in Clean Energy Investment*; Bloomberg Finance L.P.: New York, NY, USA, 2019.
7. IRENA. *World Energy Transitions Outlook: 1.5 °C Pathway*; International Renewable Energy Agency: Abu Dhabi, United Arab Emirates, 2021; pp. 1–352. Available online: www.irena.org/publications (accessed on 11 April 2024).
8. Shang, Y.; Han, D.; Gozgor, G.; Mahalik, M.K.; Sahoo, B.K. The impact of climate policy uncertainty on renewable and non-renewable energy demand in the United States. *Renew. Energy* **2022**, *197*, 654–667. [CrossRef]
9. Sweerts, B.; Longa, F.D.; van der Zwaan, B. Financial de-risking to unlock Africa's renewable energy potential. *Renew. Sustain. Energy Rev.* **2019**, *102*, 75–82. [CrossRef]
10. Anton, S.G.; Nucu, A.E.A. The effect of financial development on renewable energy consumption. A panel data approach. *Renew. Energy* **2020**, *147*, 330–338. [CrossRef]
11. Athari, S.A. Global economic policy uncertainty and renewable energy demand: Does environmental policy stringency matter? Evidence from OECD economies. *J. Clean. Prod.* **2024**, *450*, 141865. [CrossRef]
12. Alam, M.M.; Murad, M.W. The impacts of economic growth, trade openness and technological progress on renewable energy use in organization for economic co-operation and development countries. *Renew. Energy* **2020**, *145*, 382–390. [CrossRef]
13. Saliba, C.B.; Hassanein, F.R.; Athari, S.A.; Dördüncü, H.; Agyekum, E.B.; Adadi, P. The Dynamic Impact of Renewable Energy and Economic Growth on CO_2 Emissions in China: Do Remittances and Technological Innovations Matter? *Sustainability* **2022**, *14*, 14629. [CrossRef]
14. Tiwari, A.K.; Nasreen, S.; Anwar, M.A. Impact of equity market development on renewable energy consumption: Do the role of FDI, trade openness and economic growth matter in Asian economies? *J. Clean. Prod.* **2022**, *334*, 130244. [CrossRef]
15. Athari, S.A. The impact of financial development and technological innovations on renewable energy consumption: Do the roles of economic openness and financial stability matter in BRICS economies? *Geol. J.* **2024**, *59*, 288–300. [CrossRef]
16. Padhan, H.; Padhang, P.C.; Tiwari, A.K.; Ahmed, R.; Hammoudeh, S. Renewable energy consumption and robust globalization(s) in OECD countries: Do oil, carbon emissions and economic activity matter? *Energy Strategy Rev.* **2020**, *32*, 100535. [CrossRef]
17. Zhao, Z.; Gozgor, G.; Lau, M.C.K.; Mahalik, M.K.; Patel, G.; Khalfaoui, R. The impact of geopolitical risks on renewable energy demand in OECD countries. *Energy Econ.* **2023**, *122*, 106700. [CrossRef]
18. Sadeque, Z.; Rysankova, D.; Elahi, R.; Soni, R. *Scaling Up Access to Electricity: The Case of Bangladesh*; Live Wire, 2014/21; World Bank: Washington, DC, USA, 2014; Available online: https://openknowledge.worldbank.org/handle/10986/18679 (accessed on 11 April 2024).
19. Subramaniam, Y.; Masron, T.A.; Loganathan, N. Remittances and renewable energy: An empirical analysis. *Int. J. Energy Sect. Manag.* **2022**, *17*, 1034–1049. [CrossRef]
20. Acemoglu, D.; Akcigit, U.; Hanley, D.; Kerr, W. Transition to clean technology. *J. Political Econ.* **2016**, *124*, 52–104. [CrossRef]
21. Lei, W.; Xie, Y.; Hafeez, M.; Ullah, S. Assessing the dynamic linkage between energy efficiency, renewable energy consumption, and CO_2 emissions in China. *Environ. Sci. Pollut. Res.* **2022**, *29*, 19540–19552. [CrossRef] [PubMed]
22. Shafiullah, M.; Miah, M.D.; Alam, M.S.; Atif, M. Does economic policy uncertainty affect renewable energy consumption? *Renew. Energy* **2021**, *179*, 1500–1521. [CrossRef]
23. Lei, W.; Liu, L.; Hafeez, M.; Sohail, S. Do economic policy uncertainty and financial development influence the renewable energy consumption levels in China? *Environ. Sci. Pollut. Res.* **2022**, *29*, 7907–7916. [CrossRef] [PubMed]
24. Alsagr, N.; van Hemmen, S. The impact of financial development and geopolitical risk on renewable energy consumption: Evidence from emerging markets. *Environ. Sci. Pollut. Res.* **2021**, *28*, 25906–25919. [CrossRef]
25. Gozgor, G.; Lau, M.C.K.; Zeng, Y.; Yan, C.; Lin, Z. The impact of geopolitical risks on tourism supply in developing economies: The moderating role of social globalization. *J. Travel Res.* **2022**, *61*, 872–886. [CrossRef]
26. Ma, Y.; Zhao, Y.; Jia, R.; Wang, W.; Zhang, B. Impact of financial development on the energy intensity of developing countries. *Heliyon* **2022**, *8*, e09904. [CrossRef]
27. IEA. World Energy Statistics 2019, France. 2020. Available online: https://www.iea.org (accessed on 11 April 2024).
28. Global Energy Review. In *Assessing the Effects of Economic Recoveries on Global Energy Demand and CO2 Emission in 2021*; International Energy Agency: Paris, France, 2021.
29. Kirikkaleli, D.; Adebayo, T.S. Do renewable energy consumption and financial development matter for environmental sustainability? New global evidence. *Sustain. Dev.* **2021**, *29*, 583–594. [CrossRef]
30. Paramati, S.R.; Ummalla, M.; Apergis, N. The effect of foreign direct investment and stock market growth on clean energy use across a panel of emerging market economies. *Energy Econ.* **2016**, *56*, 29–41. [CrossRef]

31. Amuakwa-Mensah, F.; Näsström, E. Role of banking sector performance in renewable energy consumption. *Appl. Energy* **2022**, *306*, 118023. [CrossRef]
32. Ferrier, G.D.; Reyes, J.; Zhu, Z. Technology diffusion on the international trade network. *J. Public Econ. Theory* **2016**, *18*, 291–312. [CrossRef]
33. Swiss Agency for Development and Cooperation. Energy, Migration and the 2030 Agenda for Sustainable Development, Briefing Note. 2018. Available online: https://www.odi.org/sites/odi.org.uk/files/resource-documents/12301.pdf (accessed on 16 May 2024).
34. Sadorsky, P. Renewable energy consumption and income in emerging economies. *Energy Policy* **2009**, *37*, 4021–4028. [CrossRef]
35. Saeed, T.; Bouri, E.; Alsulami, H. Extreme return connectedness and its determinants between clean/green and dirty energy investments. *Energy Econ.* **2021**, *96*, 105017. [CrossRef]
36. Jenner, S.; Groba, F.; Indvik, J. Assessing the strength and effectiveness of renewable electricity feed-in tariffs in European Union countries. *Energy Policy* **2013**, *52*, 385–401. [CrossRef]
37. Bersalli, G.; Menanteau, P.; El-Methni, J. Renewable energy policy effectiveness: A panel data analysis across Europe and Latin America. *Renew. Sustain. Energy Rev.* **2020**, *133*, 110351. [CrossRef]
38. Johnstone, N.; Haščič, I.; Poirier, J.; Hemar, M.; Michel, C. Environmental policy stringency and technological innovation: Evidence from survey data and patent counts. *Appl. Econ.* **2012**, *44*, 2157–2170. [CrossRef]
39. Bourcet, C. Empirical determinants of renewable energy deployment: A systematic literature review. *Energy Econ.* **2020**, *85*, 104563. [CrossRef]
40. Sohail, M.T.; Xiuyuan, Y.; Usman, A.; Majeed, M.T.; Ullah, S. Renewable energy and non-renewable energy consumption: Assessing the asymmetric role of monetary policy uncertainty in energy consumption. *Environ. Sci. Pollut. Res.* **2021**, *28*, 31575–31584. [CrossRef]
41. Pommeret, A.; Schubert, K. Intertemporal emission permits trading under uncertainty and irreversibility. *Environ. Resour. Econ.* **2018**, *71*, 73–97. [CrossRef]
42. Antoniuk, Y.; Leirvik, T. Climate change events and stock market returns. *J. Sustain. Financ. Investig.* **2021**, *14*, 42–67. [CrossRef]
43. Ren, X.; Li, J.; He, F.; Lucey, B. Impact of climate policy uncertainty on traditional energy and green markets: Evidence from time-varying granger tests. *Renew. Sustain. Energy Rev.* **2023**, *173*, 113058. [CrossRef]
44. Zhou, D.; Siddik, A.B.; Guo, L.; Li, H. Dynamic relationship among climate policy uncertainty, oil price and renewable energy consumption—Findings from TVP-SV-VAR approach. *Renew. Energy* **2023**, *204*, 722–732. [CrossRef]
45. Saliba, C.; Farmanesh, P.; Athari, S.A. Does country risk impact the banking sectors' non-performing loans? Evidence from BRICS emerging economies. *Financ. Innov.* **2023**, *9*, 86. [CrossRef] [PubMed]
46. Baltagi, B.H. *Econometric Analysis of Panel Data;* John Wiley & Sons: West Sussex, UK, 2005.
47. Hsiao, C. *Analysis of Panel Data (No. 54);* Cambridge University Press: Cambridge, UK, 2014.
48. Levin, A.; Lin, C.-F.; Chu, C.-S.J. Unit root tests in panel data: Asymptotic and finite-sample properties. *J. Econ.* **2002**, *108*, 1–24. [CrossRef]
49. Im, K.S.; Pesaran, M.; Shin, Y. Testing for unit roots in heterogeneous panels. *J. Econom.* **2003**, *115*, 53–74. [CrossRef]
50. Inglesi-Lotz, R. The impact of renewable energy consumption to economic growth: A panel data application. *Energy Econ.* **2016**, *53*, 58–63. [CrossRef]
51. Yang, B.; Jahanger, A.; Khan, M.A. Does the inflow of remittances and energy consumption increase CO_2 emissions in the era of globalization? A global perspective. *Air Qual. Atmos. Health* **2020**, *13*, 1313–1328. [CrossRef]
52. Das, A.; McFarlane, A.; Carels, L. Empirical exploration of remittances and renewable energy consumption in Bangladesh. *Asia-Pac. J. Reg. Sci.* **2021**, *5*, 65–89. [CrossRef]
53. Bellakhal, R.; Ben Kheder, S.; Haffoudhi, H. Governance and renewable energy investment in MENA countries: How does trade matter? *Energy Econ.* **2019**, *84*, 104541. [CrossRef]
54. Chen, J.; Rojniruttikul, N.; Kun, L.Y.; Ullah, S. Management of green economic infrastructure and environmental sustainability in one belt and road enitiative economies. *Environ. Sci. Pollut. Res.* **2022**, *29*, 36326–36336. [CrossRef] [PubMed]
55. Hou, H.; Lu, W.; Liu, B.; Hassanein, Z.; Mahmood, H.; Khalid, S. Exploring the Role of Fossil Fuels and Renewable Energy in Determining Environmental Sustainability: Evidence from OECD Countries. *Sustainability* **2023**, *15*, 2048. [CrossRef]
56. Sadorsky, P. Renewable energy consumption, CO_2 emissions and oil prices in the G7 countries. *Energy Econ.* **2009**, *31*, 456–462. [CrossRef]
57. Li, X.; Ozturk, I.; Syed, Q.R.; Hafeez, M.; Sohail, S. Does green environmental policy promote renewable energy consumption in BRICST? Fresh insights from panel quantile regression. *Econ. Res.-Ekon. Istraživanja* **2022**, *35*, 5807–5823. [CrossRef]
58. Pesaran, M.H. *General Diagnostic Tests for Cross Section Dependence in Panels;* Cambridge Working Papers in Economics, 0435; University of Cambridge: Cambridge, UK, 2004.
59. Athari, S.A.; Adaoglu, C. Nexus between institutional quality and capital inflows at different stages of economic development. *Int. Rev. Financ.* **2019**, *19*, 435–445. [CrossRef]

60. Athari, S.A. Does the sovereign environmental, social, and governance sustainability activities jeopardize the banking sector's stability: Evidence from the Arab economies. *Sustain. Futures* **2024**, *7*, 100204. [CrossRef]
61. Athari, S.A.; Saliba, C.; Abboud, E.; El-Bayaa, N. Examining the Quadratic Impact of Sovereign Environmental, Social, and Governance Practices on Firms' Profitability: New Insights from the Financial Industry in Gulf Cooperation Council Countries. *Sustainability* **2024**, *16*, 2783. [CrossRef]

energies

MDPI

Article

Environmental Assessment of Hybrid Waste-to-Energy System in Ghana

Ekua Afrakoma Armoo [1,2,*], Theophilus Baidoo [3], Mutala Mohammed [2], Francis Boateng Agyenim [2], Francis Kemausuor [4] and Satyanarayana Narra [1,5]

1 Department of Waste and Resource Management, University of Rostock, 18051 Rostock, Germany; satyanarayana.narra@uni-rostock.de
2 Council for Scientific and Industrial Research—Institute of Industrial Research, Accra P.O. Box LG 576, Ghana; mutbaby@gmail.com (M.M.); aboffour@hotmail.com (F.B.A.)
3 Zeal Environmental Technologies, Takoradi P.O. Box TD 1395, Ghana; mobatheo@gmail.com
4 Department of Agricultural and Biosystems Engineering, Kwame Nkrumah University of Science and Technology (KNUST), Kumasi AK-039-5028, Ghana; kemausuor@gmail.com
5 German Biomass Research Centre gGmbH, 04347 Leipzig, Germany
* Correspondence: ekua.armoo2@uni-rostock.de; Tel.: +233-246-287-950

Abstract: Waste management in most parts of Africa is characterized by the disposal of mixed waste in unengineered landfills. The aim of this study is to assess the environmental impact of mixed waste received at a waste-to-energy plant in Ghana relative to the current model of landfilling. A Life Cycle Assessment was conducted using OpenLCA software version 2.3.1 based on the ReCiPe Midpoint method. For landfilling, LandGEM software version 3.03 was used. The results indicate that waste-to-energy has the potential to provide carbon savings of 3.52 tCO_2eq/ton of waste treated compared to landfilling. Pyrolysis is observed to have high avoided burden across all impact categories, with the lowest Global Warming Potential of −2.3 $kgCO_2eq$. Anaerobic digestion shows a near neutral environmental impact with the highest value of 47.56 kg 1,4DCB for Terrestrial Ecotoxicity, while Refuse-Derived Fuel and segregation processes show low environmental burdens. The net avoided burden is highest for global warming and non-carcinogenic human toxicity potential. Overall, the hybrid waste-to-energy model is concluded to be an environmentally preferred waste management option compared to conventional landfilling methods, and we recommend that decision-makers facilitate investments into it. It is also recommended for the development of local inventories and databases to encourage more country-specific environmental impact studies and to reduce uncertainty.

Keywords: life cycle analysis; waste to energy; anaerobic digestion; pyrolysis; refuse derived fuels; environmental impacts; global warming potential; openLCA; LandGEM; Ghana

1. Introduction

Developing countries continue to lag behind as the 2030 target of the Sustainable Development Goals (SDGs) draws nearer [1,2]. Waste management and energy are cross-cutting sectors with high environmental footprints mainly linked to SDGs 3, 6, 11 and 13. With a record high of 37.4 Gt energy related CO_2 emissions in 2023, greenhouse gas (GHG) emissions from these sectors are a key concern to the global agenda. Renewable energy is, in particular, crucial for reducing emissions [3,4]. The recently ended United Nations Climate Conference 2024 (COP 29) was strongly dominated by discussions of

https://doi.org/10.3390/en18030595

climate finance and carbon markets with stronger commitments to supporting developing countries [5]. Ironically, a study in South Africa, Nigeria, Cameroun and Ghana identified a lack of strong policy implementation as a challenge to renewable energy uptake and particularly waste-to-energy (WtE) projects in Ghana [6,7].

Waste is a resource material for energy generation with the potential to substitute fossil fuels and consequently reduce the environmental burden of poor waste management [8]. CEWEP estimates that in 2021 alone, the conversion of waste to energy (WtE) supplied 35,000 GWh of electricity to 21 million Europeans, together with 87,000 GWh of heat to 17 miilion residents, which translates into 22–44 million tons of CO_2 savings [9]. The end products of WtE include electricity, hydrogen, biodiesel, RDF and heat, although electricity and heat are the most common. They can generally be categorized into either energy or fuels [10].

Comparative environmental studies support key decision-making on WtE technology selection. Decisions can be made between WtE technologies and conventional/existing treatment options such as landfilling. For instance, WtE shows methane emissions reductions of about 52% for biomethane production compared to landfilling [11]. A decision can also be made on the selection of the most environmentally friendly technology or process. For instance, electricity/heat cogeneration was found to be the most environmentally friendly scenario, releasing 1.95 $kgCO_2eq/m^3{}_{biogas}$ compared to the combined end-products of electricity/biomethane [12]. In another city-scale scenario analysis, it was concluded that a thermal WtE option (incineration) provided the best GHG savings of (652,626 tCO_2eq/yr), which was more than twice that of two other scenarios: recycling/landfill gas to energy and recycling/composting [13].

Following the Paris agreement, countries like Ghana have set up a National Greenhouse Gas Inventory. Data from the directory indicate that the three major contributing sectors are energy, agriculture, forestry and other land uses (AFOLU) and waste. Between 1990 and 2019, the total national GHG emissions increased by 853.1% and 265.1% in the energy and waste sectors, respectively, with the energy sector alone contributing 45.7% in 2019. Strategies to reduce emissions as the population and economy continue to expand are critical [14]. The energy sector has been reported to be financially unsustainable due to its high dependence on externally sourced fossil fuels for power generation [15]. On the other hand, the country aims to achieve 10% renewable energy penetration by 2030 through the Intended National Determined Contribution (INDC), which is unlikely to be achieved due to low incentives [16]. Meanwhile, solar, wind and biomass (including waste) have been identified as attractive alternatives to achieve this target, considering the local abundance of these resources [17–19]. It is estimated that WtE technologies such as anaerobic digestion (AD), incineration and landfill gas to energy (LFGTE) have the potential to supplement 59% of the country's renewable energy target in 2030 [20]. Ironically, the waste management sector in Ghana is plagued by the poor implementation of sustainable waste treatment options, with an estimated 931 tons of waste sent to landfilling daily from the capital and adjoining towns alone [21].

Several WtE studies have been carried out in Ghana, mostly incorporating techno-economic assessments. Existing and operational WtE plants include the Safi Sana and HPW plants, which incorporate AD and solar PV technologies [22]. It is clear that such projects require the strict assessment of impacts, including environmental impacts and the incorporation of risk mitigation measures, to avoid adverse impacts [23]. Published studies have considered the potential for establishing WtE plants under various settings in Ghana such as rural electrification [24], university sewage treatment [25], urban municipalities [26–28], incineration at the national scale [29] and landfill sites [30]. All the studies focused on one key product, electricity to grid, and concluded on technical feasibility, considering varied conditions.

The German government has recently supported the establishment of a pilot WtE plant in Ghana in Gyankobaa in the Ashanti Region. It has been projected to generate 810 MWh of electricity per annum with other associated fuels such as RDF, plastic pellets and compost. The technical model of the plant, which involves a hybrid of solar PV, AD and plastic pyrolysis, has been proven to have the potential to be replicable in other parts of the country [31]. A detailed techno-economic assessment of this case study plant has been previously conducted and published. The results indicate that a standalone WtE plant relying only on grid connected electricity is not feasible. A business case, which involves the strong integration of solar PV at a larger capacity than the current plant, is presented [32]. In an assessment of product models, it was concluded that the conversion of generated energy into fuels and materials such as green hydrogen, biomethane, RDF and compost will potentially make the plant more feasible [33]. These studies are intended to strategically direct the business model when the plant becomes fully operational. One important element of business models is the environmental impact. Also, at the national level, WtE is generally expected to generate negative environmental burdens compared to energy generation from fossil fuels or the landfilling of waste. A research gap is therefore identified for the environmental assessment of WtE using the hybrid WtE plant as a case study.

The aim of this study is therefore to investigate the potential avoided environmental burden from the pilot plant compared to the current landfilling alternative. Also, this study will conduct a comparative assessment of the environmental performance of the subsystems at the plant. It is expected that the results will enhance investment from both the private and public sectors to support decision-making in waste management projects.

The most standardized environmental assessment technique for WtE systems is the Life Cycle Assessment (LCA) [34]. LCA is a well-developed environmental management technique used to assess the environmental impacts of technologies, scenarios or systems. The ISO 14040 and 14044:2006 provides principles and detailed guidelines for conducting an LCA [35,36]. For WtE systems, the most commonly used LCA methods are the CML impact assessment series, IPCC, ReCiPe and Eco-indicator. Global Warming Potential (GWP) is the most analyzed impact/damage category. Generally, landfilling has the highest GHG emissions compared to WtE technologies [37]. The CML method is one of the earliest and most established LCA methods and focuses on individual impact categories. IPCC is primarily considered in assessing climate change impacts such as GWP. ReCiPe, on the other hand, covers a wide range of impact categories, which is relevant for such a study [38].

2. Materials and Methods

2.1. Study Scope and System Boundaries

The study focuses on the environmental assessment of multiple technologies at a pilot WtE plant located at Atwima Nwabiagya Municipality in Ghana. The Business-as-Usual (BAU) is the current waste management model where municipal waste is collected, transported and disposed at the Kyereso landfill, which is basically an unengineered landfill/dumpsite [39]. This study uses the LCA methodology, which follows four major stages, as recommended by ISO 14040 (1997): definition of scope, inventory development, impact assessment and interpretation of results [35]. Figure 1 shows the system boundary and the process flow within the plant beginning from the gate of the treatment facility/dumpsite to the end of the treatment. Although the plant also has the capacity to treat other forms of waste through recycling and composting, this study focuses on the environmental aspects of energy/fuel production [33].

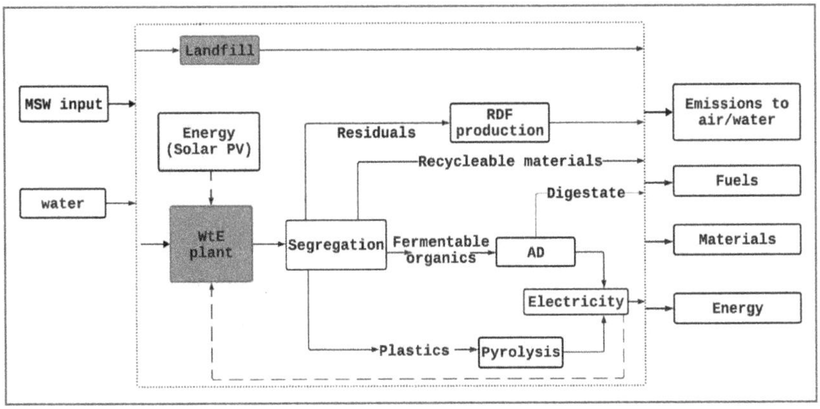

Figure 1. System boundary, technical framework and the flows for each product system for the WtE plant.

2.2. Functional Unit

Based on the goal of the study, the functional unit (FU) considered is one ton of mixed MSW waste received at the treatment facility. Since the capacity of the WtE plant is 50 t/day, it is compared with the same quantity at the landfill site over an active lifespan of 20 years.

2.3. Waste Composition and Segregation

Mixed MSW arriving at the plant or landfill is estimated to have the following characteristics according to a related study within the same municipality [39]: fermentables (37.8%), plastics (17.1%), paper (1.2%), wood (1.5%), cardboards (3.2%), metals (1.2%), glass (1.3%), textiles (4.4%), special wastes (4.2%), inerts/fine ores (16.8%) and miscellaneous materials including e-waste, sanitary materials and household repair debris (11.3%). The first stage of treatment is the four-stage manual and mechanical sorting system, similar to what is used in other Material Recovery Facilities (MRF) in Ghana, such as the Kumasi Compost and Recycling Plant (KCARP) [40,41]. After this, the various fractions are transferred to the respective processes, as shown in Figure 1.

2.4. Life Cycle Inventory Analysis (LCIA)

LCIA involves stock-taking of the inputs and outputs of materials, emissions and energy from each process within the boundaries of the product system [37]. Sources of data are primary data obtained from the plant, other related studies and calculated values based on literature. Elementary flows for emissions were estimated using emission factors from sources such as [42,43]. Pollutant emissions focus on greenhouse gases (CH_4, CO_2, N_2O) and acidic pollutants (SO_2, NO_x) [44]. The inputs and outputs of materials and energy for each process within the WtE plant were summarized in the following sections.

2.4.1. Business-As-Usual (BAU)

In BAU, mixed waste is disposed of at the dumpsite without any treatment. Landfill Gas Emissions Model (LandGEM) version 3.03 was used to estimate the emissions from 50 t/day MSW deposited at the Kyereso dumpsite [45]. The LandGEM software employs a first-order decomposition rate equation, as described in [43]. Localized values were calculated for the methane generation rate and the potential methane generation capacity to replace the default values (Table 1).

Table 1. Key input variables for landfill gas modelling in LandGEM.

Input Variable	Value	Units	Details	Reference
Precipitation	1318.34	mm/yr	Regional annual average for Ashanti Region (1990–2021)	[46]
Methane correction factor	0.8	-	Default for deep, unmanaged landfills	[47]
Temperature	26.98	°C	Regional annual average for Ashanti Region (1990–2021)	[46]
Methane fraction in landfill gas	50	%	Default	[47]

Landfill gas is generally a mixture of two major components (CH_4 and CO_2) and several trace gases. LandGEM generates default outputs for Total LFG, CH_4, CO_2 and NMOC [45]. The GWP of landfill gas for 100-year horizon (GWP_{100}) in kgCO_2eq/yr is computed as the sum of GWP of GHGs. According to IPCC, GWP_{100} for CO_2 (GWP_{100}) and non-fossil methane (GWP_{CH4}) are 1 and 27, respectively [48]. Not all methane generated in a landfill is directly emitted into the atmosphere. Losses occur through the surface cover, methane oxidation in cover soils and lateral migration or losses during storage, which is assumed to be 10% of the generated methane. Thus, mass uncontrolled methane emissions (UMP_{CH4}) from landfills without Gas Collection and Control Systems (GCCS) and subsequently (GWP_{100}) is estimated by [49]

$$UM_{CH4} = M_{CH4} \times 0.9 \tag{1}$$

$$GWP_{100} = (UM_{CH4} \times GWP_{CH4}) + (M_{CO2} \times GWP_{CO2}) \tag{2}$$

2.4.2. Anaerobic Digestion (AD)

The inventory for this study follows a similar assessment in a related publication [50]. Electricity generated is computed based on [51,52]. The source of energy for the engines is the solar PV system, while the AD receives recycled heat energy from the CHP. Sources of emission from the AD system are mainly from digester and the CHP engine. Methane losses from the digester are estimated to be 5%, and CO_2 is ignored because it is of biogenic origin, which has insignificant environmental impact [44]. Emissions from the CHP engine are estimated using the emission factor for GHG, acidifying pollutants are computed [53] and the inventory is presented in Table 2.

Table 2. Inventory for the AD process.

Flow Direction	Parameter	Units	Value
	1. Bio-digestion		
Inputs	Organic waste	tons	1.00
	Electricity (solar PV)	kWh	38.45
	Heat (CHP)	kWh	237.80
	Water (groundwater)	m^3	0.37
Outputs	Methane	m^3	62.14
	Digestate	tons	0.88
	Methane (emissions)	kg	1.34

Table 2. *Cont.*

Flow Direction	Parameter	Units	Value
	2. Cogeneration		
Input	Methane	m^3	62.14
Outputs	Electricity	kWh	216.20
	Avoided heat	kWh	237.80
	Carbon monoxide	kg	5.16×10^{-9}
	N_2O	kg	0.0012
	NO_x	kg	0.16
	SO_2	kg	0.02

2.4.3. Pyrolysis

The pyrolysis system has the capacity to process 30–50 kg/h of mixed single-use plastics (SUPs). The process is described in a recent related publication [54]. It begins with the pretreatment stage involving shredding, washing and drying, which results in a 5% loss of weight due to the elimination of impurities and moisture. The feedstock is then introduced into the reactor. The main process outputs (wt%) are bio-oil (79.38%), gas (15.22%) and char (5.4%). The bio-oil and gas are combusted in separate CHP engines to generate electricity. Electricity from bio-oil is directed for grid supply, while electricity from the gas is internally recycled to provide the process energy for the reactor. The energy output (kWh/ton) of input waste from bio-oil estimation is based on [55]. Other outputs from the system are the wastewater from washing and emissions from the CHP engine. The main source of emissions is the CHP since the reactor is specifically designed to operate under airtight conditions. Emission factors from [55] are used. The inventory analysis is summarized in Table 3 below.

Table 3. Inventory for plastic pyrolysis process.

Flow Direction	Parameter	Units	Value
	1. Pretreatment		
Inputs	Waste–plastic mixture	tons	1.00
	Water (well)	m^3	0.15
	Electricity	kWh	1.00×10^{-5}
Outputs	Waste–plastic mixture	tons	0.95
	Wastewater	m^3	0.165
	2. Pyrolysis		
Inputs	Waste–plastic mixture	tons	0.95
	Electricity (solar PV)	kWh	0.009
Outputs	Bio-oil	tons	0.75
	Char	tons	0.05
	Syngas	m^3	0.13
	3. Cogeneration		
Inputs	Bio-oil	tons	0.75

Table 3. *Cont.*

Flow Direction	Parameter	Units	Value
Outputs	Electricity	kWh	0.0033
	Heat	kWh	0.0093
	Carbon dioxide (non-fossil)	kg	2.15
	N_2O	kg	0.0003
	Methane (non-fossil)	kg	0.0001
	NO_2	kg	0.0039
	SO_2	kg	0.0209

2.4.4. RDF Production

The conversion of residuals to RDF involves three main stages: separation to remove complex waste and recyclables such as glass, metals or e-waste, cutting/shredding to homogenize waste into required particle sizes and finally compacting to reduce volume for easier transportation [56]. Energy demand for these processes has been computed using the power ratings of the equipment onsite and the 8 h of daily operating. Energy is sourced from internal solar PV system; thus, pollutant emissions are negligible. The inventory for the RDF process is specified below in Table 4.

Table 4. Inventory for RDF production.

Flow Direction	Parameter	Units	Value
Inputs	Residual materials	tons	1.00
	Electricity (solar PV)	kWh	51.58
Outputs	RDF pellets	tons	0.95
	Heat (loss)	kWh	33.0
	Trace particles	kg	0.05

2.5. Life Cycle Impact Assessment (LCA)

The OpenLCA 2.3.1 software was used to model each process following the guidelines of the ISO 14040/14044:2006 [35,36]. According to ISO 14042 [57], LCA process follows three mandatory steps: selection of characterization methods, classification and characterization. Additional optional steps are also recommended: normalization, grouping, weighing and data quality analysis [57].

The following steps were therefore followed with the aid of the OpenLCA software. Firstly, the impact categories, indicators and characterization models were selected. ReCiPe2016 midpoint (H) impact assessment method was used, which identifies 17 midpoint characterization factors that are linked to three endpoint protection areas [58,59]. This study focuses on ten midpoint impact categories due to their relevance to the scope, as is done in similar studies and this is represented in Table 5 [60].

The second stage involves assignment of LCIA results (classification). In OpenLCA, the inventory results were input for each process within the system boundary. The Ecoinvent Database v3.5 provided secondary data in the calculation of impacts. The category indicator results were then calculated by the software. Normalization is recommended in LCIA to assess the magnitude of the category indicator results relative to a reference value. This was carried out based on the World 2010 normalization method [61].

Table 5. Selected impact categories and their relevance to the study.

Impact Categories	Relevance to the Study	Link to Endpoint Impact	Unit
GWP	Risk to human health and ecosystem	Human health	$kgCO_2eq$
PMFP	Increase in respiratory diseases		kg $PM_{2.5}$ eq
HTP/C	Increased risk of carcinogenic diseases		kg 1,4-DCB
HTP/NC	Increase in other diseases		kg 1,4-DCB
WCP	Reduced access for consumptive uses		m^3
TETP	Risk to terrestrial spp.	Ecosystems	kg 1,4-DCB
FETP	Risk to freshwater spp.		kg 1,4-DCB
AP	Risk to terrestrial spp.		kg SO_2 eq
METP	Risk to marine life		kg 1,4-DCB
FFP	Increased scarcity and cost of extraction	Resource availability	kg oil eq

2.6. Interpretation of Environmental Impacts

The interpretation of results of LCIA was conducted according to ISO 14044:2006 [35]. This involves evaluation of results through checks for consistency, completeness and analysis of results. Results were then assessed through an iterative process where significant issues were searched, checked and analyzed. When significant issues were detected, corrections were made, and the simulation was re-run [57]. The characterization and normalization results were then imported from OpenLCA to be analyzed with Microsoft Excel. The characterization results were adjusted to percentage to assess the relative contribution of various impact categories. However, uncertainty results were exported from the OpenLCA software and analyzed using Python (Jupyter Notebook).

Interpretation of results includes sensitivity analysis and uncertainty analysis. This is done to check for the robustness, consistency and completeness of the data used in the analysis. A Monte Carlo simulation was conducted using 1000 iterations to perform the uncertainty analysis of the results. The data quality pedigree matrix was used to assign uncertainty values to the dataset used during the life cycle inventory stage [62].

3. Results and Discussion

3.1. Impact Assessment of Subsystems

This study focuses on an assessment of four subsystems at the WtE plant. Figure 2 shows the relative environmental impact of the various subsystems compared to the impact factors. Generally, pyrolysis shows a high negative impact across all impact categories, representing a high avoided environmental burden. The other subsystems show a positive impact, but one that is not so high. Segregation and RDF have positive values showing environmental burdens. AD, on the other hand, shows a generally neutral environmental impact. A neutral GWP was also recorded in a similar study, which assessed an AD system using different energy consumption scenarios [50]. In another study comparing AD with incineration, which is also a thermal conversion technology, AD showed near-neutral impacts, while incineration showed a high negative impact.

Figure 3, on the other hand, shows the percentage comparison of impact categories to the various subsystems. Pyrolysis for instance shows better environmental impact due to the production of new products such as char and digestate in the AD process compared to the negative impact the processes themselves generate [63]. Pyrolysis oil can also be considered a fuel similar to diesel [64]. Sources of environmental impact for the AD system include methane leakages from the biodigesters and the combustion of methane in the internal combustion engine. The highest environmental impact on soil and water in terms of toxicity may potentially originate from the digestate [65]. Overall, the magnitude of environmental burdens of the processes can be summarized as follows: RDF > segregation >AD > pyrolysis. RDF production and segregation, on the other hand, release process emissions, especially particulates [66].

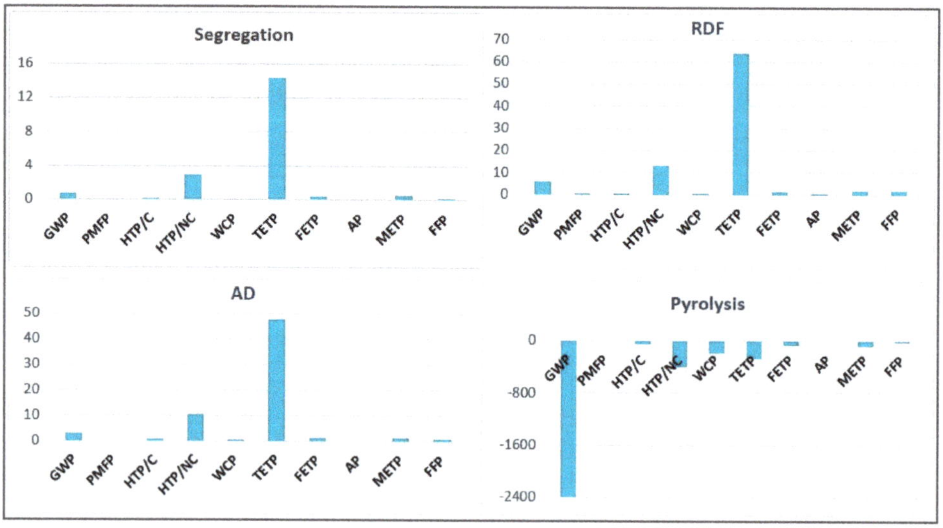

Figure 2. The impact of subsystems according to impact categories.

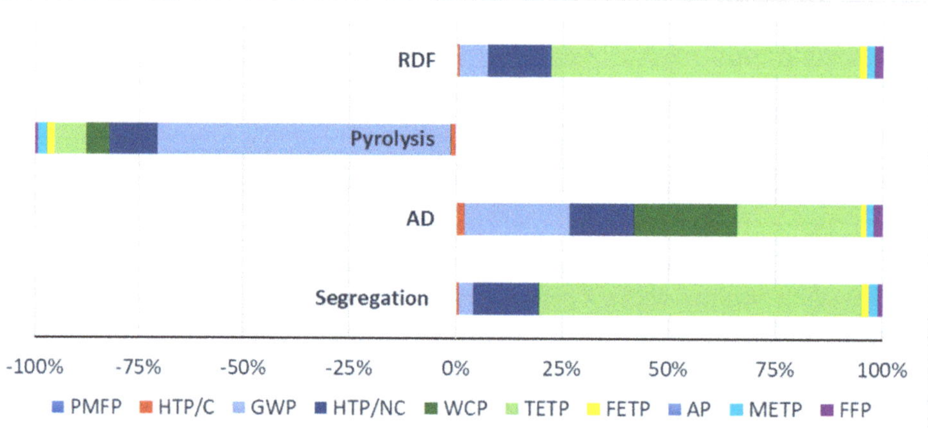

Figure 3. Percentage contribution of the relative environmental impact of subsystems at the WtE plant.

3.2. Impact Assessment According to Impact Categories

Figure 4 shows the net contribution of all processes to the impact categories. The net emissions represent a balance of positive (burdens) and negative (offsets). This highlights the actual burden after accounting for reductions. The total net GWP for the WtE plant is estimated at -2.382 tCO_2eq/FU.

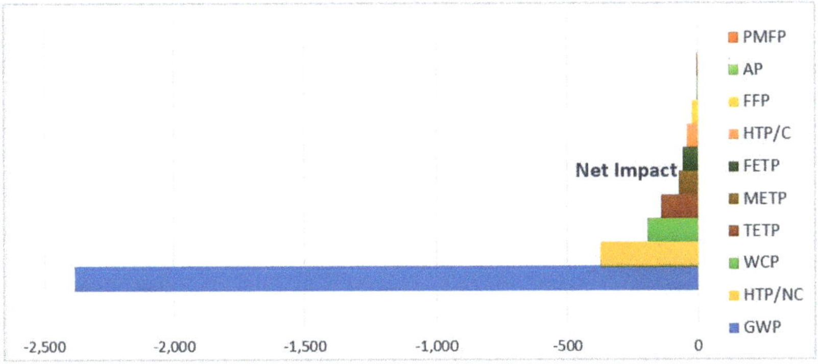

Figure 4. Net (cumulative) impact of the four subsystems according to impact categories.

Figure 5 shows the relative contribution of impact categories based on normalized results. It corroborates the results of Figure 3, where the dominant subsystems are pyrolysis and AD.

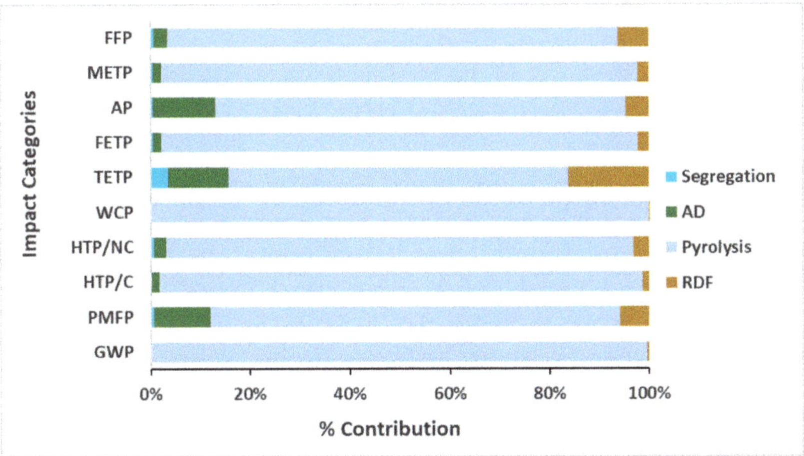

Figure 5. Percentage contribution of impact categories and processes based on the normalized values.

3.2.1. GWP

The GWP is strikingly the highest impact category with the highest avoided burden compared to the other impact categories. GWP is an important metric for environmental emissions and climate change. Pyrolysis had the lowest negative GWP value of -2.3 tCO_2eq/FU, indicating high carbon savings (Figure 3). Net emissions are important in determining the burdens.

On the other hand, the GWP_{100} is presented in Figure 6 for the same 50 t/day of waste disposed of at the landfill site. This is estimated to be 1.133 tCO_2eq/ton of waste-in-place by the end of the lifespan of the landfill. By the time of closure, the landfill would have emitted 4409.56 tons of uncontrolled methane into the atmosphere, although GHG emissions would continue for another 100 years. The total GWP over the lifespan is estimated to be 413,564.21 tCO_2eq since the landfill under consideration has no gas control infrastructure. The GWP of the landfill is based on uncontrolled methane emissions and non-biogenic CO_2.

The WtE plant reduces emissions by 3.52 tCO_2eq/FU, representing the avoided emissions. The WtE plant is expected to treat 365,000 tons of MSW over its 20-year lifespan. With the calculated net carbon reductions, this could result in carbon savings of an estimated 1.28 million tCO_2eq. This may be associated directly or indirectly with the financial feasibility of the WtE plant if such reductions are adequately verified [67]. The net emissions show that the WtE plant reduces emissions by 310%. This may, however, be subject to differences if long-term monitoring is conducted over the entire lifespan of the WtE plant compared to the long-term considerations made for the landfill. In a comparative study in Nigeria, a hybrid AD/incineration system and an LFGTE had the potential to release 75–93% and 400% less, respectively, compared to conventional landfilling [44]. A similar result was obtained in a study in the US where landfilling released 75.24% higher GHG compared to gasification systems [68]. In another study in Australia, landfilling recorded the highest environmental burden in 9 of 10 midpoint impact factors assessed, indicating that landfilling is the least environmentally friendly option [60]. Since the source of GHG at landfill sites is the biodegradable components and non-biogenic CO_2, the biochemical process can be compared to the AD system at the WtE plant, which has a GWP of 0.0032 tCO_2eq/FU. Although recent studies in Ghana have estimated that LFGTE is a feasible option, AD is even more efficient, with a lower GWP compared to LFGTE [28]. This is attributed to higher losses due to landfill cover of 10% and gas capture efficiency of 75% compared to 5% losses from the biodigesters [43,69].

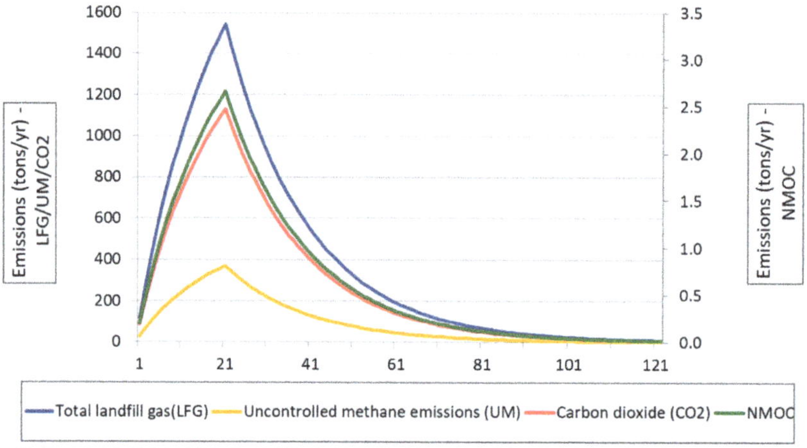

Figure 6. Landfill gas emissions based on LandGEM representing a lifespan of 20 years.

The carbon intensity of WtE technologies is linked to the source of energy [60]. Sources of emissions from the WtE plant include engines and leakages of methane from AD. Energy for the RDF and segregation equipment is sourced from internally generated electricity, solar PV systems and biogas, which has a low emission profile. Also, heat energy losses from the RDF and segregation systems add up to environmental burdens. Assuming electricity for RDF pelletizing is sourced from fossil sources (diesel), avoided emissions can be estimated using emission factors [70]. With an energy demand of 452 kWh/day at full capacity, the GWP of a diesel-fired RDF pelletizer machine is estimated at 13.84 $kgCO_2eq/ton$, which, in this study, can be considered an avoided emission. RDF produced using cleaner energy reduces the overall environmental footprint of the energy produced from it. In addition, combustibles and other non-biodegradable components of waste are more efficiently treated with thermal energy conversion technologies such as gasification and pyrolysis rather than storage in land space at a landfill.

Studies have shown that the pretreatment and dewatering of waste significantly reduce emissions [13]. The low GWP recorded for pyrolysis may be a combination of the source of energy (solar PV and syngas), incorporating pretreatment and potential carbon sequestration from char. Char produced from pyrolysis possess has the capacity to sequester carbon, leading to the negative release of carbon [63,71]. Additionally, bio-oil and syngas replace fossil fuels for the production of electricity, thus further accounting for the avoided release of stored/fossil carbon, which could potentially end up in the atmosphere.

3.2.2. Toxicity Potentials

Toxicity represents the environmental fate and accumulation of a chemical in the food chain and its associated effects [58]. The toxicity potentials analyzed in this study are Human Toxicity Potentials (HTP) and Ecotoxicity Potentials (EP) in terms of Terrestrial (TETP), Freshwater (FETP) and Marine Ecotoxicity (METP). HTP is presented in two parts by the ReCiPe method: carcinogenic (HTP/C) and non-carcinogenic (HTP/NC). Both are linked to endpoint damage to human health due to an increase in lifetime disease incidence as a result of the intake of substances such as contaminated water or food from contaminated soil [59]. Generally, AD showed a high environmental burden due to the release of digestate. Digestate is planned to be upgraded into a soil amendment product. In the case that high quantities of heavy metals and contaminants are present, digestate may potentially have an alarming disease-causing impact on humans [72].

HTP is a very important potential risk that should be further investigated due to the mixed nature of the waste. High concentrations of heavy metals such as zinc, copper, lead, chromium and nickel have been detected in waste from the same region [73]. Heavy metals and chlorines in RDF must also be avoided due to their potential risk when gasified at industrial plants [41]. HTP/NC recorded the second-highest avoided impact for pyrolysis due to the production of char. Char has the potential to adsorb such pollutants and has been effectively used in soil/wastewater remediation. This can, however, be substantial only if the char is utilized for this purpose and the sorption capacity is improved through techniques such as grinding [74].

The ecotoxicological effect measures the Potentially Affected Fraction (PAF) or the change in Potentially Disappeared Fractions (PDF) of ecological species due to exposure to unacceptable pollutant concentrations [59,75]. TETP, FETP and METP measure the impact on land-dependent, freshwater or marine organisms, respectively. EPs are notably high for AD, and the lowest avoided burdens for pyrolysis are in the order of TETP > METP > FETP. Digestate and methane losses from the AD are a major source of ET to ecological systems. Methane emissions from AD are a potent GHG, which can potentially have a greater negative impact on PAF and PDF. Marine ecosystems have been shown to be highly sensitive to changes in global climate changes from GHG. Also, the bioaccumulation of chemicals such as heavy metals in species and the food chain as a result of digestate is a source of toxicity to land, freshwater and marine environments [76]. Since this is a potential environmental burden, it is necessary to carefully monitor and control the use of digestate. In the case of pyrolysis, avoided emissions due to the substitution of internally generated energy and the avoided use of grid-powered electricity from fossil sources account for high negative impacts.

3.2.3. Water Consumption Potential

Water is an important resource for the AD and pyrolysis process that is sourced from onsite groundwater (borehole). Water consumption/depletion is assessed as a negative impact when the abstracted water evaporates or is used in a production process far from the point of abstraction, which reduces human access to the resource at the point of abstraction [61]. The pretreatment stage of pyrolysis consumes water, which is a source of environmental burden, while the RDF is insignificant. However, in this LCA model, both pyrolytic water and liquid digestate can be reused/recycled which reduces the demand for freshwater. However, the burden of AD is potentially higher if digestate is transported to another site and cannot be reused.

3.2.4. Particulate Matter Formation Potential (PMFP)

PMFP measures the endpoint impact of fine particles/aerosols with a diameter of less than 2.5 µm ($PM_{2.5}$) on human health. These particles are a complex mixture of organic and inorganic substances, which has a potential effect on the cardio-pulmonary system and mortality after long-term exposure [58,59]. This is a very important parameter because the WtE plant under study requires human operations for at least 40 h a week. Although all the PM values recorded indicated a low impact (−0.21 to 0.03 kg $PM_{2.5}$ eq), the emission factors used are only limited to the engine. Data on site-specific operational emissions are not available. Air quality measurements are highly recommended for future studies to determine the exact potential PM emissions at various waste treatment plants across Ghana and the corresponding impact on humans and the ecosystem. In a previous study to assess the concentrations of PM in the breathing zones of informal e-waste workers in Ghana, the PM 2.5, 2.5–10 concentrations were 2–3 times above a control group and also above the WHO air quality standards. The study recommended the assessment of the impact of long-term exposure [77]. In the long term, PM may also impact the ecosystem of the area, which is predominantly agricultural, thus requiring further investigation.

3.2.5. Terrestrial Acidification Potential

Acidification Potential (AP) measures the potential impact of acidifying substances such as NO_x, NH_3 and SO_2 on soils and their dependent species [58]. The values of AP were generally low, with a net avoided impact of 0.31 kg SO_2 eq, which is linked to the low emissions of these pollutants directly from the systems. The use of electricity from the solar PV system contributes to avoided emissions.

3.2.6. Fossil Resource Scarcity

Fossil resources are subject to depletion apart from their high climate footprint once combusted. The increased substitution of fossil energy for renewable energy is important. The energy mix in Ghana is composed of 68% generation from thermal combustion of fossils [78].

3.3. *Uncertainty Analysis*

Each box plot shows the distribution of uncertainty values for each impact category (Figure 7). Uncertainty in the LCA results highlights the variability and spread of data within each impact category. High uncertainty is recorded for HTP/NC, HTP/C and FETP, which indicates a high level of uncertainty in the input data, suggesting that the impact assessments for this category are highly variable and potentially influenced by many factors. FETP, METP and GWP, on the other hand, show moderate uncertainty with several outliers, indicating some level of variability. FFP, WCP, PMFP and TETP show relatively low levels of uncertainty, with smaller IQRs and fewer outliers. This suggests that the data for these impact categories are more consistent and reliable. Categories with low uncertainty provide more confidence in the results and can be used more reliably in environmental impact assessments and sustainability planning. In summary, the uncertainty plot provides valuable insights into the reliability and variability of LCA results across different impact categories.

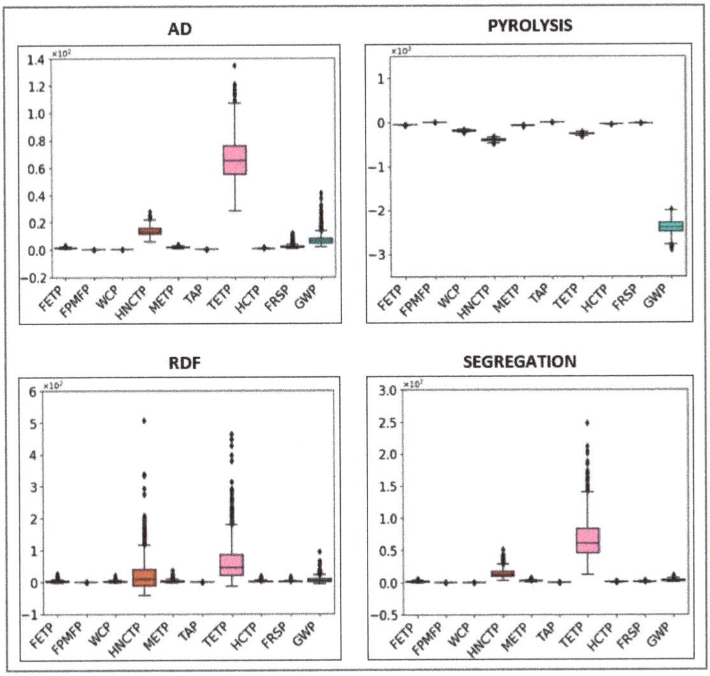

Figure 7. Uncertainty plot for the processes.

3.4. Other Potential Impacts

Other potential impacts include socio-economic and public health improvement in rural communities. According to the 2021 census data, 36.8% of rural households have no waste collection, while 57.3% use open dumps with potential health impacts. With the plant located within a municipality with 64.7% rural households, the proximity of a WtE plant will likely reduce the cost of waste collection services and encourage more residents to pay for WtE services. Also, the plant currently provides employment and income security to many residents, while compost may support agriculture at a lower cost since this is the most prevalent economic activity within the town [39]. Additionally, a buffer zone has been created around the plant site within a 1 km distance from the closest households for protection against fire outbreaks, odor and noise.

The promotion of WtE in such settings has climate mitigation impacts at both national and international levels. Ghana's NDCs aim to reduce GHGs by 64 $MtCO_2eq$ and create one million decent and green jobs by 2030. The expansion of WtE systems within a circular economy concept will support the attainment of such national targets with global impacts [79]. As of 2022, Ghana's national energy mix was made up of 1.9%, 70% and 28.1% renewable energy, fossil fuels and hydro, respectively. The sustained substitution of WtE with fossil fuels is important to reaching the set targets and protecting global resources.

3.5. Limitations and Recommendations

One major limitation of this comparative study is the lack of an ongoing monitoring system with a recent, detailed localized database on waste generation and treatment pathways across cities [38]. The Ecoinvent Database, for instance, does not contain some locally specific data for Ghana, especially for the plant site in Atwima Nwabiagya. Where locally specific data were not available, data from the rest of the world were used during the modeling in OpenLCA. This provides a limitation to the results, which may have a substantial deviation from the site-specific environmental impacts.

It is strongly proposed that a country-specific database be developed for Ghana by the government. Since Ghana is a signatory to the Paris Agreement, such a database becomes a technical resource that will enhance the accuracy of climate monitoring and auditing. The development of a database can be facilitated through the following:

- Dedicating funding to this;
- Dedicating human resources for national coordination, such as incorporating it into the Ghana Carbon Market Office;
- Future research focused on carrying out detailed and up-to-date studies on aspects such as national waste composition similar to that carried out in 2014 [80]. Other important areas of concern are emission factors and operation-related emissions from waste treatment activities;
- Facilitating training programs for waste management companies, research institutions, agencies and decision-makers to enhance monitoring and compliance.

The uncertainty in LCA results is crucial for decision-making and policy development. High uncertainty in certain impact categories, such as HTP, indicates that more research and data collection are needed to improve the reliability of these assessments. Another major source of concern is the potential impact of mixed waste on the HTP as well as ETP through bioaccumulation in terrestrial, freshwater or marine species. The waste segregation model for the socio-cultural setting in Ghana needs to be studied and implemented to avoid such impacts.

Life cycle costing is highly recommended in future studies to incorporate the cost of environmental impact in decision-making. Carbon trading can potentially stimulate more emission reduction developmental projects in local communities [81]. It is important for WtE plants to obtain international certifications such as ISO 14001, which may return positive financial benefits for the case study plant, although it requires adherence to strict requirements. Environmental assessment studies such as this to support carbon financing mechanisms that require international certification [82].

4. Conclusions

This study conducted an environmental assessment of various processes within a pilot waste-to-energy plant in Ghana. The Life Cycle Assessment was conducted using OpenLCA software version 2.3.1. Landfill emissions were considered the baseline for estimating potential emission reduction. The results indicate that the GWP has the highest potential environmental impact followed by the toxicity potentials, mainly high Human Toxicity Potentials. Pyrolysis showed high avoided impacts, while AD showed near-zero impacts. The major attributing factors are the potential sequestering of the produced char end-product and the input of renewable energy. Overall, pyrolysis technology is the most environmentally friendly. This is relevant because a comparative assessment of the GWP of landfilling showed that the WtE plant has the potential to reduce emissions by 3.52 tCO_2eq/FU, resulting in a total of 1.28 million tCO_2eq over its 20-year lifespan based on the expected capacity. This can directly or indirectly potentially improve the business case of the plant. Uncertainty assessment also points to the need for a national database with relevant localized data to improve the reliability of future studies. Also, although AD showed a near-neutral environmental impact, the release of digestate may have high human and ecotoxicity impacts when substances such as heavy metals are present. This can be a potential risk in Ghana, where the source segregation of waste is largely non-existent. As such, the implementation of AD technology must be associated with strict regulations and monitoring systems on the management of digestate. Based on these findings, it is recommended that pyrolysis of end-of-life and non-recyclable plastics be incorporated strongly into the waste-to-energy models. Decision makers have a role to play by introducing innovative approaches to encourage source segregation. It is also recommended to use strict guidelines and monitoring systems to ensure that emission control is implemented by waste-to-energy plants according to environmental compliance. With the siting of the WtE plant in a rural community, the municipal authorities are encouraged to take advantage of this to stimulate community participation in environmental protection by providing incentives such as lower waste collection costs. In the long term, this WtE plant may contribute relevant data for improving decision-making based on environmental assessments in Ghana. It may also provide a basis for carbon trading and enhance the financing profile of the WtE model studied. Future studies may be extended to include other processes such as composting and recycling.

Author Contributions: Conceptualization: E.A.A., F.B.A., F.K. and M.M.; methodology: E.A.A., F.B.A., F.K., S.N. and M.M.; Formal analysis: E.A.A., T.B. and M.M.; writing—original draft preparation: E.A.A.; Writing—review and editing: T.B. and M.M.; visualization: T.B.; supervision, S.N. and F.K. All authors have read and agreed to the published version of the manuscript.

Funding: This work was supported under the project "Waste to Energy: Hybrid Energy from Waste as Sustainable Solution for Ghana", funded by the German Federal Ministry for Education and Research (BMBF), 03SF0591A, and the Open Access Publication Fund of the University of Rostock.

Data Availability Statement: Data are contained in the article.

Conflicts of Interest: Author Theophilus Baidoo was employed by the company Zeal Environmental Technologies and Author Satyanarayana Narra was employed by the company German Biomass Research Centre gGmbH. The remaining authors declare that the research was conducted in the absence of any commercial or financial relationships that could be construed as a potential conflict of interest.

References

1. United Nations. *The Sustainable Development Goals Report 2024*; United Nations: New York, NY, USA, 2024.
2. Ndam, S.; Touikoue, A.F.; Chenal, J.; Baraka Munyaka, J.-C.; Kemajou, A.; Kouomoun, A. Urban Governance of Household Waste and Sustainable Development in Sub-Saharan Africa: A Study from Yaoundé (Cameroon). *Waste* **2023**, *1*, 612–630. [CrossRef]
3. International Renewable Energy Agency. *World Energy Transitions Outlook 2023: 1.5 °C Pathway*; International Renewable Energy Agency: Abu Dhabi, United Arab Emirates, 2023; Volume 1.
4. International Energy Agency. *CO2 Emissions in 2023*; International Energy Agency: Paris, France, 2023; Volume 24, p. 22.
5. UNFCC COP29 UN Climate Conference Agrees to Triple Finance to Developing Countries, Protecting Lives and Livelihoods. Available online: https://unfccc.int/news/cop29-un-climate-conference-agrees-to-triple-finance-to-developing-countries-protecting-lives-and (accessed on 7 December 2024).
6. Ibrahim, I.D.; Hamam, Y.; Alayli, Y.; Jamiru, T.; Sadiku, E.R.; Kupolati, W.K.; Ndambuki, J.M.; Eze, A.A. A Review on Africa Energy Supply through Renewable Energy Production: Nigeria, Cameroon, Ghana and South Africa as a Case Study. *Energy Strategy Rev.* **2021**, *38*, 100740. [CrossRef]
7. Williams, P.A.; Narra, S.; Antwi, E.; Quaye, W.; Hagan, E.; Asare, R.; Owusu-Arthur, J.; Ekanthalu, V.S. Review of Barriers to Effective Implementation of Waste and Energy Management Policies in Ghana: Implications for the Promotion of Waste-to-Energy Technologies. *Waste* **2023**, *1*, 313–332. [CrossRef]
8. International Energy Agency; International Renewable Energy Agency; United Nations Statistics Division; World Bank; World Health Organization. *Tracking SDG 7: The Energy Progress Report*; The World Bank: Washington, DC, USA, 2021.
9. CEWEP Waste-to-Energy Cycle. Available online: https://www.cewep.eu/waste-to-energy-cycle-2021/ (accessed on 7 December 2024).
10. Khan, I.; Chowdhury, S.; Techato, K. Waste to Energy in Developing Countries-A Rapid Review: Opportunities, Challenges, and Policies in Selected Countries of Sub-Saharan Africa and South Asia towards Sustainability. *Sustainability* **2022**, *14*, 3740. [CrossRef]
11. Bakkaloglu, S.; Cooper, J.; Hawkes, A. Life Cycle Environmental Impact Assessment of Methane Emissions from the Biowaste Management Strategy of the United Kingdom: Towards Net Zero Emissions. *J. Clean. Prod.* **2022**, *376*, 134229. [CrossRef]

12. El Ibrahimi, M.; Khay, I.; El Maakoul, A.; Bakhouya, M. Techno-Economic and Environmental Assessment of Anaerobic Co-Digestion Plants under Different Energy Scenarios: A Case Study in Morocco. *Energy Convers. Manag.* **2021**, *245*, 114553. [CrossRef]
13. Yaman, C. Investigation of Greenhouse Gas Emissions and Energy Recovery Potential from Municipal Solid Waste Management Practices. *Environ. Dev.* **2020**, *33*, 100484. [CrossRef]
14. Environmental Protection Agency. *Ghana's Fifth National Greenhouse Gas Inventory*; Environmental Protection Agency: Accra, Ghana, 2022.
15. Ministry of Energy. *Energy Sector Recovery Program*; Ministry of Energy: Accra, Ghana, 2019.
16. Afful-Dadzie, A.; Mallett, A.; Afful-Dadzie, E. The Challenge of Energy Transition in the Global South: The Case of Electricity Generation Planning in Ghana. *Renew. Sustain. Energy Rev.* **2020**, *126*, 109830. [CrossRef]
17. Odoi-Yorke, F.; Adu, T.F.; Ampimah, B.C.; Atepor, L. Techno-Economic Assessment of a Utility-Scale Wind Power Plant in Ghana. *Energy Convers. Manag. X* **2023**, *18*, 100375. [CrossRef]
18. Kemausuor, F.; Kamp, A.; Thomsen, S.T.; Bensah, E.C.; Stergård, H. Assessment of Biomass Residue Availability and Bioenergy Yields in Ghana. *Resour. Conserv. Recycl.* **2014**, *86*, 28–37. [CrossRef]
19. Afonaa-Mensah, S.; Majeed, I.B.; Odoi-Yorke, F. Advancing the Adoption of Net Metering: An Economic Assessment of Grid-Tied Solar Photovoltaic Systems in Urban Homes in Ghana. *Int. J. Energy Res.* **2024**, *2024*, 9919076. [CrossRef]
20. Darmey, J.; Ahiekpor, J.C.; Narra, S.; Achaw, O.W.; Ansah, H.F. Municipal Solid Waste Generation Trend and Bioenergy Recovery Potential: A Review. *Energies* **2023**, *16*, 7753. [CrossRef]
21. Abledu, E.S.; Amfo-Otu, R. Contribution of Informal Sector Recycling Workers to Sustainable Landfill Management: The Case of Kpone Landfill Site in the Greater Accra Region. *Appl. Res. J.* **2019**, *4*, 1–11.
22. Agyenim, F.B.; Dzamboe, P.D.; Mohammed, M.; Bawakyillenuo, S.; Okrofu, R.; Decker, E.; Agyemang, V.K.; Nyarko, E.H. Powering Communities Using Hybrid Solar–Biogas in Ghana, a Feasibility Study. *Environ. Technol. Innov.* **2020**, *19*, 100837. [CrossRef]
23. United Nations Environment Programme. *Africa Waste Management Outlook*; United Nations Environment Programme: Nairobi, Kenya, 2018; Volume 123, ISBN 9789280737042.
24. Odoi-Yorke, F.; Abaase, S.; Zebilila, M.; Atepor, L. Feasibility Analysis of Solar PV/Biogas Hybrid Energy System for Rural Electrification in Ghana. *Cogent Eng.* **2022**, *9*, 2034376. [CrossRef]
25. Mohammed, M.; Egyir, I.S.; Donkor, A.K.; Amoah, P.; Nyarko, S.; Boateng, K.K.; Ziwu, C. Feasibility Study for Biogas Integration into Waste Treatment Plants in Ghana. *Egypt. J. Pet.* **2017**, *26*, 695–703. [CrossRef]
26. Osei-Appiah, N.A.; Dioha, M.O. Techno-Economic Assessment of Waste-To-Energy Technologies in Ghana. *J. Sustain. Energy* **2019**, *10*, 112–121.
27. Cudjoe, D.; Nketiah, E.; Obuobi, B.; Adu-Gyamfi, G.; Adjei, M.; Zhu, B. Forecasting the Potential and Economic Feasibility of Power Generation Using Biogas from Food Waste in Ghana: Evidence from Accra and Kumasi. *Energy* **2021**, *226*, 120342. [CrossRef]
28. Obuobi, B.; Adu-Gyamfi, G.; Adjei, M.; Nketiah, E. Technologies Potential and Economic Viability Analysis of Deriving Electricity from Municipal Solid Waste in Kumasi, Ghana. *Energy Sustain. Dev.* **2022**, *68*, 318–331. [CrossRef]
29. Yakah, N.; Kwarteng, A.A.; Addy, C.; Yirenkyi, M.; Martin, A.; Simons, A. Techno-Economic Assessment of Municipal Solid Waste (MSW) Incineration in Ghana. *Processes* **2024**, *12*, 1286. [CrossRef]
30. Asante, K.; Gyamfi, S.; Amo-boateng, M. Techno-Economic Analysis of Waste-to-Energy with Solar Hybrid: A Case Study from Kumasi, Ghana. *Sol. Compass* **2023**, *6*, 100041. [CrossRef]
31. Armoo, E.A.; Narra, S.; Mohammed, M.; Boahemaa, B.; Beguedou, E.; Kemausuor, F.; Agyenim, F.B. Hybrid Waste-to-Energy Solutions within a Circular Economy Framework Directed towards Sustainable Urban Waste Management in Ghana. *Sustainability* **2024**, *16*, 4976. [CrossRef]
32. Armoo, E.A.; Narra, S.; Mohammed, M.; Beguedou, E.; Kemausuor, F. Business Case for Hybrid Waste to Energy System for Ghana. In *Innovations in Circular Economy and Renewable Energy in Africa*; Narra, M.-M., Narra, S., Eds.; Springer: Cham, Switzerland, 2024; pp. 87–104; ISBN 978-3-031-68330-5.
33. Armoo, E.A.; Mohammed, M.; Narra, S.; Beguedou, E.; Agyenim, F.B.; Kemausuor, F. Achieving Techno-Economic Feasibility for Hybrid Renewable Alternative Fuels. *Energies* **2024**, *17*, 735. [CrossRef]
34. Sharma, H.; Monnier, É.; Mandil, G.; Zwolinski, P.; Colasson, S. Comparison of Environmental Assessment Methodology in Hybrid Energy System Simulation Software. *Procedia CIRP* **2019**, *80*, 221–227. [CrossRef]
35. *ISO 14040*; Environmental Management—Life Cycle Assessment—Principles and Framework. Second Edition. International Organization for Standardization (ISO): Geneva, Switzerland, 2006.
36. *ISO 14044*; Environmental Management—Life Cycle Assessment—Requirements and Guidelines. International Organization for Standardization (ISO): Geneva, Switzerland, 2006.

37. Dastjerdi, B.; Strezov, V.; Rajaeifar, M.A.; Kumar, R.; Behnia, M. A Systematic Review on Life Cycle Assessment of Different Waste to Energy Valorization Technologies. *J. Clean. Prod.* **2021**, *290*, 125747. [CrossRef]
38. Elijah, E.U.; Oluwatobi Ojo, D.; Yakubu, E.A.; Ezeuzo, E.B.; Hampo, C.C.; Abubakar, A. Life Cycle Assessment (LCA) of Solid Waste Management Systems in African Countries: A Systematic Review. *J. Adv. Environ. Health Res.* **2023**, *11*, 194–204. [CrossRef]
39. Mohammed, M.; Dzamboe, P.D.; Rockson, M.A.D.; Ayittah, S.; Narra, S.; Agyenim, F.B.; Joshua, K.; Boakye-Danquah, E.B.; Abdul-samii, R.; Armoo, E.A.; et al. Spatial Distribution and Resource Mobilization of Waste for a Hybrid Waste-to-Energy Plant in Ghana. In *Innovations in Circular Economy and Renewable Energy in Africa*; Narra, M., Narra, S., Eds.; Springer: Cham, Switzerland, 2024; pp. 43–66; ISBN 9783031683305.
40. Kwakye, S.O.; Amuah, E.E.Y.; Ankoma, K.A.; Agyemang, E.B.; Owusu, B.G. Understanding the Performance and Challenges of Solid Waste Management in an Emerging Megacity: Insights from the Developing World. *Environ. Chall.* **2024**, *14*, 100805. [CrossRef]
41. Sarquah, K.; Narra, S.; Beck, G.; Bassey, U.; Antwi, E.; Hartmann, M.; Sarfo, N.; Derkyi, A.; Awafo, E.A.; Nelles, M. Characterization of Municipal Solid Waste and Assessment of Its Potential for Refuse-Derived Fuel (RDF) Valorization. *Energies* **2023**, *16*, 200. [CrossRef]
42. United States Environmental Protection Agency. AP 42, Fifth Edition, Volume I Chapter 2: Solid Waste Disposal, 2.4 Municipal Solid Waste Landfills. Available online: https://www.epa.gov/air-emissions-factors-and-quantification/ap-42-fifth-edition-volume-i-chapter-2-solid-waste-0 (accessed on 4 November 2024).
43. United States Environmental Protection Agency. *Background Information Document for Updating AP42 Section 2.4 for Estimating Emissions from Municipal Solid Waste Landfills*; United States Environmental Protection Agency: Washington, DC, USA, 2008.
44. Ayodele, T.R.; Ogunjuyigbe, A.S.O.; Alao, M.A. Life Cycle Assessment of Waste-to-Energy (WtE) Technologies for Electricity Generation Using Municipal Solid Waste in Nigeria. *Appl. Energy* **2017**, *201*, 200–218. [CrossRef]
45. Alexander, A.; Burklin, C.; Singleton, A. *Landfill Gas Emissions Model (LandGEM) Version 3.02*; User's Guide; United States Environmental Protection Agency: Washington, DC, USA, 2005.
46. World Bank Group Current Climate—Ghana. Available online: https://climateknowledgeportal.worldbank.org/country/ghana/climate-data-historical (accessed on 8 December 2024).
47. Intergovernmental Panel on Climate Change. *2006 IPCC Guidelines for National Greenhouse Gas Inventories*; National Greenhouse Gas Inventories Programme: Hayama, Japan, 2006.
48. Intergovernmental Panel on Climate Change. Global Warming Potential Values. In *Greenhouse Gas Protocol: AR6*; Intergovernmental Panel on Climate Change: Geneva, Switzerland, 2024.
49. United States Environmental Protection Agency. Global Methane Initiative Chapter 6. Landfill Gas Modeling. In *International Best Practices Guide for Landfill Gas Energy Projects*; U.S. Environmental Protection Agency: Washington, DC, USA, 2012; pp. 63–76.
50. Baidoo, T.; Boakye, P.; Ama, M.; Rockson, D.; Kemausuor, F.; Narra, S. A Life Cycle Assessment of Electricity Generation from the Anaerobic Digestion of Municipal Solid Waste in Ghana. In *Innovations in Circular Economy and Renewable Energy in Africa*; Narra, M.-M., Narra, S., Eds.; Springer: Cham, Switzerland, 2024; ISBN 9783031683305.
51. Addae, G.; Oduro-Kwarteng, S.; Fei-Baffoe, B.; Rockson, M.A.D.; Ribeiro, J.X.F.; Antwi, E. Market Waste Composition Analysis and Resource Recovery Potential in Kumasi, Ghana. *J. Air Waste Manag. Assoc.* **2021**, *71*, 1529–1544. [CrossRef]
52. Krou, N.M.; Baba, G.; Akpaki, O. Estimation of the Amount of Electrical Energy Available from the Biogas Produced at the Faecal Sludge Treatment Plant in the City of Sokodé. In Proceedings of the 2nd German-West African Conference on Sustainable, Renewable Energy Systems SusRES, TH Wildau, Kara, Togo, 6 April 2021; Volume 1, pp. 77–82.
53. Nielsen, M.; Nielsen, O.-K.; Thomsen, M. *Emissions from Decentralised CHP Plants 2007—Energinet.Dk Environmental Project No. 07/1882. Project Report 5—Emission Factors and Emission Inventory for Decentralised CHP Production*; Technical Report No. 07/1882; National Environmental Research Institute: Aarhus, Denmark, 2010.
54. Bassey, U.; Bowles, A.; Fowler, G.; Tom, A.O.; Beck, G.; Narra, S.; Nelles, M.; Hartmann, M. Experimental Investigation of Products from Thermal Treatment of Real-World Mixed Single-Use and Multi-Layered Waste Plastics. *Environ. Res.* **2024**, *247*, 118244. [CrossRef] [PubMed]
55. Ayodele, T.R.; Ogunjuyigbe, A.S.O.; Durodola, O.; Munda, J.L. Electricity Generation Potential and Environmental Assessment of Bio-Oil Derivable from Pyrolysis of Plastic in Some Selected Cities of Nigeria. *Energy Sources Part A Recover. Util. Environ. Eff.* **2020**, *42*, 1167–1182. [CrossRef]
56. Lee, R.D. Evaluating Uganda's Waste Management System for the Production of Refuse-Derived Fuel (RDF) and Its Potential Implementation in the Country's Growing Cement Industry. Master's. Thesis, Technical University of Wien, Vienna, Austria, 2019.
57. Guinée, J.B.; Gorrée, M.; Heijungs, R.; Huppes, G.; Kleijn, R.; de Konig, A.; van Oers, L.; Wegener Sleeswijk, A.; Udo De Haes, H.A.; de Bruijn, J.A.; et al. *Handbook on Life Cycle Assessment: Operational Guide to the ISO Standards*; Kluwer Academic Publishers: New York, NY, USA; Boston, MA, USA; Dordrecht, The Netherlands; London, UK; Moscow, Russia, 2004; ISBN 0306480557.

58. Goedkoop, M.; Heijungs, R.; Huijbregts, M.; De Schryver, A.; Struijs, J.; Zelm, R. Van ReCiPe 2008. In *A Life Cycle Impact Assessment Method Which Comprises Harmonised Category Indicators at the Midpoint and the Endpoint Level*, 1st ed.; VROM: The Hague, The Netherlands, 2009.
59. Huijbregts, M.A.J.; Steinmann, Z.J.N.; Elshout, P.M.F.; Stam, G.; Verones, F.; Vieira, M.; Zijp, M.; Hollander, A.; van Zelm, R. ReCiPe2016: A Harmonised Life Cycle Impact Assessment Method at Midpoint and Endpoint Level. *Int. J. Life Cycle Assess.* **2017**, *22*, 138–147. [CrossRef]
60. Dastjerdi, B.; Strezov, V.; Kumar, R.; He, J.; Behnia, M. Comparative Life Cycle Assessment of System Solution Scenarios for Residual Municipal Solid Waste Management in NSW, Australia. *Sci. Total Environ.* **2021**, *767*, 144355. [CrossRef]
61. Crenna, E.; Secchi, M.; Benini, L.; Sala, S. Global Environmental Impacts: Data Sources and Methodological Choices for Calculating Normalization Factors for LCA. *Int. J. Life Cycle Assess.* **2019**, *24*, 1851–1877. [CrossRef]
62. Laurent, A.; Hauschild, M. Normalisation. In *Life Cycle Impact Assessment*; Hauschild, M.Z., Huijbregts, M.A.J., Eds.; Springer: Dordrecht, The Netherlands, 2015.
63. Hospido, A.; Moreira, M.T.; Martín, M.; Rigola, M.; Feijoo, G. Environmental Evaluation of Different Treatment Processes for Sludge from Urban Wastewater Treatments: Anaerobic Digestion versus Thermal Processes. *Int. J. Life Cycle Assess.* **2005**, *10*, 336–345. [CrossRef]
64. Miandad, R.; Rehan, M.; Barakat, M.A.; Aburiazaiza, A.S.; Khan, H.; Ismail, I.M.I.; Dhavamani, J.; Gardy, J.; Hassanpour, A.; Nizami, A.S. Catalytic Pyrolysis of Plastic Waste: Moving toward Pyrolysis Based Biorefineries. *Front. Energy Res.* **2019**, *7*, 437000. [CrossRef]
65. Kumar, A.; Samadder, S.R. Assessment of Energy Recovery Potential and Analysis of Environmental Impacts of Waste to Energy Options Using Life Cycle Assessment. *J. Clean. Prod.* **2022**, *365*, 132854. [CrossRef]
66. Silva, V.; Contreras, F.; Bortoleto, A.P. Life-Cycle Assessment of Municipal Solid Waste Management Options: A Case Study of Refuse Derived Fuel Production in the City of Brasilia, Brazil. *J. Clean. Prod.* **2021**, *279*, 123696. [CrossRef]
67. Karlsson, N.P.E.; Hoveskog, M.; Halila, F.; Mattsson, M. Business Modelling in Farm-Based Biogas Production: Towards Network-Level Business Models and Stakeholder Business Cases for Sustainability. *Sustain. Sci.* **2019**, *14*, 1071–1090. [CrossRef]
68. Ouedraogo, A.S.; Frazier, R.S.; Kumar, A. Comparative Life Cycle Assessment of Gasification and Landfilling for Disposal of Municipal Solid Wastes. *Energies* **2021**, *14*, 7032. [CrossRef]
69. Ayodele, T.R.; Ogunjuyigbe, A.S.O.; Alao, M.A. Economic and Environmental Assessment of Electricity Generation Using Biogas from Organic Fraction of Municipal Solid Waste for the City of Ibadan, Nigeria. *J. Clean. Prod.* **2018**, *203*, 718–735. [CrossRef]
70. United States Environmental Protection Agency. *Emission Factor Documentation for AP-42 Section 3.1: Stationary Gas Turbines*; US United States Environmental Protection Agency: Washington, DC, USA, 2000.
71. Al-rumaihi, A.; Alherbawi, M.; Mckay, G. Assessing Plastic and Biomass-Based Biochar's Potential for Carbon Sequestration: An. *Front. Sustain.* **2023**, *4*, 1200094. [CrossRef]
72. Mishra, S.; Bharagava, R.N.; More, N.; Yadav, A.; Zainith, S.; Mani, S.; Chowdhary, P. Heavy Metal Contamination: An Alarming Threat to Environment and Human Health. In *Environmental Biotechnology: For Sustainable Future*; Sobti, R.C., Arora, N.K., Kothari, R., Eds.; Springer: Singapore, 2019; pp. 103–125; ISBN 978-981-10-7284-0.
73. Akanchise, T.; Boakye, S.; Borquaye, L.S.; Dodd, M.; Darko, G. Distribution of Heavy Metals in Soils from Abandoned Dump Sites in Kumasi, Ghana. *Sci. Afr.* **2020**, *10*, e00614. [CrossRef]
74. Sobti, C.R.; Arora, N.K.; Kothari, R. (Eds.) *Environmental Biotechnology: For Sustainable Future*; Springer: Singapore, 2019; ISBN 9789811072833.
75. Fantke, P.; Aurisano, N.; Bare, J.; Backhaus, T.; Bulle, C.; Chapman, P.M.; De Zwart, D.; Dwyer, R.; Ernstoff, A.; Golsteijn, L.; et al. Toward Harmonizing Ecotoxicity Characterization in Life Cycle Impact Assessment. *Environ. Toxicol. Chem.* **2018**, *37*, 2955–2971. [CrossRef] [PubMed]
76. Alengebawy, A.; Mohamed, B.A.; Jin, K.; Liu, T.; Ghimire, N.; Samer, M.; Ai, P. A Comparative Life Cycle Assessment of Biofertilizer Production towards Sustainable Utilization of Anaerobic Digestate. *Sustain. Prod. Consum.* **2022**, *33*, 875–889. [CrossRef]
77. Asabea, A.; Nti, A.; Arko-mensah, J.; Botwe, P.K.; Dwomoh, D.; Kwarteng, L.; Takyi, S.A.; Acquah, A.A.; Tettey, P.; Basu, N.; et al. Effect of Particulate Matter Exposure on Respiratory Health of E-Waste Workers at Agbogbloshie, Accra, Ghana. *Int. J. Environ. Res. Public Health* **2020**, *17*, 3042. [CrossRef]
78. Ghana Grid Company Limited. *2022 Electricity Supply Plan for the Ghana Power System*; Ghana Grid Company Limited: Accra, Ghana, 2022.
79. Ministry of Environment, Science, Technology and Innovation. *Ghana: Updated Nationally Determined Contribution Under the Paris Agreement (2020–2030)*; Environmental Protection Agency (EPA) and the Ministry of Environment, Science, Technology and Innovation (MESTI): Accra, Ghana, 2021.
80. Miezah, K.; Obiri-Danso, K.; Kádár, Z.; Fei-Baffoe, B.; Mensah, M.Y. Municipal Solid Waste Characterization and Quantification as a Measure towards Effective Waste Management in Ghana. *Waste Manag.* **2015**, *46*, 15–27. [CrossRef]

81. Dery, P.; Benefoh, D. *Ghana's Report on the Implementation of Article 6 of the Paris Agreement*; Ghana's Carbon Market Office: Accra, Ghana, 2023.
82. Hang, M.; Geyer-Klingeberg, J.; Rathgeber, A.W. It Is Merely a Matter of Time: A Meta-Analysis of the Causality between Environmental Performance and Financial Performance. *Bus. Strategy Environ.* **2019**, *28*, 257–273. [CrossRef]

energies

MDPI

Article

Assessing the Environmental Sustainability Corridor in South Africa: The Role of Biomass Energy and Coal Energy

Ahlam Sayed A. Salah [1,*], Serdal Işıktaş [1] and Wagdi M. S. Khalifa [2]

1 Department of Business Administration, Cyprus Health and Social Sciences University, Northern Cyprus, Mersin 10, Güzelyurt 99700, Turkey; serdal.isiktas@kstu.edu.tr

2 Department of Business Administration, University Mediterranean Karpasis, Northern Cyprus, Mersin 10, Lefkoşa 99138, Turkey; wagdi.kalifa@akun.edu.tr

* Correspondence: 230928084@kstu.edu.tr

Abstract: South Africa's national development plan has outlined aspirations to achieve a sustainable environment. However, the country remains bound for an unsustainable trajectory. Despite this ecological issue, no studies have probed how biomass and coal energy impact ecological quality. In light of this gap, this study inspects the environmental effect of political risk, coal energy, and biomass energy in South Africa. Also, this study integrates economic growth and natural resources into its framework. This study uses the load capacity factor (LC), which is a more aggregate proxy of ecological quality due to its accounting for the demand and supply aspect of the environment. This study uses the dynamic autoregressive distributive lag estimator (ARDL), which is capable of not only providing details of the influence of each determinant on LC in the long and short term but also of capturing the counterfactual shock of positive or negative exogenous variables on the LC. The kernel regularized least squares (KRLS) method is used for a robustness analysis of the dynamic ARDL approach. Furthermore, the findings of the dynamic ARDL simulation estimator disclose the negative impact of economic growth on the LC, thereby contributing to environmental deterioration by 0.552%. Natural resources and coal energy have an adverse impact on the LC, indicating a reduction in environmental sustainability by 0.037% and 0.290%, respectively. Meanwhile, biomass contributes to the LC, thereby promoting ecological quality by 0.421%. Political risk contributes to the reduction in the LC. This research provides pertinent policy considerations for policymakers and governments in South Africa, suggesting that the government of South Africa should invest in biomass energy and sustainable extraction procedures since biomass energy has a vital role in increasing ecological quality.

Keywords: load capacity factor; political risk; CO energy; natural resources; biomass energy

1. Introduction

Following the industrial revolution, the global surge in energy demand has intensified the exploitation of natural resources (NRs), driven by rapid population growth and economic development [1]. This heightened reliance on energy, particularly fossil fuels, has significantly exacerbated ecological degradation, as evidenced by numerous studies emphasizing the detrimental effects of energy consumption on environmental quality [2–4]. While energy consumption remains a critical catalyst for economic growth, its heavy dependence on nonrenewable sources, such as coal, oil, and gas, accelerates environmental deterioration. However, a substantial reduction in ecological deterioration can be achieved by shifting towards sustainable energy sources, thereby presenting a viable means to alleviate environmental harm [5,6]. Consequently, it is imperative to prioritize the pursuit of sustainable development, which can be realized through the formulation of environmental legalization. As a result, the United Nations Sustainable Development Goals (SDGs) were formulated with the goal of achieving these objectives. Sustainable development emerges as a global concern that assumes a pivotal role in safeguarding the existence of humanity [7,8]. In the contemporary milieu, the escalating utilization of fossil fuels poses a formidable impediment for nations striving to attain SDG 7, which centers on access to clean and affordable energy sources [9]. The nature and magnitude of energy production and consumption, essential for sustaining productive endeavors within countries, are widely recognized to exert a profound influence on the realization of ecological quality [8–10]. Concerning the challenges faced by most countries in meeting SDG 12, pertaining to responsible consumption and production, as well as SDG 13, addressing climate action, a primary driving factor underlying this predicament lies in the excessive depletion of resources during industrialization, which played a fundamental role in fostering nations' economic growth, and which may not exert ecological deterioration [11,12]. However, the extraction process poses formidable threats, including deforestation, biodiversity loss, and the exacerbation of global warming [13–15].

Political stability is crucial for effective ecological regulation and mitigating environmental degradation, particularly in democratic nations where people have the freedom to advocate for higher ecological standards and hold political leaders accountable [9]. Additionally, institutional quality may significantly influence environmental excellence, with stronger political institutional quality associated with lower pollution levels and higher ecological quality [16], while ineffective institutional and governance management, including corruption, can hinder the enforcement of strict environmental regulations and contribute to illegal activities and environmental deterioration [17].

The above discussion draws the conclusion that developing effective sustainable environmental policies necessitates consideration of political stability, natural resource extraction, and energy consumption patterns. Moreover, limited to no studies have been conducted on any Sub-Saharan African nation. Furthermore, the determinants of the load capacity factor (LC), which is an aggregate proxy of ecological quality due to its accounting for the demand and supply aspect of the environment, have received less attention in Sub-Saharan African nations. As a result, as a major emerging nation situated in Sub-Saharan Africa, South Africa is the obvious choice for this analysis. In addition, the nation's National Development Plan has outlined aspirations to achieve a sustainable environment, including a low-carbon economy. However, the country remains deeply rooted in an unsustainable trajectory. For instance, South Africa's ecological deficit is relatively high given that the demand aspect of the environment (ecological footprint) is 3.21 gha per capita, while the supply aspect (biocapacity) is 1.28 gha per capita [18]. Notably, this is due to South Africa's energy composition, which is heavily influenced by coal, which holds a dominant position in the overall energy supply. Coal energy dominates South Africa's energy supply, accounting for 92% of electricity and 57% of heat generation, while renewables contribute only 6.6% to the total energy supply, with biomass making up the majority at 85%, being primarily used for heat production in residential and industrial sectors [19]. Furthermore, South Africa is blessed with abundant natural resources, which is manifest in its extractive economic framework, characterized by the significant underground extraction of coal, platinum, and gold. In addition, the country has been plagued with governance problems, corruption, and inefficient administration, which contribute to South Africa's substantial political risk. Hence, having accounted for these issues, one could conclude that South Africa is an ideal candidate for this investigation.

Within this context, the objectives of this research are as follows. (i) This research ascertains the influence of economic expansion on the LC. (ii) This research inspects the influence of political risk on the LC. (iii) This research inspects the influence of natural resources on the LC in South Africa. (iv) This research ascertains the effect of coal and biomass energy on the LC. (v) This research determines the existence of a causal relationship between the LC and these variables (political risk, economic expansion, coal energy, natural resources, and biomass energy). The objectives of this study also highlight a major contribution of this study since we are unaware of any study that scrutinizes the environmental effect of political risk, economic expansion, coal energy, natural resources, and biomass energy within the framework of an emerging and Sub-Saharan African nation such as South Africa.

Another contribution of this study is the implementation of the dynamic ARDL stimulation method, which offers a comprehensive insight into both the short- and long-term impacts of key determinants, such as economic growth, political risk, and energy consumption, on the load capacity factor (LC), a nuanced proxy for ecological quality that accounts for both environmental demand and supply. This method enables the analysis to go beyond traditional techniques, allowing for a more precise evaluation of how positive or negative shocks to exogenous variables influence ecological outcomes over time. By capturing the interplay between these factors, this study provides a deeper understanding of the complex relationships underlying environmental quality in South Africa, a nation characterized by its unique mix of economic growth aspirations, energy-intensive industries, and environmental vulnerabilities. Additionally, the kernel regularized least squares (KRLS) method was used in this study as a robustness analysis for the dynamic ARDL approach.

Furthermore, this study's findings hold significant implications for shaping South Africa's energy policies and environmental strategies. This research underscores the urgent need for a diversified energy mix, emphasizing the expansion of renewable energy sources like solar and wind while reducing the dominance of coal in electricity and heat generation. Additionally, it highlights the importance of strengthening governance structures and reducing corruption to create a more stable political environment conducive to enforcing ecological regulations. These recommendations are not only critical for mitigating environmental degradation but also for aligning South Africa with the global sustainability agenda, including the achievement of SDG-7 (affordable and clean energy) and SDG-13 (climate action). Ultimately, this study's contributions extend beyond South Africa, offering valuable insights for other emerging economies facing similar challenges and reinforcing the global discourse on sustainable development and ecological preservation.

2. Literature Review

Lately, researchers have investigated the growth–energy–pollution nexus from a broad to a detailed level. Also, incorporating the role of macroeconomic variables has been increasing in recent times. However, the outcome of the literature is contradicting. As a result, the subsection that follows provides an overview of the relationship between ecological sustainability and its determinants concerning the objective of this research.

2.1. Economic Growth (GDP), Coal Energy (CO), and Ecological Degradation

The energy–growth–pollution interaction in India for the timeline from 1970 to 2017 was investigated by [20]. They employed dynamic ARDL estimators, which confirmed that GDP increases the ecological footprint (EF), whereas CO induces the ecological footprint. For South Africa, ref. [21] considered the influence of GDP and CO on CO_2 emissions for the timeline between 1965 and 2017, employing the FMOLS method. They established that EKC is valid in South Africa and CO increases CO_2 emissions, respectively, within the period of study. In a study performed on developed and developing nations, ref. [22]

reported a similar empirical result using the Augmented Anderson–Hsiao (AAH) two-step GMM estimator approach, in which GDP reduces CO_2 emissions in the entire sample and developing nations. However, in developed nations, GDP increases CO_2 emissions. Furthermore, their findings also revealed that CO_2 emissions increase as CO surges in both developed and developing nations. Utilizing the ARDL approach, ref. [23,24] confirmed that GDP and CO increase CO_2 emissions in South Africa. However, ref. [25] also probed into the effect of GDP and CO on CO_2 emissions for the period between 1966 and 2011 in India. They found that GDP and CO add to CO_2 emissions in India. Ref. [26] probed into the roles of GDP and CO on CO_2 emissions in China. They confirmed that GDP and CO promote CO_2 emissions. Ref. [27] inspected the effect of globalization on CO_2 emissions in seven selected nations. The authors established that a positive link between globalization and CO_2 emissions exists in those nations.

2.2. Biomass Energy (BM) and Ecological Degradation

In the study by [28], it was reported that biomass energy was negatively interconnected with CO_2 emissions in Germany based on an analysis of a dataset from 1965 to 2019 employing a machine learning algorithm. Utilizing the dynamic ARDL approach, ref. [29] probed into the relationship between BM and the EF in India for a dataset between 1970 and 2018; they uncovered that BM exerts an adverse effect on the EF. Meanwhile, ref. [30] assessed the effect of BM on the EF in G7 nations during the timeline from 1980 to 2016 employing the dynamic seemingly unrelated regression approach. They established a causal association between BM and the EF. The investigation by [31] in BRICS validated that the relationship between GDP and the EF is positive, but an adverse interconnection was suggested between the EF and BM over the period between 1992 and 2018. Ref. [32] considered the BM and load capacity factor interaction. The empirical outcome revealed that BM induces the load capacity factor (LC) in the USA. Also, in the research by [33] inspected the interaction between different forms of renewable energy and CO_2 emissions in China, and the empirical outcome indicated that BM reduces CO_2 emissions in China. Furthermore, the study by [34,35] probed into the relationship between BM and the EF in Malaysia and uncovered that BM exerts an adverse effect on the LC.

2.3. Natural Resources and Ecological Degradation

Due to the obvious significance of environmental issues, a majority of researchers have scrutinized the environmental effects of natural resources (NRs), which are recognized to have a major detrimental role in environmental quality. Numerous investigations have been performed to evaluate the environmental effect of NRs. For BRICS nations, ref. [36] considered the effect of NRs on CO_2, employing the MMQR, FMOLS, and DOLS approaches, which established that NRs increase CO_2 emissions. Ref. [37] inspected the linkage between NRs and the EF in 10 top emitting nations based on an analysis of a dataset from 2001 to 2021 employing the Driscoll and Kraay method and AMG, which uncovered a negative effect of NRs on the EF. However, by adopting a series of panel estimators to analyze the dataset of selected nations, ref. [38] scrutinized the effect of NRs on the LC. They uncovered a decreasing effect of NRs on the LC, thereby decreasing ecological quality. Utilizing the AMG and CCEMG method to assess a dataset of 35 South Asian economies collected between 1990 and 2020, ref. [39] evaluated the NR and EF nexus and reported that NRs exert a positive effect on the EF. For E7 nations, ref. [40] identified a decreasing effect of NRs on the EF during the period between 1993 and 2017 by employing the AMG method. However, ref. [41] assessed the NR and EF nexus and found that NRs exert a positive impact on the EF in seven selected nations. Moreover, ref. [42] appraised the relationships between NRs and the EF in BRICS and discovered a positive linkage between NRs and the

EF. However, ref. [12] confirmed that NRs decrease CO_2 emissions in BRICS nations. A similar outcome was found in the study performed in Pakistan by [43].

Conversely, the work of [44] discovered that NRs contribute to ecological degradation in African nations by utilizing the FMOLS and DOLS methods. Likewise, ref. [45] inspected the interconnection between NRs and environmental degradation and demonstrated that NRs increased environmental degradation in BRICST economies. Ref. [46] probed into the effect of NRs on CO_2 emissions in China, adopting a dataset between 2000Q1 and 2020Q4. They confirmed that NRs increase CO_2 emissions. Ref. [7] uncovered the increasing impact of NRs on CO_2 emissions in developing nations.

2.4. Political Risk and Ecological Degradation

Over time, numerous researchers have inspected the environmental role of political risk (PR); nevertheless, their studies have produced conflicting results. The effect of PR on CO_2 emissions in the Netherlands was considered by [47]. Employing the Fourier approach, they confirmed that PR decreases CO_2 emissions. Ref. [48] uncovered that political stability reduces CO_2 emissions in BRICS economies. However, another study on BRICS by [17] confirmed that political stability increases the EF. Using four selected nations as a focus of the investigation, the study by [49] found that political stability decreases CO_2 emissions. Using MENA's dataset, which spans the period between 2002 and 2018, ref. [50] identified an insignificant link between CO_2 emissions and political stability in the lower and middle quantiles, but in the upper quantiles, political stability decreased CO_2 emissions. The empirical outcome of the study by [51] revealed that PR induced ecological degradation in South Asian economies during the period from 1996 to 2019. Also, ref. [52] scrutinized the interaction between PR and the LC, and the empirical outcome revealed that PR induces the LC in Brazil. Ref. [53] studied the effect of PR on CO_2 emissions in Canada. The authors established a negative interaction between PR and CO_2 emissions.

After appraising several studies, we also discovered that the effects of CO energy, natural resources, political risk, and biomass energy on the environment are under-researched, particularly for specific Sub-Saharan African nations. Furthermore, no prior research has investigated the role of CO energy, natural resources, political risk, and biomass energy on the load capacity factor for the case of South Africa, presenting a significant void in the literature. As a result, we attempted to propose major policy guidance in this research depending on the aims stated above.

3. Data and Model Specification

3.1. Data

This study examined the role of economic growth, CO energy, biomass energy, natural resources, and political risk on the load capacity factor, in which South Africa is the focus. This study applied an annual dataset spanning between 1984 and 2022. The starting period of 1984 was selected due to data availability for political risk. The load capacity was obtained from the database of the global footprint network, whereas the datasets of natural resources and economic growth were obtained from the World Bank's database. Biomass energy was sourced from the material flow database, and, finally, the dataset of CO energy was extracted from the British Petroleum statistics database. This research converted all series into their natural logarithms to improve the analysis's reliability and normalize the data. This enabled the coefficients to be understood as elasticities. The parameters and descriptions are presented in Table 1.

Table 1. Description and sources of the variables.

Indicators	Description of the Variable Used	Metric	Sourced
LC	Load capacity factor	Hectares per capita	Global Footprint
GDP	Economic growth	Constant 2015 USD	World Bank Database
CO	CO energy	Exajoules	British Petroleum database
BM	Biomass energy	Tonnes	Material Flows Database
NRs	Natural resources	% of GDP	World Bank Database
PR	Political risk	Index	PRS Database

Based on the discussion above, economic growth, coal energy, natural resources, political risk, and biomass energy can have an adverse or positive impact on environmental sustainability. The load capacity factor was used as a metric for environmental sustainability in this study, while economic growth, CO energy, natural resources, political risk, and biomass energy were the independent variables. In line with prior studies such as [34,54–57], this study developed the economic function, which is specified as follows:

$$LCAP_t = \vartheta_0 + \vartheta_1 GDP_t + \vartheta_2 CO + \vartheta_3 BM_t + \vartheta_4 NR_t + \vartheta_5 PR_t + \varepsilon_t \quad (1)$$

where LC, CO, BM, GDP, NRs, and PR depict the load capacity factor, CO energy, biomass energy, economic growth, natural resources, and political risk. The subscript t symbolizes the period of study (1984–2022), and the error term is indicated as ε.

We anticipated a decreasing effect of GDP on the LC, i.e., $\left(\vartheta_1 = \frac{\delta LCAP}{\delta GDP} < 0\right)$. Previous studies such as refs. [58,59] agree with this perspective. We anticipated a negative connection between CO and the LC, i.e., $\left(\vartheta_2 = \frac{\delta LCAP}{\delta CO} < 0\right)$, which is echoed by previous studies [26,29]. Centered on prior studies [31,32], this research anticipated an increasing effect of BM on ecological sustainability, i.e., $\left(\vartheta_3 = \frac{\delta LCAP}{\delta BM} > 0\right)$. Concerning the effect of natural resources on the LC, we anticipated an inverse effect. This perspective was developed based on the studies by [60,61], i.e., $\left(\vartheta_4 = \frac{\delta LCAP}{\delta NR} < 0\right)$. Based on the research by [62], this research anticipated a decreasing effect of political risk on ecological sustainability, i.e., $\left(\vartheta_5 = \frac{\delta LCAP}{\delta PR} < 0\right)$.

3.2. Econometric Strategy

The conventional (PP and ADF) unit root is applied in this study to detect the stationary nature of the parameters used. Nonetheless, having received widespread criticism regarding the outcomes of conventional stationary techniques, we chose to employ the Zivot and Andrews unit root test developed by [63], since it is recognized for delivering accurate outcomes and accounting for at least one break. It is critical to remember that this approach is only employed to determine whether the series have unit root issues at various levels, while other tests cannot discriminate between structural break and unit root issues.

Furthermore, this study used the dynamic ARDL estimator to establish the long-run association. The ARDL necessitates the determination of an appropriate lag, and the potential issue of endogeneity was addressed by selecting an optimal lag period. The estimator is applicable when the series are 1(0) or 1(1) or both 1(0) and 1(1). The selection of an appropriate lag helps fix the issue of possible multicollinearity in the estimation [64,65]. Also, the ARDL can produce the effect of the LC's determinants in the long term and in the short term. Moreover, the ECT displays the convergence information. As defined in the study by [64], the ARDL equation in relation to this study's variables is shown below:

$$\Delta LCAP_t = \theta_0 + \sum_{l=1}^{p} \theta_1 \Delta LCAP_{t-1} + \sum_{i=1}^{p} \theta_2 \Delta GDP_{t-1} + \sum_{i=1}^{p} \theta_3 COAL_{t-1} + \sum_{i=1}^{p} \theta_4 BIO_{t-1} + \sum_{i=1}^{p} \theta_5 \Delta NR_{t-1} + \sum_{i=1}^{p} \theta_6 \Delta PR_{t-1} + \pi_1 LCAP_{t-1} + \pi_2 GDP_{t-1} + \pi_3 COAL_{t-1} + \pi_4 BIO_{t-1} + \pi_5 NR_{t-1} + \pi_6 PR_{t-1} + \epsilon_t \quad (2)$$

In Equation (3), $\theta_i (i = 1 \ldots 6)$ denotes the coefficient of the parameter in the short term and π_i $(i = 1 \ldots 6)$ demonstrates the long-term relationship among the parameters. t denotes lag lengths. Equation (3) is modified into Equation (4) by including the ECM in the ARDL, which is defined in the work by [66] as follows:

$$\Delta LCAP_t = \theta_0 + \sum_{l=1}^{p} \theta_1 \Delta LCAP_{t-1} + \sum_{i=1}^{p} \theta_2 \Delta GDP_{t-1} + \sum_{i=1}^{p} \theta_3 COAL_{t-1} + \sum_{i=1}^{p} \theta_4 BIO_{t-1} + \sum_{i=1}^{p} \theta_5 \Delta NR_{t-1} + \sum_{i=1}^{p} \theta_6 \Delta PR_{t-1} + \pi_1 LCAP_{t-1} + \pi_2 GDP_{t-1} + \pi_3 COAL_{t-1} + \pi_4 BIO_{t-1} + \pi_5 NR_{t-1} + \pi_6 PR_{t-1} + \pi_7 \alpha ECT_{t-1} + \epsilon_t \quad (3)$$

where the error correction term is given by ECT_t, indicating the adjustment rate of balance in the long term from a shock in the short term, and the coefficient of ECT_t is expected to be negatively significant. The ARDL approach was applied to scrutinize the linkage between CO_2 emissions and its regressors, after ascertaining the cointegrating interconnection in Equation (4). Furthermore, the findings of the dynamic ARDL stimulation method were verified using the kernel-based regularized least squares (KRLS) method.

4. Findings and Discussions

For the pre-estimation assessment in this research, we try to understand the basic nature (descriptive statistics) of the observed parameters, which is summarized in Table 2. BIO (8.161) has the highest average value next to GDP (3.722), PR (1.501), NRs (0.703), and CO (0.510), while the LC has the lowest mean value of −0.378. The range of observed variables are as follows: LC (−0.465 to −0.298), GDP (3.630 to 3.630), CO (0.405 to 0.594), BIO (8.018 to 8.236), NRs (0.395 to 1.078), and PR (1.295 to 1.580). The standard deviations of the LC, GDP, CO, BIO, NRs, and PR are 0.053, 0.058, 0.057, 0.044, 0.149, and 0.063, respectively. For skewness, the outcome shows that all series are positively skewed. Moreover, for kurtosis, we uncovered that LC, GDP, and CO are platykurtic in nature, while BIO, NRs, and PR are leptokurtic in nature.

Table 2. Descriptive statistics.

	LC	GDP	CO	BIO	NRs	PR
Mean	−0.378	3.722	0.510	8.161	0.703	1.501
Median	−0.376	3.718	0.520	8.164	0.679	1.520
Maximum	−0.298	3.796	0.594	8.236	1.078	1.580
Minimum	−0.465	3.630	0.405	8.018	0.395	1.295
Std. Dev.	0.053	0.058	0.057	0.044	0.149	0.063
Kurtosis	−0.086	−0.063	−0.191	−0.942	0.350	−1.306
Skewness	1.500	1.434	1.702	4.158	3.121	4.834

Furthermore, in this study, we utilized a battery of conventional stationary tests (PP and ADF) to uncover the integration order of the series, in which the findings are indicated in Table 3. The findings disclose that biomass energy is stationary at levels. Moreover, all series are stationary at I(1). In a scenario when the series has a break, the normal unit root test (ADF and PP) will generate erroneous estimates. Time series data are exposed to instabilities due to macroeconomic and structural phenomena, like policies that may impede the variables' stability, which pinpoint a specific economic occurrence. These issues could potentially have an impact on the outcomes of any investigation undertaken for any nation. Conventional unit root testing typically fails to take into account such disturbances during estimation. Thus, we used the Zivot and Andrew test, which was devised by [63]. Table 4 summarizes the findings, which revealed that no unit root issues exist at any level for biomass energy, coal energy, and political risk. In addition, all series also are stationary at I(1). The results of the conventional stationary tests and the Zivot and Andrew test clearly indicate that there is a mixed order of integration among the variables used.

Table 3. Conventional unit root.

	ADF		PP	
	I(0)	**I(1)**	**I(0)**	**I(1)**
LCF	−0.134	−7.702 *	−2.697	−9.412 *
GDP	−2.475	−3.894 **	−2.424	−3.850 **
CO	−2.114	−7.817 *	−2.114	−9.872 *
BM	−6.429 *	−8.094 *	−6.429 *	−8.720 *
NRs	−3.100	−8.669 *	−3.100	−8.669 *
PR	−2.643	−5.350 *	−2.866	−5.345 *

Note: $p < 1\%$ and $p < 5\%$, denote * and **.

Table 4. ZA unit root.

	I(0)	**Break**	**I(1)**	**Break**
LC	−2.032	2012	−9.986 *	2003
GDP	−2.466	2016	−5.687 *	2008
CO	−6.600 *	2008	−8.386 *	2010
BM	−7.322 *	1996	−9.211 *	1993
NRs	−4.584	2004	−9.811 *	2009
PR	−5.359 **	2009	−8.349 *	2009

Note: $p < 1\%$ and $p < 5\%$, denote * and **.

After evaluating the pattern of integration of the observed variables, we continued to assess the long-term interrelationship. We utilized the ARDL bounds test to detect the long-run interconnection. The findings of the ARDL bounds test are shown in Table 5, which disclose an indication of a long-run interrelationship between the LC and its independent variables (GDP, CO, BM, NRs, and PR) because both the T- and F-statistics are significantly higher than the critical values of Kripfganz and Schneider (LB and UB).

Upon the establishment of a cointegrating relationship between CO_2 and its determinants, this study moved on from there to assess the short-term and long-term relationships. The findings of the dynamic ARDL stimulation method are reported in Table 6. GDP exhibits a negative and significant effect on the LC, indicating that an upsurge in GDP decreases the LC, thereby increasing ecological deterioration in the long and short terms. Thus, our result shows that a 1% surge in GDP reduces the LC by 0.552% (long run) and 0.890% (short run). Moreover, this association between the LC and economic expansion specifies that South Africa is currently in the scale phase, which is experienced by emerging nations when fossil fuel energy sources underpin economic expansion. This suggests that

South Africa prioritizes economic expansion over environmental degradation. Several emerging nations have pursued this economic growth trajectory. Hence, this research's outcomes will have significant policy implications for South Africa and other emerging economies. Thus, attaining the aspirations of SDG 13 in South Africa will indeed be challenging. This is an urgent call for stakeholders and policymakers to adopt a more deliberate stance on disassociating economic expansion from pollutant emissions in the nation's energy basket. Our finding conforms with the research by [58,61] in BRICS economies, ref. [59] in Brazil, [67] in India, and [68] in OECD nations. Conversely, the research by [69] contrasts this outcome.

Table 5. Bounds ARDL cointegration.

Models	F-Statistic		T-Statistic	
	16.089 *		−10.612 *	
	Crit. Value (F-Statistic)		Crit. Value (T-Statistic)	
Significance level	**1(0)**	**1(1)**	**1(0)**	**1(1)**
10%	2.26	3.35	−2.57	−3.86
5%	2.62	3.79	−2.86	−4.19
2.5%	3.41	4.68	−3.43	−4.79
1%	3.15	4.43	−3.43	−4.99

Note: * portrays $p < 0.01$.

Table 6. ARDL.

	Long-Term			Short-Run		
Variables	**Coeffics**	***T*-Statistic**	**Prob Value**	**Coeffics**	***T*-Statistic**	**Prob Value**
GDP	−0.552 *	−4.08	0.000	−0.890 *	−3.26	0.003
BM	0.421 *	2.90	0.008	0.401 *	5.44	0.000
CO	−0.290 ***	−1.82	0.080	−0.334 **	−2.43	0.022
NRs	−0.037 **	−2.08	0.048	0.008	0.41	0.687
PR	−0.226 **	−2.67	0.013	−0.224 **	−2.56	0.017
C	−1.211	−1.37	0.183	-	-	-
ECT (−1)	-	-	-	−0.899 *	−4.72	0.000
			Diagnostic Tests			
J-B Normality			0.929 (0.628)			
χ^2 LM			1.446 (0.252)			
χ^2 RESET			1.038 (0.307)			
χ^2 BPG			0.703 (0.669)			

Note: ***, **, and * represent $p < 0.10$, $p < 0.05$, and $p < 0.01$.

In addition, the effect of CO decreases the LC, considering that the coefficient of CO is 0.290 in the long term, while for the short term, is 0.334. This result is predictable given that coal energy exerts the most carbon-intensive fossil fuel during usage, emitting 72% more climate-changing CO_2 per unit of energy than gas. Despite its target of achieving ecological sustainability, South Africa continues to consume more fossil fuel energy sources, most especially CO, to meet the increasing energy demand. Coal has been the principal energy source for South Africa's manufacturing-based economy for the past century. Over 90 percent of the electricity in South Africa is currently generated by coal because it is mined in South Africa. This result conforms with the research by [26] in China, ref. [70] in France, ref. [71] in G20 nations, and ref. [29] and ref. [20] in India.

Furthermore, our result shows that a 1% surge in BM induces the LC by 0.421% (long run) and 0.401% (short run) at the 1% significant level. Thus, biomass energy exhibits a positive interconnection with the LC in the long and short terms. This outcome indicates that biomass energy helps to achieve a sustainable environmental condition. This outcome agrees with prior studies [29,31,72–75]. This conclusion suggests that while it is unfeasible to eliminate the usage of fossil fuels, notably, coal, which seems to be the prevailing source of energy and a significant component of South Africa's economic growth, the government can enhance environmental quality by increasing its utilization of biomass energy. Thus, the use of biomass energy could support the accomplishment of sustainable development. Therefore, policymakers and stakeholders need to restructure their energy basket by incorporating more greener energy. Furthermore, other prior studies, such as [28] in Germany, ref. [33] in China, and ref. [32] in the USA, also give credence to the positive role of BM on ecological quality. Meanwhile, the study by [30] offers a contradicting outcome by confirming the detrimental effect of BM on ecological quality in G7 nations.

This study's findings show that natural resources exacerbate the degradation of the environment in South Africa. Based on these findings, when all other variables are held constant, an increase in NRs by 1% will lead to a reduction in the LC by 0.037% in the long term, indicating that the practices of natural resource extraction in South Africa have not been sustainable. It is not surprising to observe this outcome, as the pursuit of economic development in South Africa propels the nation towards industrialization and fosters the excessive exploitation of its natural resources. Consequently, this process depletes the availability of natural resources and accelerates ecological degradation, as overdependence on these resources hampers their regeneration, which diminishes biocapacity. Furthermore, the extensive coal mining in South Africa engenders profound ecological degradation within delicate natural ecosystems, including forests, and possesses the capacity to inflict irreparable scars upon the landscape, thereby resulting in detrimental consequences for humans, animals, and plants through the destruction of habitats and the contamination of the environment. This result conforms with the investigations by [76–79], positing that the escalating global demand for natural resources, driven by population expansion and economic progress, frequently outpaces the earth's capability to regenerate them. As a result, resources are exploited at an unsustainable rate, causing depletion and environmental damage.

Our results reveal that a 1% upsurge in PR mitigates the LC by 0.226% (long run) and 0.224% (short run). Thus, this outcome highlights the adverse effect of PR on the LC in South Africa. This adverse association suggests that the current political climate hinders ecological sustainability in South Africa. Political instability engenders a circumstance wherein the integrity of institutions and the efficacy of governmental administration are compromised, thereby precipitating environmental degradation. Notably, corruption serves as a tangible exemplification of the diminished quality of institutions, as it has the potential to undermine the stringency of environmental regulations. This, in turn, fosters a rise in illegal manufacturing and extraction activities, thereby contributing to ecological deterioration. In addition, this result conforms with the investigations by [17,51], which reported the detrimental effect of PR on ecological quality in South Asian nations and BRICS nations, respectively. However, the positive role of PR on ecological quality was observed in the studies by [52] in Brazil, ref. [53] in Canada, and [47] in the Netherlands.

Furthermore, in Table 6, the results of the post-estimation analysis are disclosed, which suggest that the residuals of the model exhibit a normal distribution and do not suffer from serial correlation, misspecification, or heteroscedasticity issues. Figure 1 shows that the residual of this model is stable at a 5% level of significance.

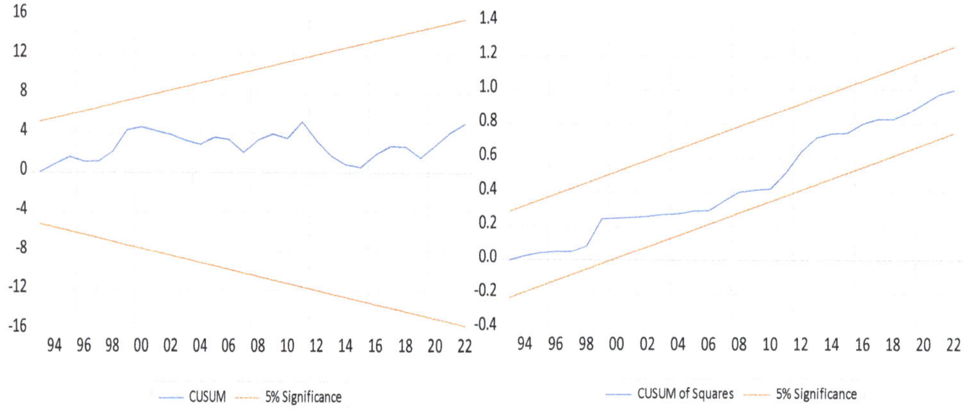

Figure 1. Stability test.

The dynamic ARDL stimulation method has the ability to automatically generate insightful impulse–response plots, which reveal the counterfactual shock of 1% positive or negative change in the exogenous variables on the LC. The outcome of the dynamic ARDL stimulation method is presented in Figures 2–6. Figure 2 shows the responses of the LC to a change in GDP. For the positive shock in GDP (see Figure 2A), a 1% positive change in GDP will lead to a decrease in the LC. Meanwhile, a 1% negative shock in GDP will induce the LC (see Figure 2B). Furthermore, the responses of the LC to ±1% changes in biomass energy are presented in Figure 3. A 1% surge in BM will lead to a rise in the LC (see Figure 3A), while a 1% decrease in BM will cause a decrease in the LC (see Figure 3B). In addition, Figure 4 shows the responses of the LC to ±1% changes in natural resources. Figure 4A shows that the LC declines due to a 1% rise in the LC. Likewise, Figure 4B shows that a 1% decline in NRs will cause a reduction in NRs. Furthermore, Figure 5 reveals the positive and negative changes in NRs on the LC. Figure 5A indicates that a 1% positive change in NRs reduces the LC. Conversely, Figure 5B depicts that a 1% decline in NRs decreases the LC. The responses of the LC to ±1% changes in political risk are showcased in Figure 6. A 1% positive shock in PR leads to an adverse impact on the LC, as highlighted in Figure 6A, while Figure 6B shows that a 1% negative change in PR mitigates the LC.

For the robustness analysis of the result of the dynamic ARDL method, this study conducted the kernel regularized least squares (KRLS) method. Table 7 shows the results of the KRLS method. It shows that CO has an adverse effect on the LC, wherein a 1% surge in CO will lead to an average decline in the LC by 0.288%. Likewise, a 1% surge in GDP will cause an average drop in the LC by 0.261%. Thus, GDP decreases the LC in South Africa. However, natural resources have a negative and insignificant effect on the LC. Furthermore, PR causes a decline in the LC, wherein a 1% rise in PR will lead to an average decline in the LC by 0.112. Conversely, a 1% surge in BM will cause a rise in the LC by 0.241%. Thus, the KRLS method shows that CO energy, GDP, natural resources, and political risk exert an adverse influence on the LC, while biomass energy increases the LC in South Africa. The outcome from the KRLS method provides support for the result of the dynamic ARDL stimulation method.

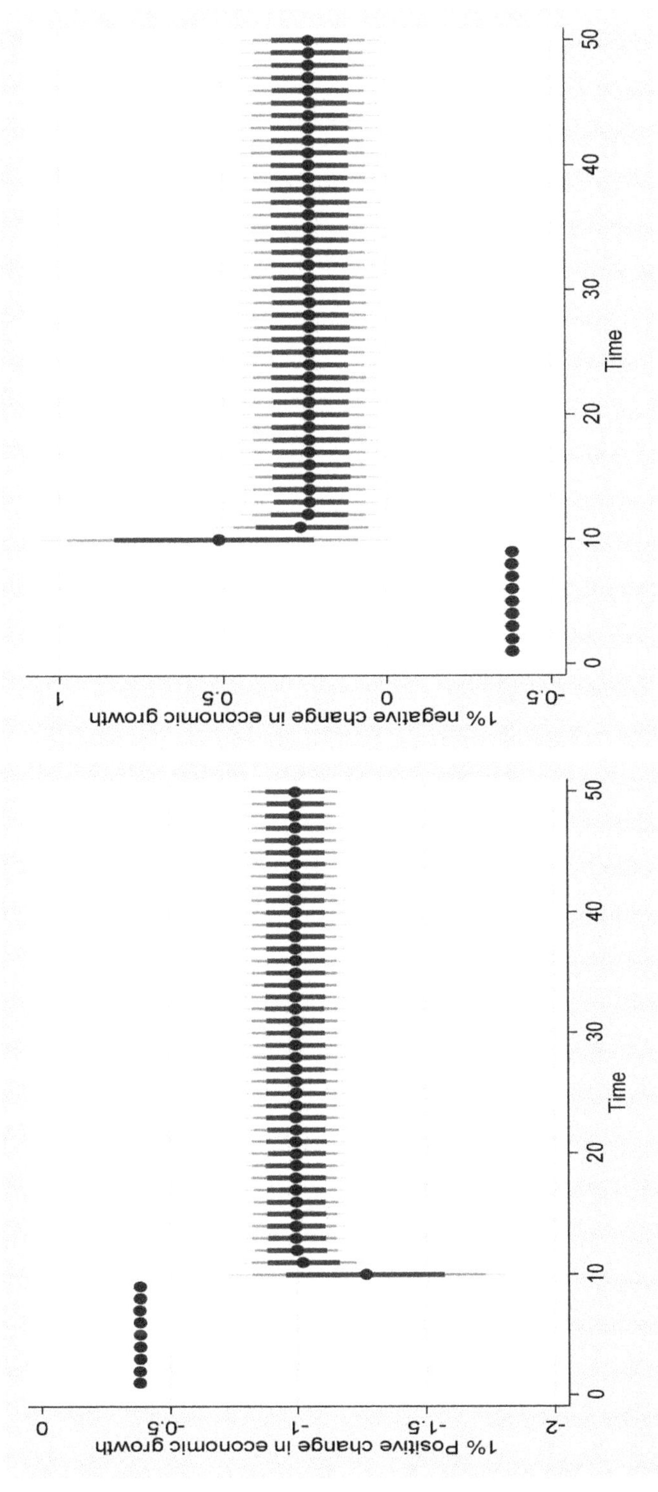

Figure 2. Response of the LC to changes in GDP. **Note:** (**A**) shows a positive change in economic growth, while (**B**) indicates a negative change in economic growth.

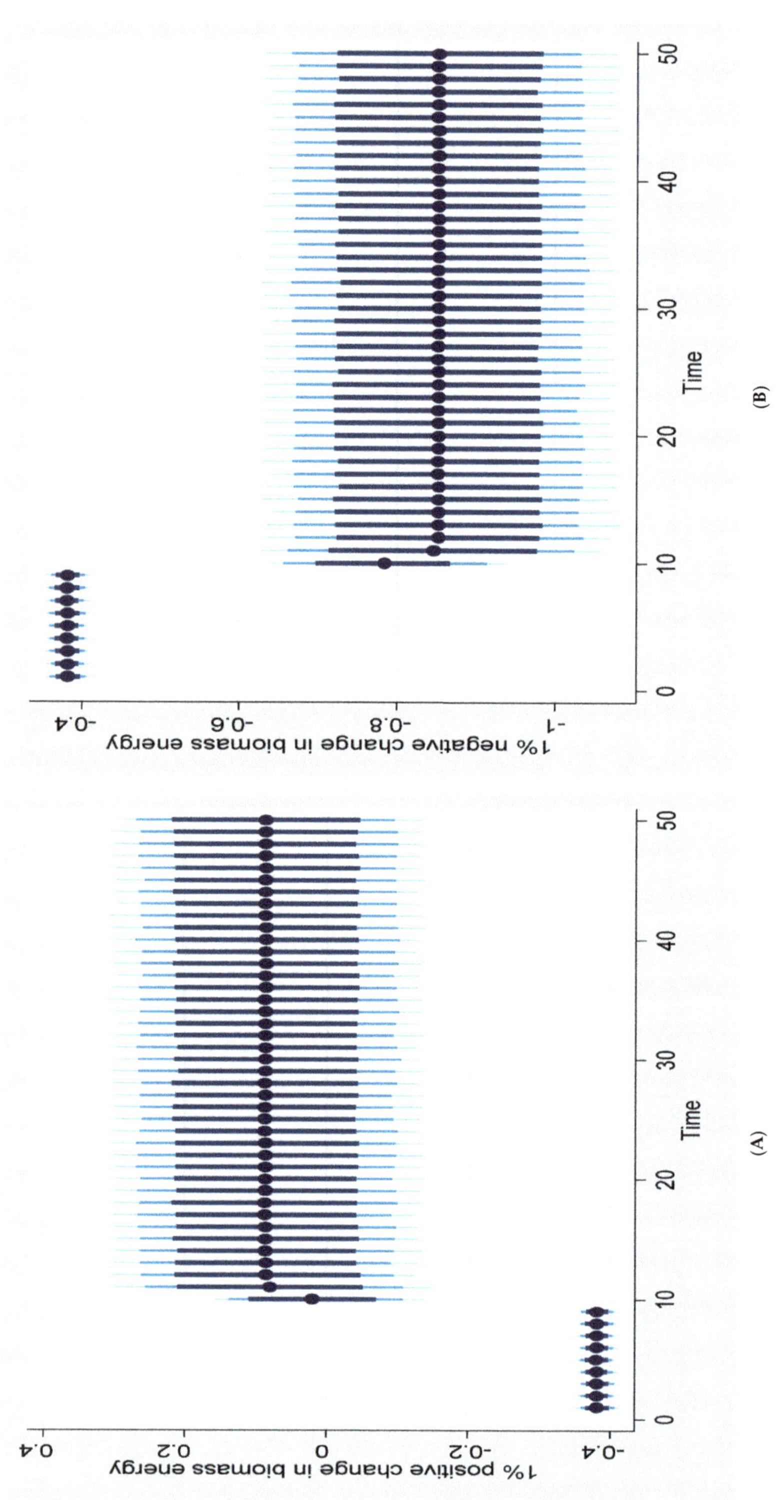

Figure 3. Response of the LC to changes in biomass energy. **Note**: **(A)** shows a positive change in biomass, while **(B)** indicates a negative change in biomass.

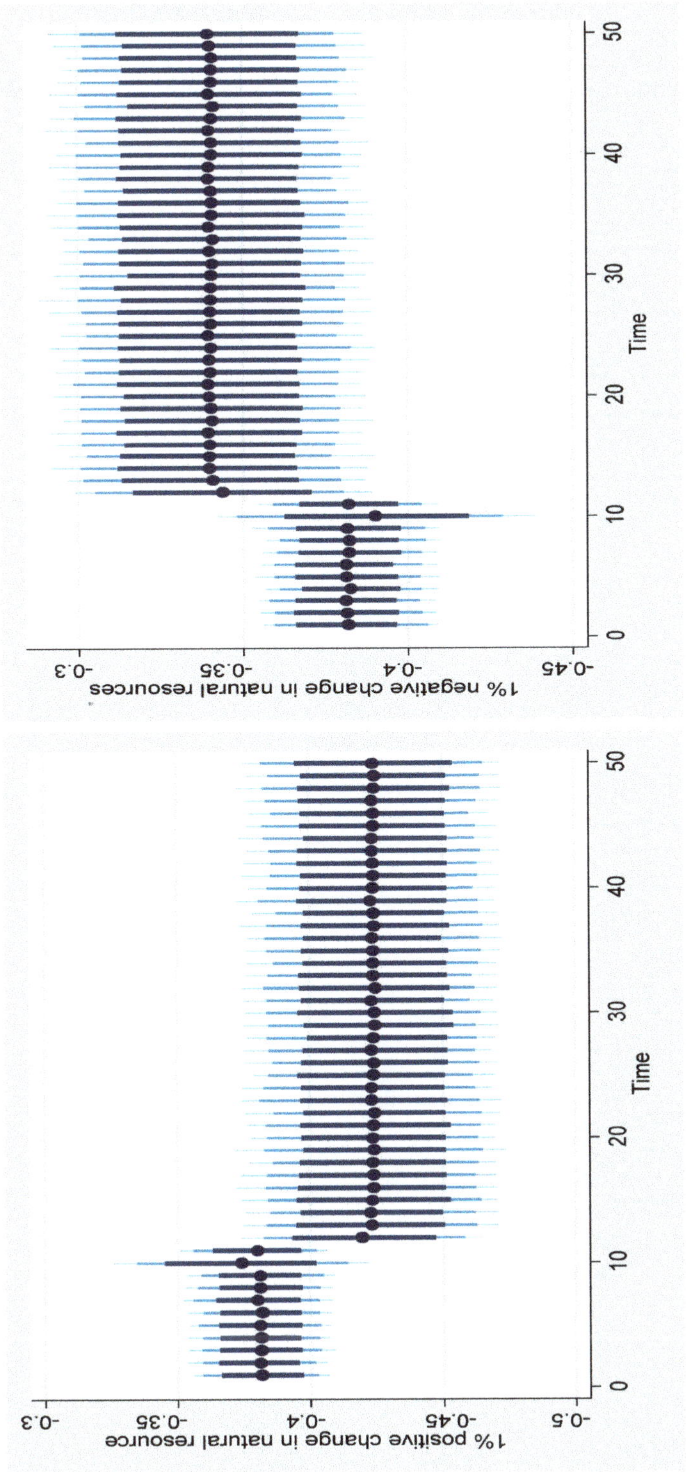

(A) (B)

Figure 4. Response of the LC to changes in natural resources. **Note**: (**A**) shows a positive change in natural resources, while (**B**) indicates a negative change in natural resources.

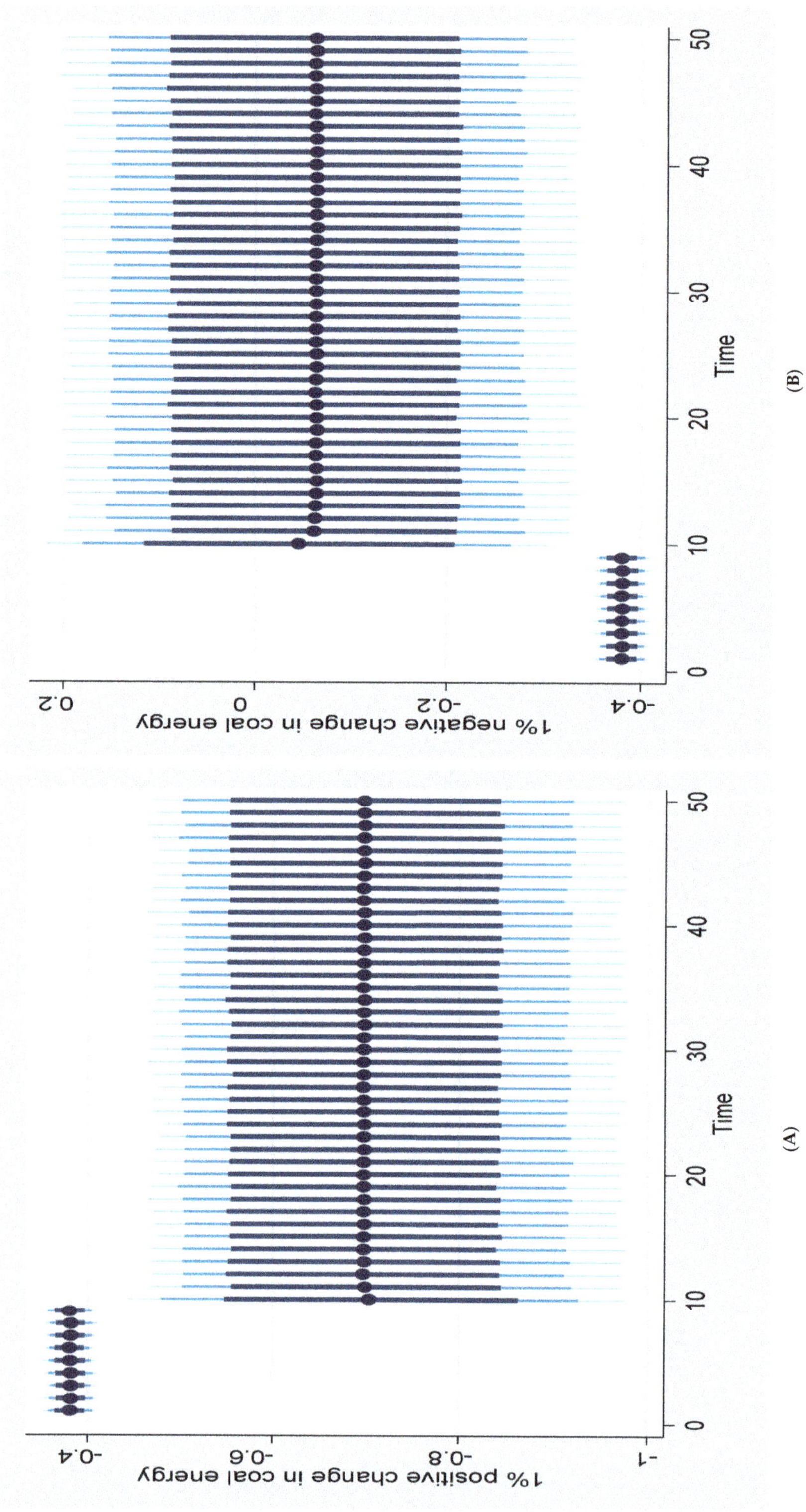

Figure 5. Response of the LC to changes in CO energy. **Note:** **(A)** shows a positive change in CO energy, while **(B)** indicates a negative change in CO energy.

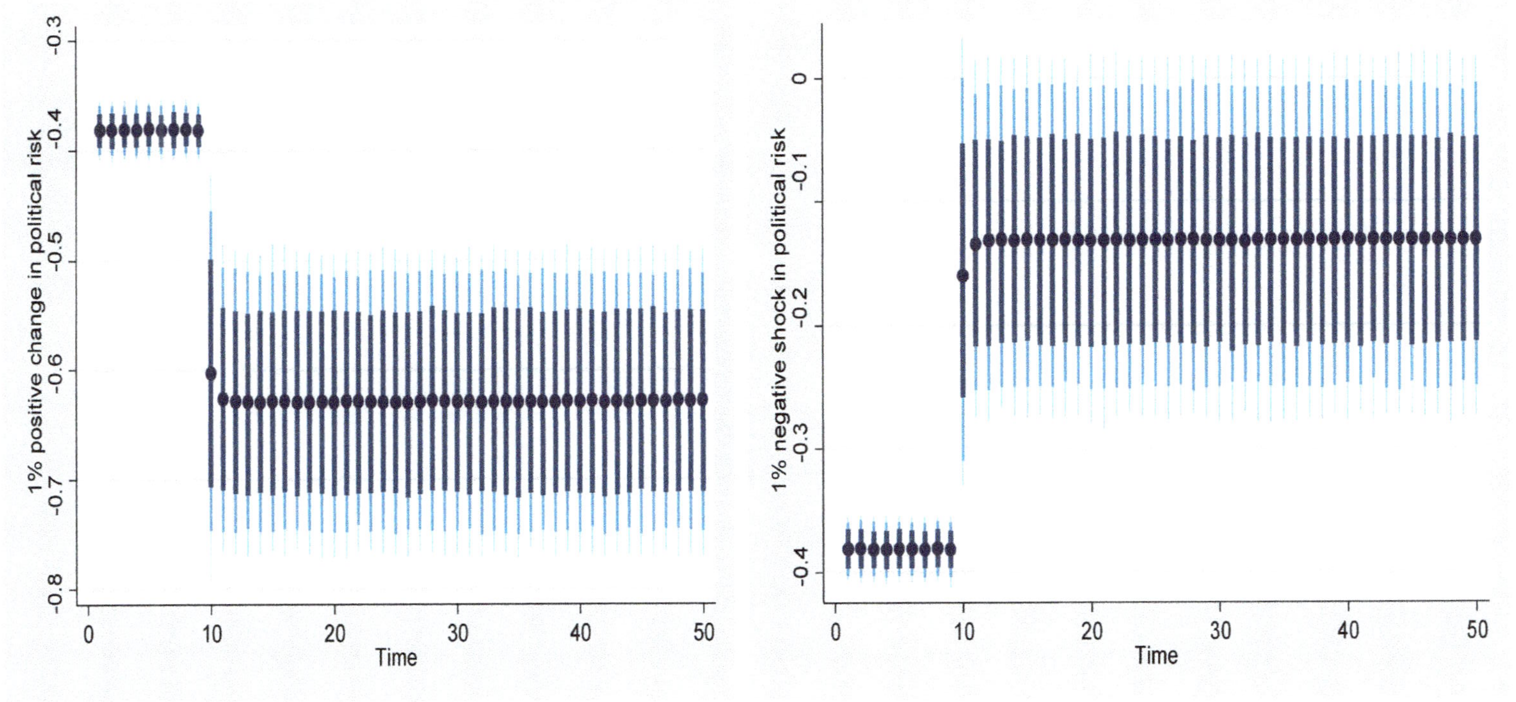

(A) (B)

Figure 6. Response of the LC to changes in political risk. **Note**: **(A)** shows a positive change in political risk, while **(B)** indicates a negative change in political risk.

Table 7. Result of the KRLS methods.

	Average	SE	t-Value	p-Value	P25	P50	P7
CO	−0.288 *	0.481	−5.987	0.000	−0.392	−0.298	−0.145
BM	0.241 *	0.053	4.555	0.000	0.053	0.155	0.447
GDP	−0.261 *	0.038	−6.758	0.000	−0.429	−0.258	−0.098
NRs	−0.007	0.012	−0.549	0.586	−0.039	−0.007	0.025
PR	−0.112 *	0.035	−3.162	0.003	−0.195	−0.119	−0.007

Note: Significance at the 1% is indicated with * symbols.

5. Conclusions and Policy Implication

Identifying the effect of economic growth, coal energy, biomass energy, natural resources, and political risk on the load capacity factor is the major objective of this study. This study focused on South Africa, in which the dataset covers the timeline between 1984 and 2022. For the stationarity nature of the observed parameters, the mix order of integration was confirmed by the conventional unit root tests (ADF and PP) and the ZA unit root. Moreover, this study also used the Bound test approach, along with [80]'s critical approximation, which confirmed a long-term equilibrium interconnection between the LC and its regressors. Furthermore, the findings of the dynamic ARDL simulation estimator disclosed that a 1% surge in GDP reduces the LC by 0.552% (long run) and 0.890% (short run), thereby suggesting that economic growth contributes to environmental deterioration. The results further reported a decline in the LC by 0.290 in the long run due to a unit increase in CO. The result of natural resources has an adverse impact on the LCAP, indicating a reduction in environmental sustainability by 0.037%. Meanwhile, biomass contributes to the LC, thereby promoting ecological quality by 0.421%. Political risk contributes to a reduction in the LC by 0.226% (long run) and 0.224% (short run). The KRLS method also corroborated the impact of each regressor on the LC, except for natural resources.

The results of this study offer credence for South African authorities to increase R&D spending on biomass technologies. South Africa's commitment to achieving ecological sustainability and producing a sustainable ecosystem by expanding the quantity of biomass energy in its energy market may be mainly ascribed to this milestone. Furthermore, firms specializing in producing and assembling biomass energy sources need to be eligible for incentives such as subsidies, price control regulations, and tax reliefs. This will empower these firms to increase their output level, thereby making products more affordable and available to all citizens. Furthermore, natural resources undermine environmental quality; therefore, the implementation of sustainable strategies for the effective assessment and oversight of resources is of critical importance for South Africa. Prioritizing a sustainable extraction and production process is a pivotal strategy, thereby positioning South Africa to effectively promote biodiversity, resource recycling, and enhance the overall ecological state while mitigating the depletion of biocapacity.

The establishment of a more robust institutional framework must be implemented by the South African authorities with the aim of mitigating the bureaucracy that hampers the efficient implementation of ecological policies. The institutional framework must be meticulously designed to facilitate the effective implementation and administration of ecological regulations. South African authorities can address the obstacles undermining the effective implementation of ecological regulations by cultivating robust cross-sectoral collaboration, optimizing administrative mechanisms, and reinforcing governance structures to ensure cohesive and efficient regulatory enforcement. Empowering regulatory authorities with the essential knowledge and expertise to proficiently address complex ecological challenges represents a strategic investment that ensures the successful realization of their objectives. This approach significantly mitigates risks associated with ineffective implementation, thereby facilitating the seamless execution of ecological regulations.

Prior to drawing any conclusions from this study, it is crucial to identify that no policy structure can encapsulate all appropriate policy issues within the borders of a specific framework, and even this investigation is not different. The model was designed by considering only five variables, which is this study's drawback. Subsequent research should replicate similar investigations in other emerging and developing countries by examining the asymmetric influence of these variables or incorporating other socioeconomic indicators. Another limitation of this study is the coverage of the data adopted by this study. As a result, future studies can extend the period of study. Furthermore, another limitation of this research is that it relied on time series analysis using the ARDL estimator. However, future studies could adopt other time series approaches such as the quantile-on-quantile approach, wavelet tools, Fourier ARDL, and quantile ARDL.

Author Contributions: Conceptualization, A.S.A.S., S.I. and W.M.S.K.; methodology, W.M.S.K.; software, W.M.S.K.; validation, A.S.A.S., S.I. and W.M.S.K.; formal analysis, A.S.A.S. and S.I.; investigation, W.M.S.K.; resources, A.S.A.S.; data curation, A.S.A.S. and S.I.; writing—original draft preparation, A.S.A.S., S.I. and W.M.S.K.; writing—review and editing, A.S.A.S., S.I. and W.M.S.K.; visualization, A.S.A.S., S.I. and W.M.S.K. All authors have read and agreed to the published version of the manuscript.

Funding: This research received no external funding.

Data Availability Statement: The data are readily available upon request from the corresponding author.

Conflicts of Interest: The authors declare no conflicts of interest.

Nomenclature

ARDL	autoregressive distributive lag estimator
AAH	Augmented Anderson–Hsiao
LC	load capacity factor
GDP	economic growth
CO	coal energy
BM	biomass energy
NRs	natural resources
PR	political risk
EF	ecological footprint

References

1. Alola, A.A.; Adebayo, T.S. Analysing the waste management, industrial and agriculture greenhouse gas emissions of biomass, fossil fuel, and metallic ores utilization in Iceland. *Sci. Total Environ.* **2003**, *887*, 164115. [CrossRef] [PubMed]
2. Díaz, A.; Marrero, G.A.; Puch, L.A.; Rodríguez, J. Economic growth, energy intensity and the energy mix. *Energy Econ.* **2019**, *81*, 1056–1077. [CrossRef]
3. Miao, Y.; Razzaq, A.; Adebayo, T.S.; Awosusi, A.A. Do renewable energy consumption and financial globalisation contribute to ecological sustainability in newly industrialized countries? *Renew. Energy* **2022**, *187*, 688–697. [CrossRef]
4. Ayobamiji, A.A.; Somoye, O.A. How does the shock in technological innovation and hydroelectricity consumption influence the pursuit of carbon neutrality in Colombia? *Clean Technol. Environ. Policy* **2024**, *26*, 2379–2394. [CrossRef]
5. Wang, X.; Chen, G.; Afshan, S.; Awosusi, A.A.; Abbas, S. Transition towards sustainable energy: The role of economic complexity, financial liberalization and natural resources management in China. *Resour. Policy* **2023**, *83*, 103631. [CrossRef]
6. Mu, H.; Ramzan, M.; Nassani, A.A.; Awosusi, A.A.; Olanrewaju, V.O. Pathway towards decarbonization: How do renewable energy, natural gas consumption trade and monetary policy influence this pursuit? *Int. J. Sustain. Dev. World Ecol.* **2025**, *34*, 1–16. [CrossRef]

7. Khan, Y.; Hassan, T. Promoting sustainable development: Evaluating the influence of natural resources, high-tech export and corruption on CO_2 emissions in developing economies. *Resour. Policy* **2024**, *88*, 104511. [CrossRef]
8. Ramzan, M.; Razi, U.; Quddoos, M.U.; Adebayo, T.S. Do green innovation and financial globalization contribute to the ecological sustainability and energy transition in the United Kingdom? Policy insights from a bootstrap rolling window approach. *Sustain. Dev.* **2023**, *31*, 393–414. [CrossRef]
9. Adebayo, T.S.; Özkan, O. Investigating the influence of socioeconomic conditions, renewable energy and eco-innovation on environmental degradation in the United States: A wavelet quantile-based analysis. *J. Clean. Prod.* **2024**, *434*, 140321. [CrossRef]
10. Awosusi, A.A.; Ozdeser, H.; Seraj, M.; Abbas, S. Can green resource productivity, renewable energy, and economic globalization drive the pursuit of carbon neutrality in the top energy transition economies? *Int. J. Sustain. Dev. World Ecol.* **2023**, *30*, 745–759. [CrossRef]
11. Zang, X.; Adebayo, T.S.; Oladipupo, S.D.; Kirikkaleli, D. Asymmetric impact of renewable energy consumption and technological innovation on environmental degradation: Designing an SDG framework for developed economy. *Environ. Technol.* **2023**, *44*, 774–791. [CrossRef] [PubMed]
12. Adebayo, T.S.; Ullah, S.; Kartal, M.T.; Ali, K.; Pata, U.K.; Ağa, M. Endorsing sustainable development in BRICS: The role of technological innovation, renewable energy consumption, and natural resources in limiting carbon emission. *Sci. Total Environ.* **2023**, *859*, 160181. [CrossRef] [PubMed]
13. Razzaq, A.; Wang, S.; Adebayo, T.S.; Al-Faryan, M.A.S. The potency of natural resources on ecological sustainability in PIIGS economies. *Resour. Policy* **2022**, *79*, 102941. [CrossRef]
14. Forster, P.M.; Smith, C.; Walsh, T.; Lamb, W.F.; Lamboll, R.; Hall, B.; Hauser, M.; Ribes, A.; Rosen, D.; Gillett, N.P.; et al. Indicators of Global Climate Change 2023: Annual update of key indicators of the state of the climate system and human influence. *Earth Syst. Sci. Data* **2024**, *16*, 2625–2658. [CrossRef]
15. Wang, Z.; Xue, W.; Li, K.; Tang, Z.; Liu, Y.; Zhang, F.; Cao, S.; Peng, X.; Wu, E.Q.; Zhou, H. Dynamic combustion optimization of a pulverized coal boiler considering the wall temperature constraints: A deep reinforcement learning-based framework. *Appl. Therm. Eng.* **2025**, *259*, 124923. [CrossRef]
16. Su, Z.-W.; Umar, M.; Kirikkaleli, D.; Adebayo, T.S. Role of political risk to achieve carbon neutrality: Evidence from Brazil. *J. Environ. Manage.* **2021**, *298*, 113463. [CrossRef]
17. Kirikkaleli, D.; Altuntaş, M.; Awosusi, A.A.; Adebayo, T.S. Role of technological innovation and globalization in BRICS economies: Policy towards environmental sustainability. *Int. J. Sustain. Dev. World Ecol.* **2022**, *29*, 593–610. [CrossRef]
18. Global Footprint Network. Ecological Footprint—Global Footprint Network. Available online: https://www.footprintnetwork.org/our-work/ecological-footprint/ (accessed on 11 October 2021).
19. IEA Bioenergy. Implementation of Bioenergy in South Africa—2021, Update 2021. Available online: https://www.ieabioenergy.com/wp-content/uploads/2021/11/CountryReport2021_SouthAfrica_final.pdf (accessed on 27 February 2024).
20. Akadiri, S.S.; Adebayo, T.S.; Altuntaş, M.; Awosusi, A.A. Environmental effects of structural change, hydro and coal energy consumption on ecological footprint in India: Insights from the novel dynamic ARDL simulation. *Environ. Dev. Sustain.* **2022**, *25*, 14309–14332. [CrossRef]
21. Magazzino, C.; Bekun, F.V.; Etokakpan, M.U.; Uzuner, G. Modeling the dynamic Nexus among coal consumption, pollutant emissions and real income: Empirical evidence from South Africa. *Environ. Sci. Pollut. Res.* **2020**, *27*, 8772–8782. [CrossRef]
22. Alhassan, A.; Ozturk, I.; AL-Zyoud, M.F.; Bekun, F.V. Coal consumption-environmental sustainability nexus in developed and developing major coal-consuming economies. *Heliyon* **2024**, *10*, e25619. [CrossRef]
23. Bekun, F.V.; Adebayo, T.S.; Awosusi, A.A.; Altuntaş, M. Coal energy consumption beat renewable energy consumption in South Africa: Developing policy framework for sustainable development. *Renew. Energy* **2021**, *175*, 1012–1024. [CrossRef]
24. Shahbaz, M.; Tiwari, A.K.; Nasir, M. The effects of financial development, economic growth, coal consumption and trade openness on CO_2 emissions in South Africa. *Energy Policy* **2013**, *61*, 1452–1459. [CrossRef]
25. Tiwari, A.K.; Shahbaz, M.; Hye, Q.M.A. The environmental Kuznets curve and the role of coal consumption in India: Cointegration and causality analysis in an open economy. *Renew. Sustain. Energy Rev.* **2013**, *18*, 519–527. [CrossRef]
26. Adebayo, T.S. Trade-off between environmental sustainability and economic growth through coal consumption and natural resources exploitation in China: New policy insights from wavelet local multiple correlation. *Geol. J.* **2023**, *58*, 1384–1400. [CrossRef]
27. Akadiri, S.S.; Akpan, U.; Aladenika, B.; Adebayo, T.S. Asymmetric effect of financial globalization on carbon emissions in G7 countries: Fresh insight from quantile-on-quantile regression. *Energy Environ.* **2022**, *34*, 1285–1304. [CrossRef]
28. Magazzino, C.; Mele, M.; Schneider, N.; Shahbaz, M. Can biomass energy curtail environmental pollution? A quantum model approach to Germany. *J. Environ. Manag.* **2021**, *287*, 112293. [CrossRef]
29. Ding, C.; Awosusi, A.A.; Abbas, S.; Ojekemi, O.R. Formulating ecological sustainability policies for India within the coal energy, biomass energy, and economic globalization framework. *Environ. Sci. Pollut. Res.* **2023**, *30*, 112758–112772. [CrossRef]

30. Wang, Z.; Bui, Q.; Zhang, B.; Pham, T.L.H. Biomass energy production and its impacts on the ecological footprint: An investigation of the G7 countries. *Sci. Total Environ.* **2020**, *743*, 140741. [CrossRef]
31. Awosusi, A.A.; Adebayo, T.S.; Altuntaş, M.; Agyekum, E.B.; Zawbaa, H.M.; Kamel, S. The dynamic impact of biomass and natural resources on ecological footprint in BRICS economies: A quantile regression evidence. *Energy Rep.* **2022**, *8*, 1979–1994. [CrossRef]
32. Pata, U.K.; Kartal, M.T.; Adebayo, T.S.; Ullah, S. Enhancing environmental quality in the United States by linking biomass energy consumption and load capacity factor. *Geosci. Front.* **2023**, *14*, 101531. [CrossRef]
33. Wang, C.; Raza, S.A.; Adebayo, T.S.; Yi, S.; Shah, M.I. The roles of hydro, nuclear and biomass energy towards carbon neutrality target in China: A policy-based analysis. *Energy* **2022**, *262*, 125303. [CrossRef]
34. Zhang, S.; Ramzan, M.; Awosusi, A.A.; Eweade, B.S.; Ojekemi, O.S. Unraveling causal dynamics: Exploring resource efficiency and biomass utilization in Malaysia's context. *Renew. Energy* **2024**, *226*, 120368. [CrossRef]
35. Awosusi, A.A.; Eweade, B.S.; Ojekemi, O.S. Analyzing the environmental role of resource efficiency, economic globalization, and biomass usage in Malaysia: A time-varying causal approach. *Environ. Dev. Sustain.* **2024**, *30*, 1–24. [CrossRef]
36. Zhou, H.; Awosusi, A.A.; Dagar, V.; Zhu, G.; Abbas, S. Unleashing the asymmetric effect of natural resources abundance on carbon emissions in regional comprehensive economic partnership: What role do economic globalization and disaggregating energy play? *Resour. Policy* **2023**, *85*, 103914. [CrossRef]
37. Ashraf, M.Z.; Wei, W.; Usman, M.; Mushtaq, S. How can natural resource dependence, environmental-related technologies and digital trade protect the environment: Redesigning SDGs policies for sustainable environment? *Resour. Policy* **2024**, *88*, 104456. [CrossRef]
38. Zhang, Y.; Radmehr, R.; Ali, E.B.; Samour, A. Natural resources, financial globalization, renewable energy, and environmental quality: Novel findings from top natural resource abundant countries. *Gondwana Res.* **2024**, *140*, 60–70. [CrossRef]
39. Qing, L.; Usman, M.; Radulescu, M.; Haseeb, M. Towards the vision of going green in South Asian region: The role of technological innovations, renewable energy and natural resources in ecological footprint during globalization mode. *Resour. Policy* **2024**, *88*, 104506. [CrossRef]
40. Uzar, U. The dynamic effect of income distribution, natural resources, and freedom of press on ecological footprint: Theory and empirical evidence for emerging economies. *Resour. Policy* **2024**, *89*, 104682. [CrossRef]
41. Satrovic, E.; Cetindas, A.; Akben, I.; Damrah, S. Do natural resource dependence, economic growth and transport energy consumption accelerate ecological footprint in the most innovative countries? The moderating role of technological innovation. *Gondwana Res.* **2024**, *127*, 116–130. [CrossRef]
42. Udemba, E.N.; Shah, S.A.R.; Philip, L.D.; Zhao, G. The mediating role of green energy and environmental policies in sustainable development for BRICS economies: A tripartite impact of entrepreneurial activities, urban development and economic growth on ecological footprint. *Sustain. Dev.* **2024**, *32*, 4649–4670. [CrossRef]
43. Iqbal, S.; Wang, Y.; Shaikh, P.A.; Maqbool, A.; Hayat, K. Exploring the asymmetric effects of renewable energy production, natural resources, and economic progress on CO_2 emissions: Fresh evidence from Pakistan. *Environ. Sci. Pollut. Res.* **2022**, *29*, 7067–7078. [CrossRef] [PubMed]
44. Erdogan, S. On the impact of natural resources on environmental sustainability in African countries: A comparative approach based on the EKC and LCC hypotheses. *Resour. Policy* **2024**, *88*, 104492. [CrossRef]
45. Li, S.; Sun, H.; Sharif, A.; Bashir, M.; Bashir, M.F. Economic complexity, natural resource abundance and education: Implications for sustainable development in BRICST economies. *Resour. Policy* **2024**, *89*, 104572. [CrossRef]
46. Yu, Z. Do natural resources promote carbon neutrality: The role of green finance. *Resour. Policy* **2024**, *88*, 104424. [CrossRef]
47. Kartal, M.T.; Kirikkaleli, D.; Depren, S.K. Investigating political stability effect on the environment in the Netherlands by Fourier-based approaches. *Int. J. Sustain. Dev. World Ecol.* **2024**, *31*, 125–136. [CrossRef]
48. Guan, L.; Ali, Z.; Uktamov, K.F. Exploring the impact of social capital, institutional quality and political stability on environmental sustainability: New insights from NARDL-PMG. *Heliyon* **2024**, *10*, e24650. [CrossRef]
49. Han, Y.; Bao, M.; Niu, Y.; Rehman, J.U. Driving towards net zero emissions: The role of natural resources, government debt and political stability. *Resour. Policy* **2024**, *88*, 104479. [CrossRef]
50. Bilgili, F.; Alsanusi, M.; Kabir, M.M.; Awan, A. Quantile dynamics of control of corruption, political stability, and renewable energy on environmental quality in the MENA region. *Environ. Dev. Sustain.* **2024**, *30*, 1–21. [CrossRef]
51. Asif, K.; Sabir, S.; Qayyum, U. Corruption, Political Instability, and Environmental Degradation in South Asia: A Comparative Analysis of Carbon Footprint and Ecological Footprint. *J. Knowl. Econ.* **2023**, *15*, 4072–4096. [CrossRef]
52. Kirikkaleli, D.; Adebayo, T.S. Political risk and environmental quality in Brazil: Role of green finance and green innovation. *Int. J. Finance Econ.* **2022**, *29*, 1205–1218. [CrossRef]
53. Adebayo, T.S. Renewable Energy Consumption and Environmental Sustainability in Canada: Does Political Stability Make a Difference? *Environ. Sci. Pollut. Res.* **2022**, *29*, 61307–61322. [CrossRef] [PubMed]

54. Sun, X.Q.; Awosusi, A.A.; Han, Z.; Uzun, B.; Öncü, E. Racing towards environmental sustainability: A synergy between economic complexity, political stability, and energy transition: Policy insight from a bootstrap time varying causality approach. *Int. J. Sustain. Dev. World Ecol.* **2024**, *31*, 206–221. [CrossRef]
55. Ojekemi, O.S.; Ağa, M.; Magazzino, C. Towards Achieving Sustainability in the BRICS Economies: The Role of Renewable Energy Consumption and Economic Risk. *Energies* **2023**, *16*, 5287. [CrossRef]
56. Ojekemi, O.S.; Ağa, M. In the era of globalization, can renewable energy and eco-innovation be viable for environmental sustainability in BRICS economies? *Environ. Sci. Pollut. Res.* **2023**, *36*, 85249–85262. [CrossRef]
57. Awosusi, A.A.; Akadiri, S.S.; Olanrewaju, V.O.; Rjoub, H.; Ozdeser, H.; Ojekemi, O. The role of energy, political stability, and real income on achieving carbon neutrality: Asymmetric evidence. *Environ. Sci. Pollut. Res.* **2023**, *30*, 83302–83318. [CrossRef]
58. Adebayo, T.S.; Samour, A. Renewable energy, fiscal policy and load capacity factor in BRICS countries: Novel findings from panel nonlinear ARDL model. *Environ. Dev. Sustain.* **2023**, *26*, 4365–4389. [CrossRef]
59. Xu, D.; Salem, S.; Awosusi, A.A.; Abdurakhmanova, G.; Altuntaş, M.; Oluwajana, D.; Kirikkaleli, D.; Ojekemi, O. Load Capacity Factor and Financial Globalization in Brazil: The Role of Renewable Energy and Urbanization. *Front. Environ. Sci.* **2022**, *9*, 823185. Available online: https://www.frontiersin.org/article/10.3389/fenvs.2021.823185 (accessed on 11 February 2022). [CrossRef]
60. Ni, Z.; Yang, J.; Razzaq, A. How do natural resources, digitalization, and institutional governance contribute to ecological sustainability through load capacity factors in highly resource-consuming economies? *Resour. Policy* **2022**, *79*, 103068. [CrossRef]
61. Yang, M.; Magazzino, C.; Awosusi, A.A.; Abdulloev, N. Determinants of Load capacity factor in BRICS countries: A panel data analysis. *Nat. Resour. Forum* **2023**, *25*, 525–548. [CrossRef]
62. Ahmed, Z.; Le, H.P. Linking Information Communication Technology, trade globalization index, and CO_2 emissions: Evidence from advanced panel techniques. *Environ. Sci. Pollut. Res.* **2021**, *28*, 8770–8781. [CrossRef]
63. Zivot, E.; Andrews, D.W.K. Further Evidence on the Great Crash, the Oil-Price Shock, and the Unit-Root Hypothesis. *J. Bus. Econ. Stat.* **2002**, *20*, 25–44. [CrossRef]
64. Ayobamiji, A.A.; Kalmaz, D.B. Reinvestigating the determinants of environmental degradation in Nigeria. *Int. J. Econ. Policy Emerg. Econ.* **2020**, *13*, 52–71. [CrossRef]
65. Phillips, P.C.B.; Hansen, B.E. Statistical Inference in Instrumental Variables Regression with I(1) Processes. *Rev. Econ. Stud.* **1990**, *57*, 99–125. [CrossRef]
66. Pesaran, M.H.; Shin, Y.; Smith, R.J. Bounds testing approaches to the analysis of level relationships. *J. Appl. Econom.* **2001**, *16*, 289–326. [CrossRef]
67. Wei, Y.; Ramzan, M.; Uzun, B.; Khudoykulov, K. Racing toward attaining sustainable development in India: Probing the asymmetric effect of country risk and coal energy. *Nat. Resour. Forum* **2024**, *45*, 1–20. [CrossRef]
68. Guloglu, B.; Caglar, A.E.; Pata, U.K. Analyzing the determinants of the load capacity factor in OECD countries: Evidence from advanced quantile panel data methods. *Gondwana Res.* **2023**, *118*, 92–104. [CrossRef]
69. Somoye, O.A.; Ayobamiji, A.A. Can energy intensity, clean energy utilization, economic expansion, and financial development contribute to ecological progress in Iceland? A quantile-on-quantile KRLS analysis. *Nat. Resour. Forum* **2024**, *44*, 1–20. [CrossRef]
70. Kartal, M.T.; Pata, U.K.; Depren, S.K.; Depren, Ö. Effects of possible changes in natural gas, nuclear, and coal energy consumption on CO_2 emissions: Evidence from France under Russia's gas supply cuts by dynamic ARDL simulations approach. *Appl. Energy* **2023**, *339*, 120983. [CrossRef]
71. Waris, A.; Khan, S.; Hronec, M.; Suplata, M. The impact of hydro-biofuel-wind-solar energy consumption and coal consumption on carbon emission in G20 countries. *Environ. Sci. Pollut. Res.* **2023**, *30*, 72503–72513. [CrossRef]
72. Kim, G.; Choi, S.-K.; Seok, J.H. Does biomass energy consumption reduce total energy CO_2 emissions in the US? *J. Policy Model.* **2020**, *42*, 953–967. [CrossRef]
73. Naqvi, S.A.A.; Hussain, M.; Hussain, B.; Shah, S.A.R.; Nazir, J.; Usman, M. Environmental sustainability and biomass energy consumption through the lens of pollution Haven hypothesis and renewable energy-environmental kuznets curve. *Renew. Energy* **2023**, *212*, 621–631. [CrossRef]
74. Xin, Y.; Li, H.; Adebayo, T.S.; Awosusi, A.A. Asymmetric linkage between biomass energy consumption and ecological footprints in top ten biomass-consuming nations. *Econ. Res.-Ekon. Istraz.* **2022**, *36*, 2147567. [CrossRef]
75. Yasmeen, R.; Zhaohui, C.; Shah, W.U.H.; Kamal, M.A.; Khan, A. Exploring the role of biomass energy consumption, ecological footprint through FDI and technological innovation in B&R economies: A simultaneous equation approach. *Energy* **2021**, *244*, 122703. [CrossRef]
76. Adebayo, T.S.; Samour, A.; Alola, A.A.; Abbas, S.; Ağa, M. The potency of natural resources and trade globalisation in the ecological sustainability target for the BRICS economies. *Heliyon* **2023**, *9*, e15734. [CrossRef]
77. Zhao, W.-X.; Samour, A.; Yi, K.; Al-Faryan, M.A.S. Do technological innovation, natural resources and stock market development promote environmental sustainability? Novel evidence based on the load capacity factor. *Resour. Policy* **2023**, *82*, 103397. [CrossRef]

78. Xu, P.; Hussain, M.; Ye, C.; Wang, J.; Wang, C.; Geng, J.; Liu, Y.; Chen, J. Natural resources, economic policies, energy structure, and ecological footprints' nexus in emerging seven countries. *Resour. Policy* **2022**, *77*, 102747. [CrossRef]
79. Appiah, M.; Li, M.; Sehrish, S.; Abaji, E.E. Investigating the connections between innovation, natural resource extraction, and environmental pollution in OECD nations; examining the role of capital formation. *Resour. Policy* **2023**, *81*, 103312. [CrossRef]
80. Kripfganz, S.; Schneider, D.C. ARDL: Stata Module to Perform Autoregressive Distributed Lag Model Estimation. 2016. Available online: https://www.stata.com/meeting/chicago16/slides/chicago16_kripfganz.pdf (accessed on 26 October 2024).

energies

MDPI

Article

Energy Entrepreneurship in the Emerging New Globalization: A Macro–Meso–Micro Perspective with Evidence from a Less-Developed Regional Ecosystem

Dimos Chatzinikolaou [1,2,3] and Charis Michael Vlados [1,2,3,*]

1 Department of Economics, Democritus University of Thrace, 69100 Komotini, Greece
2 Knowledge Management, Innovation and Strategy Center (KISC), University of Nicosia, P.O. Box 24005, Nicosia CY-2417, Cyprus
3 School of Business, University of Nicosia, P.O. Box 24005, Nicosia CY-2417, Cyprus
* Correspondence: cvlados@econ.duth.gr

Abstract: This study aims to analyze the shifting focus and emerging themes in contemporary energy entrepreneurship research, alongside the challenges and opportunities faced by select energy entrepreneurs in a rapidly transforming global landscape that is driven by sustainability imperatives, resilience, and systemic energy transitions. Employing a semi-systematic and critical literature review, 238 relevant scientific articles from the Web of Science database were identified and analyzed. We then conducted focused case studies of energy entrepreneurs in a less-developed regional ecosystem. The findings reveal two distinct "generations" of energy entrepreneurship research, marked by a clear shift towards sustainability themes, innovative energy business models, and corporate responsibility. Additionally, we introduce the "Energy Innovation Scorecard" (EN.I.SCORE) framework—a comprehensive macro–meso–micro guideline designed to support energy entrepreneurs. Applying this framework to a sample of 89 surveyed and 8 interviewed firms reveals that, especially among microfirms, there is weak integration of strategy–technology–management, limited energy innovation, and poor financial performance prevail—challenges typical of underdeveloped ecosystems. By examining the origins, evolution, and holistic transformation of energy entrepreneurship, and by investigating the selected regional case, this research potentially helps bridge critical gaps in understanding the dynamics of this field.

Keywords: energy entrepreneurship; new globalization; energy transition; sustainability; macro–meso–micro

1. Introduction

In the context of the transformational crisis within the energy sector—an integral feature of the latest phase of global capitalism characterized by the gradual emergence of a new wave of restructured globalization—the priorities of stakeholders worldwide have undergone a significant shift [1–5]. The era of unrestrained growth and unchecked exploitation of energy resources has been replaced by more complex and often conflicting developmental imperatives across the production spectrum, with a pronounced focus on energy production, distribution, and consumption [6]. In this evolving landscape, a "new model of entrepreneurship" in the energy sector can play a critical role in facilitating the transition towards sustainable energy sources and practices. Historically, innovative entrepreneurs, alongside supportive state policies, have been pivotal in revitalizing socioeconomic systems and propelling capitalism into new phases of development [7]. In line with this tradition, innovative entrepreneurship in the energy sector has the potential to significantly contribute to the ongoing energy transition. By establishing new ventures in renewable energy or challenging established utilities to adapt to the demands of this emerging era, such entrepreneurship can serve as a key driver of transformative change [8,9].

Recent research has increasingly explored the multifaceted dimensions of entrepreneurship within the energy sector. However, systematic approaches remain relatively scarce. A notable contribution in this area is the now-classic *Handbook of Research on Energy Entrepreneurship*, which outlines key research directions and compiles a broad spectrum of contributions [10]. A significant insight from this handbook is the evolving role of energy customers—transitioning from passive consumers to active producers, thus giving rise to a new category of "prosumers." This term, introduced by Toffler, describes individuals or entities capable of generating energy, reflecting the convergence of production and consumption roles in a mature market [11,12]. This shift has significant implications for all energy organizations, particularly incumbent ones, which must innovate and adapt to the industry's evolving energy structures. A notable trend is the increasing view of energy as a commodity. In financial markets, energy has been treated as a commodity in derivative trading for decades, and its prominence in such contexts only continues to grow. This marks a sharp departure from its traditional status as a public service, a shift that demands strategic agility from industry players [13]. Furthermore, recent studies highlight the evolving role of entrepreneurial actors in energy transition processes, emphasizing their impact on sustainability, technological innovation, and policy adaptation [8,9,14,15].

This necessitates further exploration into the sophisticated dimensions of energy entrepreneurship. Currently, it remains unclear whether microlevel entrepreneurial issues have been adequately addressed in the field. Moreover, it is uncertain how the landscape of entrepreneurship research has evolved recently and the ways in which energy entrepreneurs can navigate the new challenges of this era. This paper aims to answer the following questions:

1. How has the focus and thematic emphasis of contemporary entrepreneurship research in the energy sector evolved over the last few decades?
2. According to recent research, how can energy entrepreneurs successfully navigate the challenges and leverage the opportunities of the evolving socioeconomic environment?
3. How do select energy entrepreneurs innovate, what deficiencies do they exhibit, and how might these be addressed?

To address these questions, this paper investigates the evolution of energy entrepreneurship research using a macro–meso–micro analytical framework. This approach enables us to unravel the complex layers of the energy ecosystem and examine critical factors affecting the development of vulnerable socioeconomic systems in terms of their energy balance. First, we conducted a critical semi-systematic literature review, analyzing 238 scientific articles from the Web of Science (WoS) database. This analysis revealed recurring patterns in the literature, their evolution over time, and their defining characteristics. A key finding is the shift in post-2010 literature compared to earlier works, reflecting changes in perceptions, preferences for energy sources, a more integrated understanding of energy ecosystems and communities, and a heightened emphasis on innovation, sustainability, and responsibility—all indicative of the emerging new globalization. Next, we present an empirical investigation comprising a survey of 89 energy firms and in-depth interviews with 8 microfirms. This investigation examines the strategic, technological, and managerial strengths and weaknesses of these organizations and explores why certain firms appear more developed than others.

The remainder of the paper is structured as follows. Section 2 delineates the research field and provides a detailed background on pertinent energy entrepreneurship research. Section 3 describes the methodology used to critically analyze the selected literature, explains how we developed a new framework for supporting energy firms, and details the case selection process to illustrate these developments. In Section 4, we present our findings by identifying key themes in the literature—categorized at the macro, meso, and microlevels—and trace their evolution across different generations of energy entrepreneurship research. This section also introduces a framework designed to help energy entrepreneurs navigate the new era, supported by case studies of selected energy firms. Section 5 critically discusses these findings, and finally, Section 6 concludes the paper with final remarks and suggestions for future research avenues.

2. Theoretical Background

Energy economics and entrepreneurship in the energy sector have undergone significant transformations, shaped by historical events, technological advancements, and the demands of globalization. This section delves into the theoretical foundations of these shifts, focusing first on the evolving paradigms of energy economics (2.1) and then on the emergence of energy entrepreneurship as a critical driver of innovation and sustainability in the globalized era (2.2).

2.1. Research Generations of Energy Economics and the Emergence of a New Era of Globalization

Energy economics explores the societal dynamics of energy supply, production, distribution, storage, and consumption, delving into the costs and benefits of energy services that facilitate energy delivery to end-use applications [6]. Over time, the evolution of energy economics has closely mirrored the shifts in global capitalism, particularly since World War II (postwar). This relationship is extensively documented in globalization literature, which emphasizes transformations in political–economic structures [1,3].

The development of global capitalism since the postwar era can be categorized into periods that represent distinct phases in global economic and political structures characterized by shifts in priorities such as industrial reconstruction, energy security, market liberalization, and sustainability, driven by changing technological, geopolitical, and economic conditions [2,16]. These periods—reconstruction and growth (1945–1973), energy crises and diversification (1973–1980), liberalization and climate awareness (1980–2008), and energy transition and digitalization (post-2008)—illustrate significant structural changes in global priorities (Figure 1).

Parallel to these shifts, energy economics has evolved through broader periods that are distinct phases defined by dominant paradigms in energy production, consumption, and policy, reflecting the interaction between political-economic forces and technological advancements in the energy sector [17–21]. These broader periods include the rebound effect (1860s–1930s), energy security and expansion (1940s–1970s), energy efficiency gap (1980s–1990s), and green nudges (1990s–present), each reflecting a unique focus on efficiency, security, or sustainability:

- Rebound effect (1860s–1930s): This period coincides with the rapid industrialization of economies during the late 19th and early 20th centuries. The rebound effect refers to the phenomenon where improvements in energy efficiency lower energy costs, leading to increased energy consumption that offsets the initial efficiency gains [22]. The focus during this era was on maximizing production output without consideration for environmental or sustainability concerns.
- Energy security and expansion (1940s–1970s): The postwar period saw a global emphasis on energy security and infrastructure expansion. Nations prioritized securing

reliable energy supplies to support industrial reconstruction and geopolitical stability. Investments in large-scale energy systems, including oil, coal, and nuclear power, were central to this phase. The oil crises of the 1970s brought energy vulnerability to the forefront, compelling policymakers to consider diversification strategies [23,24]. This period bridged the gap between industrial expansion and the emerging awareness of resource limitations.

- Energy efficiency gap (1980s–1990s): By the 1980s, globalization and market liberalization shaped energy policies that prioritized cost-effective energy use. The energy efficiency gap describes the persistent under-adoption of economically viable energy efficiency measures, attributed to market failures, behavioral barriers, and policy gaps [25,26]. During this phase, energy economics began integrating environmental considerations, particularly as the impacts of fossil fuel dependence on climate change became evident.
- Green nudges (1990s–present): The most recent period reflects a global shift toward sustainability and renewable energy systems. Green nudges refer to policy and behavioral interventions designed to promote environmentally friendly energy consumption practices. This phase is characterized by decentralized energy systems, digital innovation, and a focus on consumer empowerment to achieve climate goals [19,27]. The transition to renewable energy has also driven advancements in energy entrepreneurship and community-based energy ecosystems.

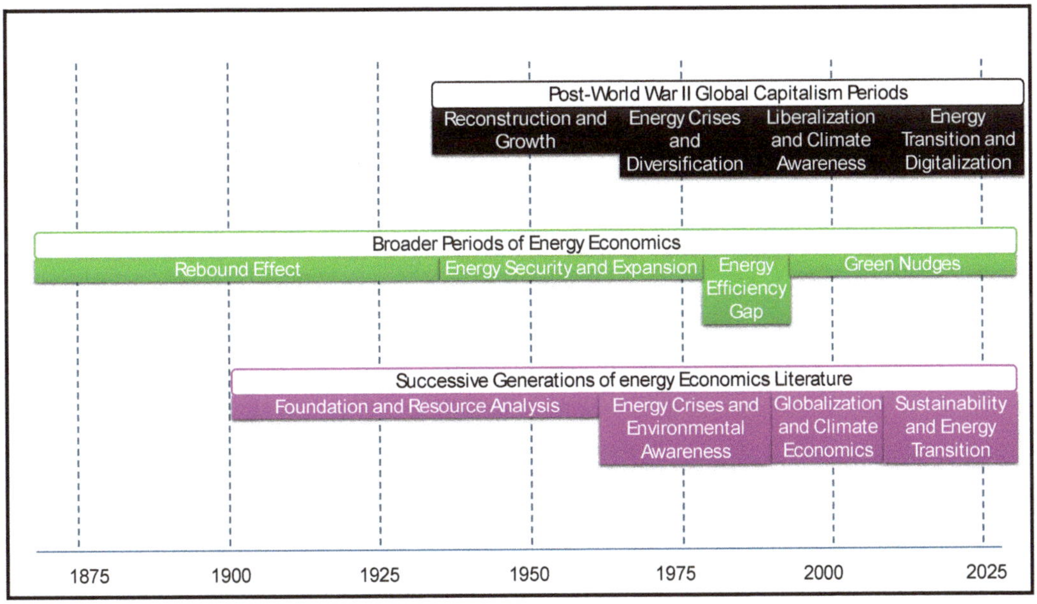

Figure 1. Evolution of energy economics and global capitalism: key periods and paradigms.

These broader periods align with successive generations of energy economics literature, which refer to distinct phases in academic inquiry and research focus shaped by the evolving priorities of energy systems and policies, including resource pricing, environmental awareness, climate change economics, and sustainability. Early research focused on resource and pricing analysis in the mid-20th century, emphasizing the economics of fossil fuels like coal and oil [28,29]. In the 1970s and 1980s, attention shifted to energy crises and environmental awareness, driven by the oil crises and the need for energy diversification [30–32]. From the 1990s onward, globalization and climate economics emerged, incorporating market liberalization, carbon pricing, and the integration of renewable energy systems [33–35]. The current phase emphasizes sustainability and energy transition, with a focus on digitalization, decentralized systems, and ecosystem-based approaches [8,9,14,15].

The interconnected evolution of postwar global capitalism periods, broader periods of energy economics, and successive generations of energy economics literature illustrates the dynamic interplay between global economic transformations and energy systems. These frameworks highlight how changes in global economic priorities have both influenced and been influenced by shifts in energy policies, technologies, and consumption patterns. Distinguishing between these categories is crucial because they serve different yet complementary roles: postwar global capitalism periods provide a macrolevel context of economic and political restructuring, broader periods of energy economics focus on the paradigms shaping real-world energy practices, and successive generations of energy economics literature capture the evolving intellectual frameworks developed in response to these transformations. For example, the rebound effect period aligns with industrial expansion and the prioritization of production efficiency during the late 19th and early 20th centuries, reflecting a focus on maximizing output without sustainability considerations. In contrast, the green nudges period corresponds to the modern transition toward renewable energy, characterized by digital innovation, decentralized systems, and a growing focus on consumer empowerment and environmental consciousness. Recognizing these distinctions allows for a more nuanced understanding of how global economic transitions, energy systems, and academic thought mutually influence and shape one another.

Some recent approaches increasingly advocate for an integrative "macro–meso–micro" framework. The integrated evolutionary macro–meso–micro synthesis in social sciences is a framework that examines societal changes across large-scale (macrolevel), intermediate (mesolevel), and individual (microlevel) levels through an evolutionary lens [36]. This approach seeks to understand the interplay between broader societal structures, various aggregational dynamics, and individual behaviors in the context of social evolution and adaptation [37]. Such an evolutionary perspective offers valuable insights for analyzing energy sector ecosystems, encompassing the interactions among energy suppliers, producers, distributors, and consumers [38–46]. It underscores the importance of examining microlevel roles of individual energy firms within mesolevel local and trans-spatial ecosystems, as well as macrolevel—national and international—systemic dynamics [47,48].

Nonetheless, despite notable advancements in contemporary energy economics, microlevel entrepreneurial issues seemingly require further investigation. Reflecting on Schumpeter's [7] emphasis on innovative entrepreneurs as key drivers of innovation and creators of new economic sectors, understanding entrepreneurship's role in energy economics becomes crucial, especially in the context of new globalization. This underscores the imperative to foster innovation, resilience, and sustainability in energy entrepreneurship—a theme expanded upon in the following section, which emphasizes its pivotal role in shaping the new globalization era.

2.2. *Fostering Energy Entrepreneurship in the New Globalization*

The "living organization" perspective is seemingly one of the most intriguing and relatively recent approaches to entrepreneurship. It understands business elements like culture, mission, and vision differently than traditional business literature, suggesting these elements evolve organically over time [49,50]. This perspective, rooted in biological analogies, posits innovation as an evolutionary process, which is, among other factors, crucial in the emerging global era [51,52].

Energy entrepreneurship is seemingly an emerging field, whose trajectory in the new globalization context remains largely under-researched. A recent notable handbook nevertheless delineates promising research avenues within energy entrepreneurship [10]. The highlighted themes encompass institutional adaptation, entrepreneurial marketing, renewable energy and cleantech innovations, political implications, infrastructure constraints, and developing nations' prospects. Within these, the prosumer framework stands out, illustrating the merged roles of producers and consumers in the new era, necessitating flexibility for energy businesses to thrive [13].

Digital technologies are also spearheading the shift towards decentralized energy and prosuming platforms, enabling households to manage renewable electricity production, consumption, storage, and exchange, both locally and on a larger grid scale [53]. The relevant literature appears to lack clear guidelines for fostering the development of prosuming and other energy organizations, underscoring the necessity of further investigating frameworks that could assist with this [10].

One such framework is the Stra.Tech.Man approach, which evolutionarily intertwines strategy, technology, and management. It asserts that the effective combination of these elements drives and renews innovation potential [54]. This synthesis is encapsulated within a "business physiology" perspective, which classifies organizations based on their development of philosophies and procedures within these interconnected spheres [55]. Existing research suggests that Greek firms—and other socioeconomic contexts of comparable developmental scope—often exhibit a "monad-centered" organizational physiology [39,56,57]. Strategically, they typically rely on an instinctive approach, while their technological approach tends to be sporadic, and their management style is rooted in day-to-day practical experience. By contrast, more "flexible" physiologies, which entail cutting-edge technological approaches, dialectical strategic thinking, and participative management, remain relatively scarce in Greece and similarly structured economies [39,56,57]. Building upon this, the Stra.Tech.Man Scorecard (Stra.Tech.Man.SC) serves as a tool for organizations to chart their evolutionary trajectory through introspective analysis (Figure 2). This analysis highlights vulnerabilities and promotes remediation through intra-organizational consultation, scrutinizing additional forces that impinge upon the organization's strategic, technological, and managerial approaches, alongside its innovativeness and financial performance [55].

In response to contemporary global challenges, this scorecard approach has incorporated a "RASI" analysis—the dimensions of resilience, adaptability, sustainability, and inclusiveness—facilitating organizations in navigating the demands of the rising new globalization [58]. Moreover, the "Green Stra.Tech.Man.SC-RASI" signifies a substantial advancement, urging organizations to critically assess their green strategies, technologies, and management practices in alignment with the new globalization imperatives [59].

In conclusion, situating energy entrepreneurship within the broader framework of the "living organization" and a biological and ecosystemic analytical perspective amplifies its importance, particularly in the context of current demands for fostering organizational innovation and advancing digital energy technologies amidst complex, dynamic interactions. Moreover, innovative entrepreneurship is crucial in addressing the challenges arising from the new globalization and the transition towards sustainable energy era. While the Green Stra.Tech.Man.SC-RASI offers a roadmap, the future course of energy entrepreneurship is still ambiguous, highlighting the need to formulate recommendations that support energy entrepreneurs in navigating this shifting landscape.

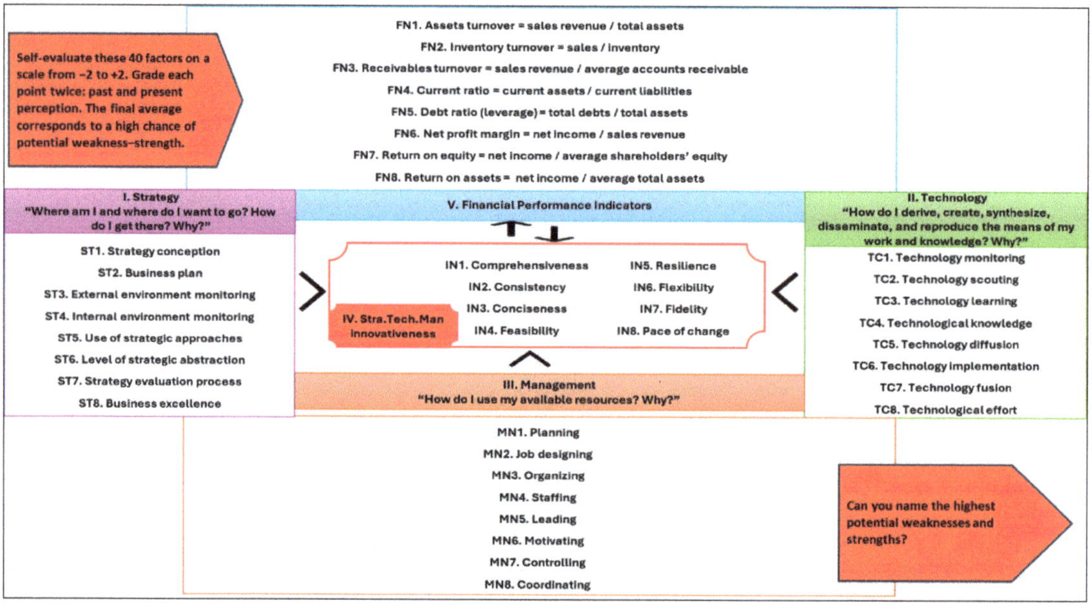

Figure 2. The Stra.Tech.Man.SC approach based on Chatzinikolaou and Vlados [55].

3. Methodology

To address the study's research questions, we employed a twofold methodological design. First, we conducted a semi-systematic literature review of 238 scientific articles to examine the evolution and key thematic shifts in energy entrepreneurship research. Second, we performed an empirical investigation of energy firms located in a less-developed Greek region (Eastern Macedonia and Thrace, ReMTh). This dual approach aimed to (a) establish a critical theoretical backdrop on the phenomenon of energy entrepreneurship, (b) illuminate how contemporary energy firms perceive and implement strategic, technological, and managerial (Stra.Tech.Man) dimensions of innovation, and (c) integrate these findings to propose guidelines that may support energy entrepreneurs in less-developed business ecosystems.

3.1. Semi-Systematic Literature Review

Following recent guidelines on semi-systematic literature reviews in energy sector management research [60–62], we focused on identifying publications examining entrepreneurship in the energy field while ensuring a reproducible and transparent process [63,64]. In October 2023, we searched the Web of Science (WoS)—a database covering over 10,000 academic journals—using a concise set of energy-related keywords (e.g., "energy sector", "energy system", "energy ecosystem") combined with "entrepreneurship" or "business".

The results were refined by selecting WoS categories likely to yield relevant research in economics, management, and sustainability. Non-English publications, conference proceedings, books, and book chapters were excluded. After discarding incomplete entries or items with no available full text, 238 peer-reviewed journal articles remained (refer to Supplementary Files S1 and S2). A bibliometric analysis was then carried out using VOSviewer to track publication patterns, journal distribution, and co-occurrence of authors' keywords. In addition, a word-frequency analysis was conducted by splitting the sample into two periods (1995–2009 and 2010–2023) to detect key terms and thematic evolution, and a critical thematic synthesis grouped the final articles into macrolevel, mesolevel, and microlevel perspectives to identify the main research trends in energy entrepreneurship (Appendix A). This process revealed a surge in publications after 2010 and a shift in emphasis from energy efficiency and liberalization to sustainability, corporate social responsibility (CSR), and innovative business models, as well as limited but emerging research addressing the strategic, technological, and managerial integration required for energy entrepreneurs.

3.2. Case Studies

Second, we collected empirical data from the ReMTh region—a NUTS 2 Greek region characterized by comparatively low competitiveness, structural weaknesses, and less-developed entrepreneurship [65,66]. We aimed to understand how (and to what extent) local energy firms perceive and implement innovative practices and respond to global sustainability imperatives, thereby complementing our literature-based insights.

Registered energy-related firms in the ReMTh were compiled by scanning the Greek Business Registry (ΓΕΜΗ) for businesses with at least one energy-related activity code, based on Bradford's [6] comprehensive energy-sector definition. From an initial set of 2237 entities, those under suspension or lacking any online presence were excluded, resulting in 1025 active firms. In March 2024, each of these firms was invited via email to complete an online survey, yielding 89 valid responses, which represents approximately an 8.7% response rate.

The survey featured 15 five-point Likert-scale questions and one open-ended question on local development (Appendix B). These questions were derived in part from the newly proposed "Energy Innovation Scorecard" (EN.I.SCORE) presented later in the study and were adapted for large-scale use. Collected data were processed using IBM SPSS 29.0.0.0 for descriptive analyses and Pearson's correlation tests. At the end of the survey, respondents were asked if they wished to participate in more extensive in-depth interviews; eight microfirm owners agreed. These interviews, conducted between May and June 2024, lasted approximately one to two hours each and employed a more detailed questionnaire that combined Likert-type items with open-ended inquiries (Appendix C). This approach enriched the quantitative data with nuanced qualitative insights through word frequency analysis and "thick descriptions" [39,67,68].

By triangulating literature-based insights with the survey and interview findings, we aimed to explain why some energy entrepreneurs in less-developed regions exhibit monad-centered strategic conduct, limited managerial sophistication, and weak technological adoption. Throughout our study, we hypothesized that microfirms in less-developed ecosystems often face persistent innovation and performance challenges, fueling significant developmental gaps [57,69].

In overall terms, the scientific hypothesis of this study follows the explanatory and predictive scheme that can be captured as follows: If "X, Y, Z..." in conditions "A, B, C...", then "K, L, M..." for the reasons "N1, N2, N3..." [67,70]. Based on this explanatory mechanism, we can arrive at predictions as conditions change over time—predictions produced specifically from the collection and analysis of primary data. In line with this research perspective, a central hypothesis is formulated as follows: If energy firms in the ReMTh region—particularly microfirms—remain less competent at effectively integrating strategy, technology, and management, they will likely face inadequate innovation and financial performance, thus perpetuating the developmental lag in less competitive areas.

By designing a mixed-method approach that combines a semi-systematic literature review with survey and in-depth interviews, the study aims to test this hypothesis. In other words, our goal is to investigate whether and how smaller energy firms in the ReMTh—confronted with the conditions of a relatively underdeveloped regional ecosystem—struggle with strategy–technology–management integration and, consequently, exhibit diminished innovative and financial capacities.

4. Results

Section 4.1 explores the evolution of energy entrepreneurship research by categorizing it into macrolevel, mesolevel, and microlevel dimensions. This categorization reveals how themes like globalization, energy transition, energy ecosystems, and business model innovation have evolved over two distinct generations of literature, reflecting shifts in research priorities and societal needs. Section 4.2 introduces the "EN.I.SCORE" framework, a tool designed to integrate macro–meso–micro perspectives to support innovation and sustainability in energy organizations. Section 4.3 presents the empirical findings from the ReMTh study, highlighting how energy firms in this region perceive their strategic, technological, and managerial performance, and how they adapt or fail to adapt to the area's developmental constraints.

4.1. Evolution of Macro, Meso, and Micro Energy Dimensions in Entrepreneurship Research

The literature review highlights a significant acceleration in the development of the broader thematic area since 2010. In light of the emerging new globalization era, the 238 studies can be classified into two generations of entrepreneurship research in the energy sector: the first generation comprises 26 studies from 1995 to 2009, while the second consists of 212 studies from 2010 to 2023.

From 1995, there is a significant growth in publications, starting from just 1 and rising to 12 by 2023—a staggering 1200% increase over almost three decades. The peak in research interest is in 2019, with 29 publications.

In the second generation, when analyzed with VOSviewer, several author keywords with an entrepreneurial focus emerge, with a minimum occurrence of five times. These keywords predominantly center on articulating business models, contextualized within the larger frame of the energy transition. Additionally, the intersection of sustainability and CSR with climate change is explored. It is worth noting that the first generation displayed no keyword co-occurrences, suggesting that the field of energy entrepreneurship was still nascent.

A closer look at the journals publishing this research reveals some interesting facts. The journal "*Energy Policy*" leads the pack, witnessing a rise from 18 articles in 1995–2009 to 77 articles in 2010–2023. Similarly, "*Energy Economics*" shows a growth from 2 publications in the former period to 12 in the latter, highlighting the burgeoning interest in merging energy and entrepreneurship themes.

However, the majority of articles adopt a perspective either in terms of the macrolevel or mesolevel, with fewer articles focusing on microlevel issues. This trend becomes evident when examining the frequency of certain terms in the literature (as shown in Figure 3).

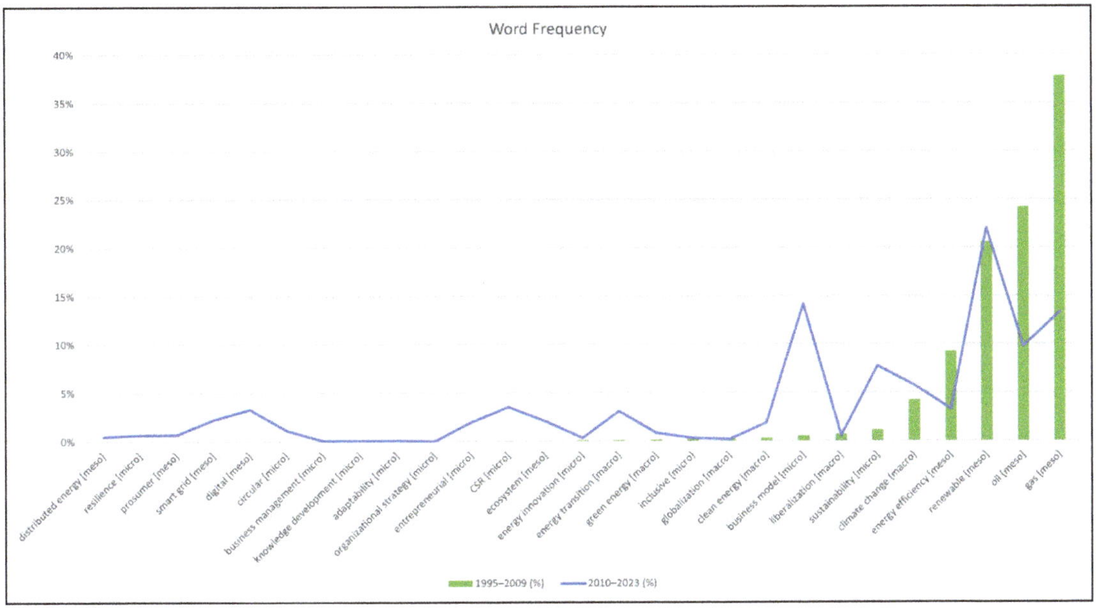

Figure 3. Frequency of macrolevel, mesolevel, and microlevel terms in the two generations.

4.1.1. Changing Perceptions of Globalization and Liberalization in the Energy Sector at the Macrolevel

In the early era, the relevant literature primarily viewed globalization as a unifying force for a global market, combined with the significant impacts from the ongoing liberalization of the electricity sector. This liberalization arises largely from massive foreign direct investments, mainly through privatizing public sector assets [71]. In the later period, some scholars argued that globalization is entering a prolonged restructuring in favor of local production due to increasing transportation costs. Some contended that reimagined CSR forms can offset the potential negative effects [72,73]. A standout that needs mention in this new era is recognizing that the liberalization of the electricity sector results in the industry's concentration under a few major (state-owned) players [9]. The energy sector is now generally viewed not just as a public service but also as a commodity. However, due to ongoing regulatory uncertainties, some utilities are decreasing their R&D investments [74]. The usage frequency of the terms "globalization" and "liberalization" remains consistent across both periods, which may stem from the literature's dominant mesolevel perspective.

4.1.2. Evolution of Clean-Green Energy Transition, Climate Change, and Related Policies at the Macrolevel

Historically, energy policies leaned towards ensuring stable, affordable energy, often from carbon-heavy sources to fuel growing economies and maintain energy security, with clean, green energy being a secondary concern or seen as a luxury [75]. Today, policies prioritize clean and renewable energy, aiming for a sustainable, low-carbon future. This shift highlights a gradual departure from previous government passivity in energy policies, as some scholars now stress the government's pivotal role in this change [76,77]. While the term "energy transition" does not appear at all in the past generation, in contemporary discussions it rises to prominence because of increasing climate sustainability concerns. This evolution is evident in the increasing prevalence of terms such as "climate change", "clean energy", and "green energy" in relevant research themes. The term "energy policies" continues to dominate the literature, as demonstrated by the prominence of the renowned journal "*Energy Policy*" in the field. Nevertheless, scholars have traditionally approached this topic from a mesolevel and sectoral perspective. Only recently have they begun to explore broader and more holistic issues, such as energy transition and climate change, signaling a shift in research focus.

4.1.3. Rise of Renewable Energy Communities at the Mesolevel

The current generation of literature is more directly concerned with localized energy communities, whereas earlier works primarily focus on the traditional energy efficiency of production and distribution at the national level. Larger energy utilities, component manufacturers, and other organizations with the expertise to coordinate extensive data flows are now aiming to establish both real and virtual energy communities, or to aggregate smaller communities on a larger scale [8]. Energy communities enhance renewable energy capacity, mobilize private capital, and empower consumers [9]. The growing focus on renewable energy communities and trans-spatial networks is reflected in the declining frequency of terms such as "oil" and "gas" related to simple production, while the use of "renewable" has risen significantly. Additionally, the reduced mention of "energy efficiency" indicates a shift away from traditional discussions on efficiency, highlighting a growing interest in fostering resilient and sustainable energy communities, local institutions, and regional energy subsystems.

4.1.4. The Gradual Emergence of the Energy Ecosystems Concept at the Mesolevel

This perspective provides new insights and helps us understand the role that supporting organizations have in the emergence and growth of energy communities. It aids in analyzing the roles that ecosystem actors play by looking at the kinds of support they offer [33]. Previous research in this field often lacked an ecosystem-oriented perspective, failing to explicitly incorporate concepts relevant to "biological" or "symbiotic" evolution. It was linear and did not fully acknowledge the interconnectedness among different actors and organizations crucial for sustaining the entire energy system. The term "ecosystem" only emerges in the second generation of literature, as previously noted.

4.1.5. The Era of Smart Grids, Digital-Distributed Energy and Prosuming at the Mesolevel

In the current era, scholars examine the smart grid as an advanced electricity network infrastructure featuring a two-way flow of information and often, a two-way flow of electricity. It does not stem from a single company; instead, it is a complex system of interconnected products and niche technologies, including storage and peer-to-peer trading platforms [78]. This distributed energy system primarily uses renewable sources and relies on small-scale energy generation technologies [53]. At the same time, global communities of prosumers are emerging. These prosumers not only produce electricity or heat for their own use but also contribute surplus power to the electric grid or share excess heat through district heating networks [79]. This emerging trend is underscored by the exclusive use of terms such as "distributed energy", "prosumer", "smart grid", and "digital" in the present era, as opposed to the past. Unlike previous research, which focused primarily on the unidirectional energy flow from centralized power stations to consumers, the current generation emphasizes dynamic interactions with prosumers.

4.1.6. Navigating the Evolution of Strategy, Technology, and Management in Energy Organizations at the Microlevel

When analyzing organizational strategy, technology, and management in energy organizations, only a few approaches emerge as particularly noteworthy. Regarding the strategy for energy organizations, some researchers argue that success is more likely when setting diverse strategic goals based on forms of organizational ambidexterity. This refers to the balance between exploration and exploitation skills within organizations [80]. Others advocate for renewable energy organizations to shift strategies from transactional stakeholder approaches (win–win) to relational ones. These relational strategies aim to foster stronger stakeholder relationships and understanding, often demanding flexibility and compromise in their implementation [81].

Regarding the valorization of technology in energy organizations, some scholars recommend a proactive approach to knowledge, tapping into technical guidelines, standardized products, technology reputation, and support programs. In these programs, it becomes essential to create relevant knowledge resources [82]. As for general management in energy organizations, some researchers analyze large-scale utilities and advise segmenting them into distinct operational activities and businesses [83]. Others underscore broader guidelines, such as addressing the looming environmental, economic, and human crises. From these crises, a new business management model, sustainable development, emerges, linking economic, environmental, and human domains [84].

Though guidelines exist in the literature for strategy, technology, and management of energy organizations, specific findings for microfirms or SMEs in these areas are scarce. Most scholars primarily focus on aspects of larger utilities and seldom explore strategy, technology, and management in a holistic manner. For example, management in an energy organization, particularly an SME, should integrate operational activities instead of segmenting them, which contrasts with the prevailing literature. The limited mention of terms such as "organizational strategy", "knowledge development", and "business management" in more recent literature is also a point of interest, whilst earlier literature does not address these terms at all. Furthermore, terms like "organizational technology" and "organizational management" are absent in all relevant studies. Consequently, it seems challenging to derive concrete recommendations for enhancing these areas in energy organizations based on the examined literature.

4.1.7. Business Model Innovations Within Energy Organizations at the Microlevel

Innovation in energy organizations is often explored from a business model perspective [85–90]. In the present era, scholars identify energy innovations that support decarbonization, digitalization, and decentralization of the energy system [91]. Among the many approaches to business model generation and reinforcement, certain methods stand out. Green business models emphasize the flexibility to build upon innovation and sustainability foundations, essential for thriving in the dynamic environments that new market mechanisms introduce [92]. Moreover, the vital role of entrepreneurship in shaping and growing a circular business is recognized, as entrepreneurs proactively pursue innovative solutions to environmental challenges while maintaining profitability [93]. Some propose a customer-centric business model for future energy suppliers, where they act both as an energy supply contractor (via third-party ownership) and as an aggregator of energy assets for their clients [94].

Most approaches on energy innovations in the current literature tend to offer general guidelines and focuses on broad business model typologies. Such sources seldom delve into strategic, technological, and managerial microlevel issues, occasionally attempting to propose a more integrated approach, as discussed in the previous section. In essence, this business model literature explicitly endorses a microlevel viewpoint, even though it implicitly views its examined issues from a mesolevel perspective. Regarding the frequency of pertinent terms, it is notable that many relevant terms do not emerge in the first generation of related research. However, terms like "circular", "entrepreneurial", "energy innovation", and "business model" become prominent in the second generation, with "business model" seeing the most significant surge relative to all other terms examined.

4.1.8. Advancements in CSR and RASI Dimensions in Energy Organizations at the Microlevel

The energy sector is under intense pressure from both public and private actors to undergo radical transformation. Simultaneously, like in other industries, energy companies must address CSR demands, especially since they rank among the primary sources of greenhouse gas emissions. This positions them as pivotal contributors to the sustainable development goals (SDGs) of the United Nations' 2030 Agenda [95,96]. As profit-driven entities—constituting the majority globally—energy companies need to harness resources and produce energy while demonstrating a profound commitment to society and emphasizing CSR, which influences corporate financial performance [72].

Concerning RASI principles (resilience–adaptability–sustainability–inclusiveness) in energy organizations, some researchers offer relevant definitions. The synthesis of these definitions is as follows:

- Resilience describes the capability to bounce back and sustain operations in the face of unexpected disruptions, ensuring a stable energy supply [97].
- Adaptability highlights an energy organization's ability to evolve and adapt to shifts in energy-related technology and regulations, ensuring its continued relevance and effectiveness amid environmental strains [98].
- Sustainability commits to eco-friendly practices, resource optimization, and minimizing negative impacts on both the environment and socioeconomic systems. This ensures organizational longevity while reducing ecological and social footprints [99].
- Inclusiveness emphasizes efforts to guarantee equal energy access and opportunities within the socioeconomic system, addressing energy poverty and promoting diverse stakeholder involvement [100].

Regarding the CSR concept, a rising interest in its nuances exists, with significant research coming primarily from the journal "*Corporate Social Responsibility and Environmental Management*", which takes primarily a mesolevel approach to the energy sector. Similarly to other microlevel aspects discussed in the preceding sections, discussions on the various RASI dimensions do not explicitly explore these areas since they seem to be in their early stages. In the reviewed literature, terms such as "resilience", "adaptability", "CSR", and "sustainability" are largely absent in the first generation of research but gain prominence in the second generation. Among these aspects, CSR and sustainability concepts dominate, while adaptability and inclusiveness receive less focus. This pattern might exist because CSR and sustainability are now more mainstream, but all dimensions are equally critical in tandem.

4.1.9. Synthesis: Shifting Perspectives in Energy Entrepreneurship

The literature review reveals the emergence of distinct and autonomous energy entrepreneurship research in the 1990s—since 1995 in the examined sample. Dividing the 238 studies across two generational periods reveals a notable increase in focus on this sub-theme starting in 2010. The second generation of research demonstrates significant differences from the first in several key areas:

Macrolevel:

(a) Early literature portrayed globalization as a homogenizing force and celebrated the liberalization of the electricity sector. In contrast, recent research identifies global industry consolidation under a few oligopolistic major players, alongside dynamic local, national, and international disparities. The perspective has shifted from viewing energy solely as a public service to recognizing it as a commodity with incremental innovative potential.

(b) Policy discussions have moved away from reliance on carbon-intensive energy sources, emphasizing clean, renewable alternatives, unified distribution channels, and advanced storage technologies. This shift underscores the increasingly active role of governments in mitigating climate change.

Mesolevel:

(c) The second generation highlights a growing interest in renewable energy communities. Larger energy entities are working to establish or consolidate these communities, moving away from traditional energy efficiency models and emphasizing the development of resilient, pluralistic, and sustainable local energy systems.

(d) An ecosystem-oriented perspective has emerged, offering a more interconnected understanding of the roles various actors play in the energy sector. This holistic and evolutionary approach replaces earlier linear perspectives.

Microlevel:

(e) Recent research emphasizes the bidirectional flow of electricity and information in smart grids, emphasizing the critical role of prosumers in these digitally advanced networks.

(f) The focus has shifted from broad, generic strategies to more nuanced examinations of organizational ambidexterity, stakeholder relationships, and knowledge resources, primarily within larger utilities. However, strategies for microfirms and SMEs remain underexplored.

(g) Modern energy organizations explore innovative business models that support decarbonization, digitalization, and decentralization. Innovative entrepreneurship plays a critical role in developing ecologically friendly and socioeconomically sustainable solutions.

(h) CSR has become a focal point due to the energy sector's environmental impact. The emerging RASI principles—resilience, adaptability, sustainability, and inclusiveness—are gaining prominence, emphasizing their integration within energy organizations and the local systems that support them.

In conclusion, contemporary energy entrepreneurship research has evolved significantly from earlier studies, reflecting shifts in perceptions, priorities, and approaches. The focus has transitioned to cleaner energy sources, a more interconnected view of the energy ecosystem, and increased emphasis on innovation, sustainability, and social responsibility. These changes align with broader global trends and challenges, particularly regarding climate change and sustainable development. Notably, these shifts are especially relevant for less-developed countries and regions, which face unique challenges in achieving energy equity and sustainability.

4.2. The Energy Innovation Scorecard (EN.I.SCORE) Framework

Many existing approaches to entrepreneurship within the energy sector tend to focus predominantly on just one of the three dimensions: macro, meso, or micro. This often results in an incomplete understanding, as the dialectic interplay of these macro–meso–micro dimensions is overlooked [36,37]. Such a limited perspective potentially hampers the development of comprehensive guidelines that address the evolutionary variations of entrepreneurship in the energy sector, as its implications—both internal and external—span all analytical levels. To address this need, we introduce the "Energy Innovation Scorecard" framework, henceforth referred to as "EN.I.SCORE".

The integration of macrolevel, mesolevel, and microlevel perceptions in the proposed questions underscores the interconnected nature of energy innovation within a firm, as outlined in the previous section. At the macrolevel, the questions examine broad systemic dynamics, such as the influence of globalization and national policies on energy innovation, highlighting how external forces shape strategic priorities. At the mesolevel, the focus shifts to the organization's role within energy ecosystems, investigating collaborations, partnerships, and the development of renewable energy communities that drive localized and sectoral energy transitions. At the microlevel, the questionnaire delves into internal organizational processes, including strategies for leveraging technology, fostering innovative business models, and addressing CSR and RASI dimensions. This tripartite approach ensures a comprehensive evaluation of energy innovation, synthesizing macro–meso–micro perspectives to capture how global trends, industry-level networks, and internal capabilities converge to shape a firm's innovation potential and strategic decision-making.

The questionnaire employs a Likert-based scale to measure perceptions across two timeframes—present and past—allowing for the identification of trends in organizational satisfaction [101]. An open-ended section complements the quantitative data by capturing qualitative justifications, enabling respondents to explain their scores and offer deeper insights. This design positions the questionnaire as a self-assessment tool akin to strategic audit methodologies [102–105] and promotes firm-wide introspection and action-oriented research [106]. By involving all senior management members, the process ensures a comprehensive evaluation of organizational dynamics. Aggregated responses, recommended to be collected on an annual basis, serve as a foundation for discussions focused on areas of divergence and significant interest, fostering opportunities for collaborative problem-solving and innovation. This iterative approach seamlessly integrates the precision of quantitative analysis with the depth of qualitative exploration, establishing it as a powerful tool for continuous organizational improvement and alignment with the evolving challenges of the energy sector.

Additionally, we recommend integrating the questionnaire with the Stra.Tech.Man.SC framework [55] previously introduced in this paper, thereby serving as a foundation for a comprehensive scorecard methodology. Drawing upon the eight key findings outlined earlier—presented as points "a" through "h" and discussed in Sections 4.1.1–4.1.8, respectively—we propose the following series of self-assessment questions:

Macrolevel:

(a) How does your organization interpret the implications of globalization within the energy sector, especially in the context of the ongoing consolidation under major industry players and the shift in viewing energy as a commodity?
(b) What initiatives is your organization undertaking to transition from carbon-reliant sources towards sustainable energy, and how are these aligned with governmental policies on climate change mitigation?

Mesolevel:

(c) In what manner does your organization collaborate with and foster the development or consolidation of renewable energy communities?
(d) How is your organization positioning itself within the evolving energy ecosystem, promoting partnerships and synergies with various industry stakeholders?

Microlevel:

(e) What strategies does your organization employ to capitalize on the bidirectional flow of electricity and data within smart grids, and how are you integrating and supporting prosumers?

(f) How does your organization navigate the balance between exploration and exploitation in its strategy, leverage knowledge resources in terms of organizational technology, and integrate its operations for efficient management amid a broad spectrum of stakeholders in the energy landscape?
(g) In which ways does your organization transcend the boundaries with innovative business models that accentuate decarbonization, digital transformation, and decentralization?
(h) How does your organization emphasize CSR and integrate the principles of resilience, adaptability, sustainability, and inclusiveness into its operations?

By building on the foundation established by the Stra.Tech.Man.SC, the EN.I.SCORE could serve as both a complement and an extension. This fusion could culminate in the creation of a "Stra.Tech.Man.SC.EN.I.SCORE" framework. This extended framework could not only encapsulate the core elements of the original but also act as a guiding tool for energy organizations of all sizes. Its primary objective would be to identify vulnerabilities among energy entrepreneurs and recommend corrective actions through evidence-based, methodical self-evaluation. This approach emphasizes systematic, longitudinal internal assessment and performance tracking, grounded in the interpretations and evaluations of the organization's own leadership.

4.3. Empirical Analysis: Evaluating Energy Entrepreneurship in the ReMTh

4.3.1. Survey Results

Between April and June 2024, 89 valid responses were collected. Each respondent took approximately 15–20 min to complete the questionnaire. This relatively extended duration hints at the respondents' willingness to reflect on their practices, possibly also because they were familiar with (and interested in) the region's major public university in which the survey was conducted. The sample demographics (Table 1) indicate that 25.8% of firms are less than three years old, 27% have been in operation for 10 to 20 years, and 19.1% have been operating for more than 20 years. In terms of location, 30.3% of the firms are based in Evros, 22.5% in Xanthi, 19.1% in Drama, 16.9% in Kavala, 10.1% in Rodopi, and 1.1% in Thasos. Regarding employment, 64% of the firms have between one and five employees, while 21.3% are operated on by a single individual, and only 5.6% have more than ten employees. The largest revenue group (25.8%) reports annual earnings between EUR 50,000 and EUR 250,000, although just 5.6% exceed EUR 1 million, and 20.2% chose not to disclose their revenues. Overall, most surveyed firms are microfirms facing structural competitiveness hurdles. Their energy-related activities vary across production, supply of raw materials, distribution, and usage, with many focusing on small-scale production, often of photovoltaic energy, or on energy usage.

Table 1. Survey identity (frequencies).

Years of Operation	
<3	25.8%
3–5	12.4%
5–10	15.7%
10–20	27%
>20	19.1%
Headquarters (regional unit–city)	
Drama	19.1%
Evros	30.3%
Thassos	1.1%

Table 1. *Cont.*

Years of Operation	
Kavala	16.9%
Xanthi	22.5%
Rodopi	10.1%
Number of permanent employees	
0	21.3%
1–5	64%
6–10	9%
11–50	2.2%
51–250	3.4%
>250	0%
Activity in the energy sector	
Energy distributor	5.6%
Energy distributor and user	1.1%
Energy producer	30.3%
Energy producer and supplier	5.6%
Energy producer and user	7.9%
Energy supplier	11.2%
Energy supplier and energy user	2.2%
Energy supplier, distributor, and user	3.4%
Energy user	32.6%
Annual revenue	
<EUR 10,000	6.7%
EUR 10,000–50,000	24.7%
EUR 50,000–250,000	25.8%
EUR 250,000–1,000,000	16.9%
EUR 1,000,000–10,000,000	5.6%
>EUR 10,000,000	0%
I do not wish to report	20.2%

Regarding the Likert-scale items (EN1–EN15), the survey results indicated that items related to external forces, such as oligopolistic consolidation (EN2) and government policies (EN15), received comparatively low scores (Table 2). This reflects a general dissatisfaction among respondents with factors perceived as beyond their control. In contrast, items capturing internal capabilities, such as resilience (EN10), adaptability (EN11), sustainability (EN12), and financial performance (EN14), tended to receive moderate or high satisfaction ratings. However, some paradoxical findings emerged; for instance, although many respondents claimed high adaptability, the correlation between "crisis adaptation" (EN1) and policy (EN15) was notably weak (0.20), suggesting that entrepreneurs do not clearly associate their crisis responses with broader policy changes.

Table 2. Percentage frequencies of Likert scales in the survey.

	EN1	**EN2**	**EN3**	**EN4**	**EN5**	**EN6**	**EN7**	**EN8**	**EN9**	**EN10**	**EN11**	**EN12**	**EN13**	**EN14**	**EN15**	**AVR**
Too little (1)	15.7	16.9	10.1	6.7	16.9	20.2	13.5	20.2	14.6	10.1	10.1	6.7	9.0	13.5	13.5	13.2
A little (2)	11.2	36	9.0	12.4	10.1	18	5.6	9.0	5.6	7.9	3.4	6.7	7.9	7.9	49.4	13.3
Moderate (3)	24.7	24.7	25.8	34.8	37.1	38.2	32.6	32.6	32.6	21.3	24.7	28.1	32.6	28.1	18	29.1
A lot (4)	29.2	18	32.6	33.7	23.6	19.1	33.7	29.2	30.3	42.7	39.3	36	31.5	39.3	15.7	30.2
Very much (5)	19.1	4.5	22.5	12.4	12.4	4.5	14.6	9	16.9	18.0	22.5	22.5	19.1	11.2	3.4	14.2

Correlation analyses (Figure 4) revealed significant positive relationships among resilience, adaptability, and sustainability, indicating that where entrepreneurs view one of these dimensions as strong, they tend to rate the others highly. Conversely, variables like crisis adaptation (EN1) and policy support (EN15) remain relatively uncorrelated, implying a limited sense of synergy between governmental frameworks and firm-level transformations. Overall, the data suggest that while microentrepreneurs view themselves as resilient, they are critical of what they regard as structurally unfair market conditions and inadequate policy support.

	EN1: Crisis	EN2: Consolidation	EN3: Transition	EN4: Cooperation	EN5: Communities	EN6: Smart grid	EN7: Stra.Tech.Man	EN8: Innovation	EN9: CSR	EN10: Resilience	EN11: Adaptability	EN12: Sustainability	EN13: Inclusiveness	EN14: Financials	EN15: Policy
EN1: Crisis	1.00	0.37	0.48	0.33	0.13	0.07	0.20	0.32	0.27	0.20	0.32	0.26	0.27*	0.46	0.20
EN2: Consolidation	0.37	1.00	0.32	0.39	0.28	0.30	0.22*	0.22*	0.15	0.09	0.13	0.19	0.14	0.15	0.36
EN3: Transition	0.48	0.32	1.00	0.45	0.29	0.36	0.35	0.29	0.34	0.32	0.38	0.40	0.19	0.36	0.09
EN4: Cooperation	0.33	0.39	0.45	1.00	0.41	0.47	0.39	0.48	0.40	0.34	0.40	0.29	0.30	0.32	0.16
EN5: Communities	0.13	0.28	0.29	0.41	1.00	0.41	0.30	0.29	0.29	0.20	0.13	0.16	0.11	0.14	0.23*
EN6: Smart grid	0.07	0.30	0.36	0.47	0.41	1.00	0.33	0.23*	0.18	0.29	0.16	0.11	0.07	0.17	0.14
EN7: Stra.Tech.Man	0.20	0.22*	0.35	0.39	0.30	0.33	1.00	0.46	0.47	0.40	0.51	0.46	0.25*	0.46	0.11
EN8: Innovation	0.32	0.22*	0.29	0.48	0.29	0.23*	0.46	1.00	0.27	0.29	0.30	0.30	0.24	0.37	0.17
EN9: CSR	0.27	0.15	0.34	0.40	0.29	0.18	0.47	0.27	1.00	0.32	0.45	0.34	0.41	0.36	0.17
EN10: Resilience	0.20	0.09	0.32	0.34	0.20	0.29	0.40	0.29	0.32	1.00	0.65	0.64	0.34	0.48	0.25
EN11: Adaptability	0.32	0.13	0.38	0.40	0.13	0.16	0.51	0.30	0.45	0.65	1.00	0.79	0.50	0.68	0.20
EN12: Sustainability	0.26*	0.19	0.40	0.29	0.16	0.11	0.46	0.30	0.34	0.64	0.79	1.00	0.43	0.55	0.22
EN13: Inclusiveness	0.27*	0.14	0.19	0.30	0.11	0.07	0.25*	0.24*	0.41	0.34	0.50	0.43	1.00	0.44	0.30
EN14: Financials	0.46	0.15	0.36	0.32	0.14	0.17	0.46	0.37	0.36	0.48	0.68	0.55	0.44	1.00	0.30
EN15: Policy	0.20	0.36	0.09	0.16	0.23*	0.14	0.11	0.17	0.17	0.25*	0.20	0.22*	0.30	0.30	1.00

* The correlation is significant at the 0.05 level.

1.00 / 0.50 / 0.00

Figure 4. Pearson correlation matrix of the survey. The right-hand legend indicates the strength of correlation from 0 (light blue) to 1 (dark blue) to observe how strong each pairwise correlation is.

4.3.2. In-Depth Interviews

In addition to the survey, eight microfirm owners with zero to five employees volunteered for detailed semi-structured interviews. Their feedback was both quantified through Likert-type items and qualitatively analyzed for recurring themes. Descriptive statistics were computed across six categories—strategy (ST), technology (TC), management (MN), strategy–technology–management-based innovation (IN), energy-related innovation (EN), and financial performance (FN)—revealing several notable outlier values (Table 3).

Table 3. Some variables that show outliers in the descriptive statistics of in-depth interviews.

	ST1	ST2	ST6	ST7	ST8	TC1	TC2	TC3	TC8	MN5	MN8	IN6	IN7	EN2	EN3	EN10	FN6
Mean	4.00	3.88	4.38	3.75	3.50	4.50	4.38	4.13	4.38	2.88	3.38	2.88	3.63	2.63	3.75	4.50	3.50
Std. Error of Mean	0.33	0.35	0.32	0.37	0.27	0.27	0.26	0.30	0.26	0.13	0.26	0.30	0.26	0.63	0.56	0.27	0.27
Median	4.00	3.50	5.00	3.00	3.00	5.00	4.50	4.00	4.50	3.00	3.00	3.00	3.50	2.00	4.50	5.00	3.00
Mode	3.00	3.00	5.00	3.00	3.00	5.00	5.00	4.00	5.00	3.00	3.00	3.00	3.00	1.00	5.00	5.00	3.00
Std. Deviation	0.93	0.99	0.92	1.04	0.76	0.76	0.74	0.83	0.74	0.35	0.74	0.83	0.74	1.77	1.58	0.76	0.76
Variance	0.86	0.98	0.84	1.07	0.57	0.57	0.55	0.70	0.55	0.13	0.55	0.70	0.55	3.13	2.50	0.57	0.57
Coefficient of Variation (%)	0.23	0.20	0.39	0.22	0.21	0.32	0.17	0.22	0.15	0.15	0.22	0.12	0.31	0.22	0.22	0.22	0.15
Skewness	0.00	0.31	−1.00	0.64	1.32	−1.32	−0.82	−0.28	−0.82	−2.83	1.95	−1.69	0.82	0.55	−0.90	−1.32	1.32
Std. Error of Skewness	0.75	0.75	0.75	0.75	0.75	0.75	0.75	0.75	0.75	0.75	0.75	0.75	0.75	0.75	0.75	0.75	0.75
Normality of Skewness	0.00	0.41	−1.33	0.86	1.76	−1.76	−1.10	−0.37	−1.10	−3.76	2.59	−2.25	1.10	0.73	−1.20	−1.76	1.76
Kurtosis	−2.10	−2.36	−1.04	−2.24	0.87	0.88	−0.15	−1.39	−0.15	8.00	3.20	4.97	−0.15	−1.80	−0.69	0.87	0.88
Std. Error of Kurtosis	1.48	1.48	1.48	1.48	1.48	1.48	1.48	1.48	1.48	1.48	1.48	1.48	1.48	1.48	1.48	1.48	1.48
Normality of Kurtosis	−1.42	−1.59	−0.70	−1.51	0.59	0.59	−0.10	−0.94	−0.10	5.40	2.16	3.36	−0.10	−1.21	−0.47	0.59	0.59
Sum	32	31	35	30	28	36	35	33	35	23	27	23	29	21	30	36	28

Regarding some key observations, high ratings often surfaced in technology (TC1–TC3, TC8) and financial (FN6) questions. Interviewees cited routine machinery upgrades and stable (though modest) profit margins. They perceived continuous monitoring of technology as crucial but seldom embraced formal R&D processes. Lower ratings arose in management (especially MN5 on leadership) and innovation (IN6 on innovation flexibility). Here, entrepreneurs openly admitted a lack of modern leadership skills, organizational processes, or a "culture of continuous innovation."

Figure 5 shows that the most frequently mentioned word in the interviews was "improvement", indicating that many believe conditions are gradually improving. However, the contradictory phrase "has not changed" was often used to describe internal managerial systems—suggesting that small-scale operations may experience incremental gains without undergoing fundamental restructuring. The interviews further revealed that most owners lacked formal training in strategic planning. Many had launched photovoltaic-based ventures primarily in response to government subsidies rather than through proactive opportunity-seeking. As these subsidies waned, the owners found it difficult to pivot toward new business models, such as storage or aggregator roles.

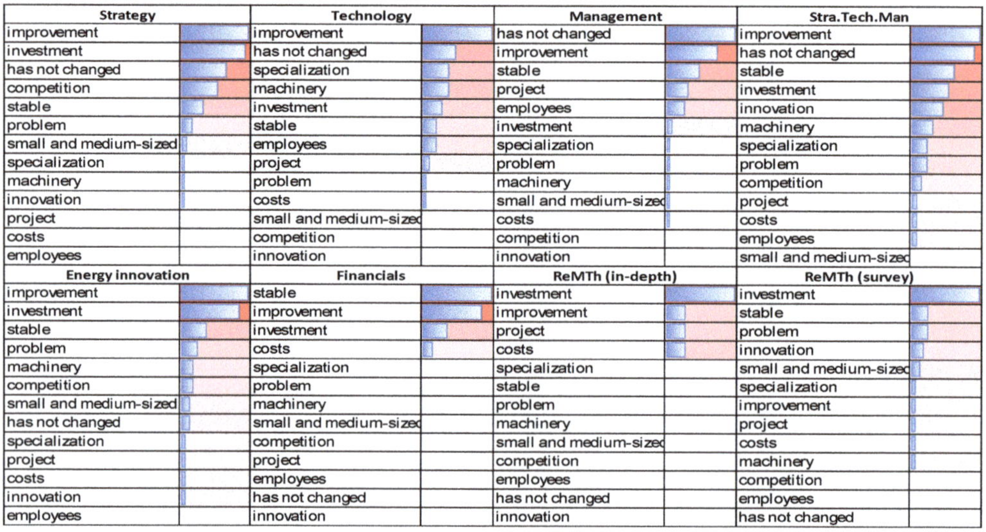

Figure 5. Frequency of occurrence of related words in interviews. The color-coded bars (in red and blue) visually represent how frequently each word was used in the interviews, where larger bars and darker red tones indicate higher occurrence.

In terms of a thick description analysis and comparing with broader evidence, patterns echo previous findings on "monad-centered" business physiology in Greece [54–56]. Microfirms typically emphasize daily survival over strategic repositioning. This approach, while understandable in a region with subdued demand, undermines deeper market differentiation or advanced integrative practices.

Overall, the combined survey and interview data confirm that many ReMTh energy microfirms exhibit limited capacity to dynamically integrate the strategic–technological–managerial nexus—a critical factor stifling innovation and financial growth. They seldom leverage the potential of distributed energy, smart grids, or meaningful collaborations within energy ecosystems. Yet they do see themselves as adaptable and financially stable, suggesting a paradox: they recognize certain threats (oligopolistic consolidation, limited policy support) but hesitate to pursue advanced (and riskier) innovation pathways.

5. Discussion

Our findings indicate a significant shift in energy entrepreneurship research over the past two decades. Notably, recent years have seen increased interest, driven by the gradual emergence of a new era of globalization that mirrors broader transformations in both the global economy and the energy sector [107–109]. While the patterns identified indicate increasing emphasis on sustainability and innovation, further studies are needed to confirm whether these trends are consistent across all regional and sectoral contexts. Before 2010, most literature addressed liberalization, large-scale efficiency, and static energy markets.

Conversely, since 2010, attention turns to macrolevel imperatives (e.g., sustainability, climate policies), mesolevel ecosystems (renewable energy communities, prosumers), and microlevel entrepreneurship (innovation-based business models, CSR, RASI principles). This shift corresponds with broader dynamics in the global energy transition and with calls for strategic reorientation amid pressing sustainability challenges [1,8,10].

However, our review also highlights gaps concerning microlevel organizational integration of strategy, technology, and management—areas that appear underexplored or addressed only in fragmented business model frameworks [85,88]. The proposed EN.I.SCORE framework (Section 4.2) responds to this gap by offering an integrated approach that situates energy entrepreneurship within a macro–meso–micro dialectic, underscoring how entrepreneurs can better navigate the interplay of global energy markets, local networks, and internal processes.

Compared to seminal insights from Wüstenhagen and Wuebker [10], our approach expands their broad agenda into an explicit and integrated macro–meso–micro framework. Rather than simply mapping energy entrepreneurship themes, our analysis identifies specific correlations—such as the paradox of perceived adaptability coexisting with limited real innovation—and suggests that systematic introspection via the EN.I.SCORE questionnaire can help entrepreneurs diagnose and address strategic, technological, and managerial weaknesses.

In conclusion, the findings confirm that most microfirms in the ReMTh region demonstrate "monad-centered" organizational physiologies, typified by intuitive strategic approaches, sporadic technology adoption, and empirically based management. These observations strongly support our central hypothesis: namely, when microfirms lack systematic strategy–technology–management integration, they tend to underperform in innovation and financial outcomes, thereby reinforcing the structural weaknesses often seen in less-competitive regions. In this sense, the hypothesis is validated in the specific context examined.

Comparing these results with the broader literature [1–5,8–10] suggests that improving local energy entrepreneurship hinges on targeted, integrative interventions that align with macro (policy), meso (ecosystem), and micro (organizational) levels. On the one hand, policymakers could foster resource accessibility and reduce market-entry barriers for microfirms, promoting collaborative networks and capacity-building. On the other, microfirm owners themselves might adopt systematic self-assessment tools—such as the proposed EN.I.SCORE—to regularly audit their strategic, technological, and managerial orientations and to catalyze incremental improvements.

In sum, by examining smaller energy entrepreneurs under real constraints, this study demonstrates how insufficient Stra.Tech.Man integration can perpetuate underdevelopment. Conversely, structured capacity building and ecosystemic coordination can enable even small-scale firms to transition toward cleaner, smarter, and more resilient energy models. Strengthening these dimensions appears vital for moving beyond a necessity-based mode of survival and fulfilling the sustainability imperatives of the emerging new globalization era.

6. Conclusions

In addressing the evolution of the thematic focus in contemporary research on energy entrepreneurship, we observe a pronounced shift toward sustainability, energy transition, CSR, and innovative business models after 2010, moving away from earlier emphases on deregulation and unidirectional efficiency. When considering how energy entrepreneurs can navigate the evolving socioeconomic environment, it becomes evident that integrated macro–meso–micro strategies—incorporating climate policies, local energy communities, and consistent internal Stra.Tech.Man interplay—can enable a transition to green, digital, and decentralized models. Finally, regarding the innovation practices of selected energy entrepreneurs, the microfirms in the ReMTh display limited strategic planning, sporadic technological adoption, and rudimentary management processes, indicating that targeted frameworks such as the "Stra.Tech.Man.SC.EN.I.SCORE" can facilitate ongoing self-assessment, guide capacity building, and foster more robust business ecosystems supportive of energy transitions.

These results reveal specific organizational deficiencies in real-world entrepreneurial contexts and illustrate how local, smaller-scale energy actors can innovate under constraints. Policymakers, practitioners, and scholars can use the new scorecard approach to better diagnose gaps in strategy–technology–management integration, thus informing targeted policy measures and capacity-building programs.

While this study enriches our understanding of energy entrepreneurship, several avenues for further exploration remain. First, future research could expand the literature search beyond WoS by incorporating additional databases and literature sets to ensure a more comprehensive review of global research trends. Second, empirical studies could be conducted in diverse geographical and economic contexts to examine the generalizability of the findings primarily in less-developed regional ecosystems and beyond. Third, a longitudinal approach could be employed to analyze how energy entrepreneurs adapt over time in response to shifting policy and market dynamics. Additionally, future research could refine and empirically validate the EN.I.SCORE framework by testing it in other samples and cross-sectoral comparisons. This could help assess its effectiveness as a diagnostic and strategic tool for firms of various sizes and developmental scopes. Finally, more studies could explore the role of emerging digital technologies, such as blockchain, artificial intelligence, and smart grids, in reshaping energy entrepreneurship models, particularly within decentralized and prosumer-driven energy systems.

This challenge highlights the potential value of a policy approach that enhances these activities. Recent research [110] suggests that energy policy must shift from its traditional focus on energy security and market liberalization toward a holistic, integrated strategy addressing sustainability and adaptability at multiple systemically interconnected levels—macrolevel (macrosocial policy and regulation), mesolevel (regional ecosystems and communities), and microlevel (firm-level strategies and technological innovation). Such a framework should prioritize support for smaller firms and underrepresented regions, particularly in less-developed areas, by promoting decentralized renewable energy projects, fostering collaboration among local stakeholders, and reducing bureaucratic barriers. As a result, it moves away from fragmented analyses toward a unified policy that adapts to our evolving socioeconomic and technological landscape, ultimately reinforcing energy strategies that are resilient, inclusive, and capable of addressing contemporary environmental challenges while encouraging active consumer participation in energy production and management.

Nonetheless, the findings suggest a pivotal insight: the strategic alignment of macrolevel, mesolevel, and microlevel forces could help shift energy entrepreneurship toward more resilient, sustainable, and inclusive outcomes. By adopting a comprehensive approach, energy entrepreneurs—even in less-developed regions—might progress from "necessity-based" survival modes to genuinely innovative roles in the global energy transition.

Supplementary Materials: The following supporting information can be downloaded at: https://www.mdpi.com/article/10.3390/en18061323/s1.

Author Contributions: Conceptualization, D.C. and C.M.V.; methodology, D.C. and C.M.V.; validation, D.C. and C.M.V.; formal analysis, D.C. and C.M.V.; investigation, D.C. and C.M.V.; resources, D.C. and C.M.V.; writing—original draft preparation, D.C. and C.M.V.; writing—review and editing, D.C. and C.M.V.; visualization, D.C. and C.M.V. All authors have read and agreed to the published version of the manuscript.

Funding: This research received no external funding.

Institutional Review Board Statement: Ethical review and approval were waived for this study due to local practice and in accordance with the applicable data protection regulations (the Greek Personal Data Law as amended to comply with the GDPR), non-interventional research that does not involve the collection of identifiable data is generally considered exempt from formal Ethics Committee or Institutional Review Board approval.

Informed Consent Statement: Informed consent was obtained from all subjects involved in the study.

Data Availability Statement: The original contributions presented in this study are included in the article/Supplementary Materials. Further inquiries can be directed to the corresponding author.

Conflicts of Interest: The authors declare no conflicts of interest.

Appendix A. Specific Bibliometric Results

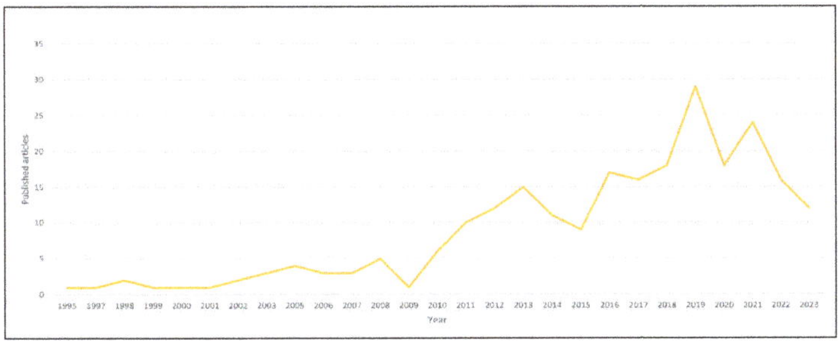

Figure A1. Timeline of the 238 articles examined.

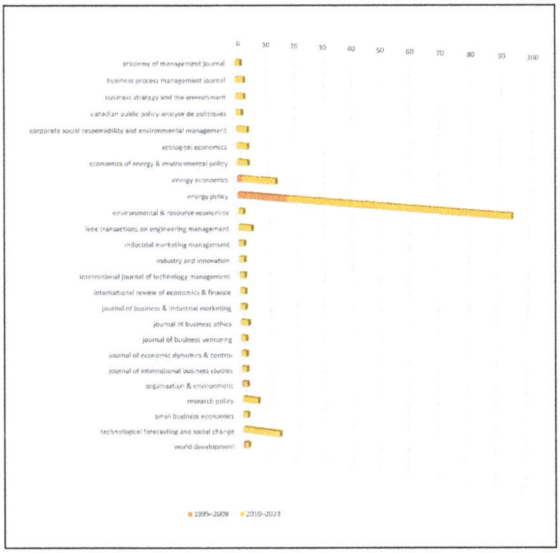

Figure A2. Distribution per journal (2 occurrences or more) of the 238 articles examined.

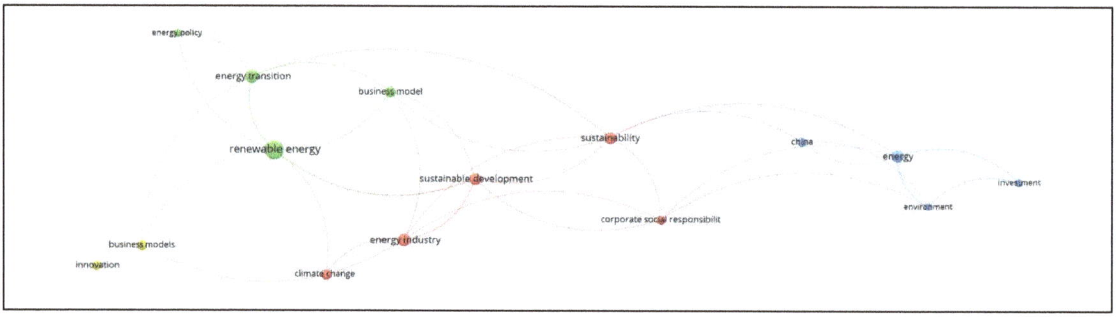

Figure A3. Author keywords (5 occurrences) in VOSviewer.

Appendix B. Survey Research

Respondents are prompted to rate 15 questions on a 5-point scale from 1 to 5, indicating the degree to which their satisfaction has decreased or increased, respectively. The questions are labeled "EN" followed by a sequential number. The rating scale is as follows: "Too little" (1), "A little" (2), "Moderate" (3), "A lot" (4), and "Very much" (5). The final question, labeled "ReMTh", concerns the study region and is open-ended. Note that, in the main text, this questionnaire appears in a slightly adapted form as "EN.I.SCORE".

1. General Information
    - Years in operation (<3, 3–5, 5–10, 10–20, >20)
    - Headquarters (Drama, Evros, Thassos, Kavala, Xanthi, Rhodope)
    - Number of permanent employees (0, 1–5, 6–10, 11–50, 51–250, >251)
    - Activity in the energy sector (supplier of raw materials, energy producer, energy distributor, energy user)
    - Annual revenue (EUR 10,000, EUR 10,000–50,000, EUR 50,000–250,000, EUR 250,000–1,000,000, EUR 1,000,000–10,000,000, >EUR 10,000,000 and "I do not wish to report")
2. Energy-Sector Innovation (EN)
    - EN1: Are you satisfied with how your firm has adapted to the effects of the global energy crisis?
    - EN2: Are you satisfied with how your firm has adapted to the ongoing oligopolistic consolidation of the sector?
    - EN3: Are you satisfied with your firm's energy-transition initiatives?
    - EN4: Are you satisfied with your firm's collaboration with suppliers of raw materials, producers, distributors, and energy users?
    - EN5: Are you satisfied with your firm's initiatives to promote renewable energy communities?
    - EN6: Are you satisfied with how your firm supports smart grids and prosumer models?
    - EN7: Are you satisfied with your firm's strategy, technology, and management processes in the energy domain?
    - EN8: Are you satisfied with your firm's processes for articulating an innovative business model in the energy domain?
    - EN9: Are you satisfied with the emphasis your firm place on CSR?
    - EN10: Are you satisfied with your firm's resilience?
    - EN11: Are you satisfied with your firm's adaptability?
    - EN12: Are you satisfied with your firm's sustainability?
    - EN13: Are you satisfied with your firm's inclusiveness?
    - EN14: Are you satisfied with your firm's financial performance?

- EN15: Are you satisfied with national energy policy and your firm's adjustment to it?

3. Local Development
    - ReMTh: Do you believe that Greece's energy-transition efforts, specifically in the ReMTh, involve significant opportunities and threats? What do you consider to be the region's comparative strengths and weaknesses, and why?

Appendix C. In-Depth Interview

Respondents are prompted to rate questions on a 5-point scale from 1 to 5, indicating the degree to which their satisfaction has decreased or increased, respectively. The questions are labeled "ST" for strategy, "TC" for technology, "MN" for management, "IN" for Stra.Tech.Man-based innovativeness, "EN" for energy-related innovation, and "FN" for financials. Each label is followed by a sequential number. There are 54 Likert-scale questions in total. The rating scale is as follows: "Too little" (1), "A little" (2), "Moderate" (3), "A lot" (4), and "Very much" (5). Respondents must explain each choice in detail, noting key problems, gaps, or difficulties, and cite any relevant actions and changes made. The final question, labeled "ReMTh", concerns the study region and is open-ended. Note that this questionnaire integrates the Stra.Tech.Man.SC with EN.I.SCORE. Definitions of the items are provided in prior published research that introduced and implemented the Stra.Tech.Man.SC [55,111].

1. Strategy (ST)
    - ST1: Are you satisfied with the overall strategic conception in your firm?
    - ST2: Are you satisfied with your firm's business plan?
    - ST3: Are you satisfied with your firm's processes for monitoring the external environment?
    - ST4: Are you satisfied with your firm's processes for monitoring the internal environment?
    - ST5: Are you satisfied with your firm's processes for generating and leveraging alternative strategic approaches?
    - ST6: Are you satisfied with your firm's strategic abstraction processes?
    - ST7: Are you satisfied with your firm's processes for evaluating its strategy?
    - ST8: Are you satisfied with your firm's processes for pursuing excellence?
2. Technology (TC)
    - TC1: Are you satisfied with your firm's processes for tracking technological developments?
    - TC2: Are you satisfied with your firm's processes for identifying useful new technology?
    - TC3: Are you satisfied with your firm's processes for learning and assimilating new technology?
    - TC4: Are you satisfied with your firm's processes for creating technological knowledge?
    - TC5: Are you satisfied with your firm's processes for disseminating technology and expertise?
    - TC6: Are you satisfied with your firm's processes for applying technology?
    - TC7: Are you satisfied with your firm's processes for synthesizing new technological data?
    - TC8: Are you satisfied with your firm's processes for overseeing the firm's overall technological endeavor?
3. Management (MN)

- MN1: Are you satisfied with your firm's planning processes?
- MN2: Are you satisfied with your firm's processes for designing job roles?
- MN3: Are you satisfied with your firm's processes for effective organization?
- MN4: Are you satisfied with your firm's staffing processes?
- MN5: Are you satisfied with your firm's leadership processes?
- MN6: Are you satisfied with your firm's motivational process?
- MN7: Are you satisfied with your firm's processes for control?
- MN8: Are you satisfied with your firm's coordination processes?

4. Stra.Tech.Man Innovativeness (IN)

- IN1: Are you satisfied with the completeness of your firm's innovation?
- IN2: Are you satisfied with the coherence of your firm's innovation?
- IN3: Are you satisfied with the simplicity of your firm's innovation?
- IN4: Are you satisfied with the feasibility of your firm's innovation?
- IN5: Are you satisfied with the resilience of your firm's innovation?
- IN6: Are you satisfied with the flexibility of your firm's innovation?
- IN7: Are you satisfied with the precision of your firm's innovation?
- IN8: Are you satisfied with the pace of your firm's innovative change?

5. Energy-Sector Innovation (EN)

- EN1: Are you satisfied with how your firm has adapted to the effects of the global energy crisis?
- EN2: Are you satisfied with how your firm has adapted to the ongoing oligopolistic consolidation of the sector?
- EN3: Are you satisfied with your firm's energy-transition initiatives?
- EN4: Are you satisfied with your firm's collaboration with suppliers of raw materials, producers, distributors, and energy users?
- EN5: Are you satisfied with your firm's initiatives to promote renewable energy communities?
- EN6: Are you satisfied with how your firm supports smart grids and prosumer models?
- EN7: Are you satisfied with your firm's strategy, technology, and management processes in the energy domain?
- EN8: Are you satisfied with your firm's processes for articulating an innovative business model in the energy domain?
- EN9: Are you satisfied with the emphasis your firm place on CSR?
- EN10: Are you satisfied with your firm's resilience?
- EN11: Are you satisfied with your firm's adaptability?
- EN12: Are you satisfied with your firm's sustainability?
- EN13: Are you satisfied with your firm's inclusiveness?
- EN14: Are you satisfied with your firm's financial performance?
- EN15: Are you satisfied with national energy policy and your firm's adjustment to it?

6. Financial Performance Indicators (FN)

- FN1: Asset turnover ratio = turnover/total assets
- FN2: Inventory turnover ratio = cost of sales/inventories
- FN3: Receivables turnover ratio = turnover/receivables
- FN4: Current ratio = total current assets/total short-term liabilities
- FN5: Financial leverage ratio = total assets/total equity
- FN6: Net profit margin = net profits for the fiscal year/turnover
- FN7: Return on equity = net profits for the fiscal year/total equity
- FN8: Return on assets = (net profits for the fiscal year/total assets) $\times$ 100

7. Local Development
    - ReMTh: Do you believe that Greece's energy-transition efforts, specifically in Eastern Macedonia and Thrace (ReMTh), involve significant opportunities and threats? What do you consider to be the region's comparative strengths and weaknesses, and why?

References

1. Baldwin, R. *The Great Convergence: Information Technology and the New Globalization*; Belknap Press: Cambridge, MA, USA, 2016; Volume 18.
2. Chatzinikolaou, D.; Vlados, C. The World System's Mutational Crisis and the Emergence of the New Globalization. In *Proceedings of the Économie Sociale et Crises du XXIe Siècle*; Lamotte, B., Ed.; L'Harmattan: Grenoble, France, 2022; pp. 163–192.
3. Strohmer, M.F.; Easton, S.; Eisenhut, M.; Epstein, E.; Kromoser, R.; Peterson, E.R.; Rizzon, E. Introduction. In *Disruptive Procurement: Winning in a Digital World*; Strohmer, M.F., Easton, S., Eisenhut, M., Epstein, E., Kromoser, R., Peterson, E.R., Rizzon, E., Eds.; Springer International Publishing: Cham, Switzerland, 2020; pp. 1–17. ISBN 978-3-030-38950-5.
4. Quitzow, R.; Bersalli, G.; Eicke, L.; Jahn, J.; Lilliestam, J.; Lira, F.; Marian, A.; Süsser, D.; Thapar, S.; Weko, S.; et al. The COVID-19 Crisis Deepens the Gulf between Leaders and Laggards in the Global Energy Transition. *Energy Res. Soc. Sci.* **2021**, *74*, 101981. [CrossRef]
5. Peet, R.; Robbins, P.; Watts, M. (Eds.) *Global Political Ecology*; Routledge: London, UK; New York, NY, USA, 2011; ISBN 978-0-415-54814-4.
6. Bradford, T. *The Energy System: Technology, Economics, Markets and Policy*; MIT Press: Cambridge, MA, USA; London, UK, 2018; ISBN 978-0-262-03752-5.
7. Schumpeter, J. *Capitalism, Socialism and Democracy*; Edition Published in the Taylor&Francis e-Library, 2003; Harper & Brothers: New York, NY, USA; London, UK, 1942; ISBN 0-203-20205-8.
8. Plewnia, F.; Guenther, E. The Transition Value of Business Models for a Sustainable Energy System: The Case of Virtual Peer-to-Peer Energy Communities. *Organ. Environ.* **2021**, *34*, 479–503. [CrossRef]
9. Vernay, A.-L.; Sebi, C.; Arroyo, F. Energy Community Business Models and Their Impact on the Energy Transition: Lessons Learnt from France. *Energy Policy* **2023**, *175*, 113473. [CrossRef]
10. Wüstenhagen, R.; Wuebker, R. An Introduction to Energy Entrepreneurship Research. In *Handbook of Research on Energy Entrepreneurship*; Wüstenhagen, R., Wuebker, R., Eds.; Edward Elgar Publishing: Cheltenham, UK, 2011; ISBN 978-0-85793-122-1.
11. Kubli, M.; Loock, M.; Wüstenhagen, R. The Flexible Prosumer: Measuring the Willingness to Co-Create Distributed Flexibility. *Energy Policy* **2018**, *114*, 540–548. [CrossRef]
12. Toffler, A. *The Third Wave*, 1st ed.; Morrow: New York, NY, USA, 1980; ISBN 978-0-688-03597-6.
13. Heiskanen, E.; Matschoss, K. Consumers as Innovators in the Electricity Sector? Consumer Perceptions on Smart Grid Services. *Int. J. Consum. Stud.* **2016**, *40*, 665–674. [CrossRef]
14. Sovacool, B.K. How Long Will It Take? Conceptualizing the Temporal Dynamics of Energy Transitions. *Energy Res. Soc. Sci.* **2016**, *13*, 202–215. [CrossRef]
15. Creutzig, F.; Agoston, P.; Goldschmidt, J.C.; Luderer, G.; Nemet, G.; Pietzcker, R.C. The Underestimated Potential of Solar Energy to Mitigate Climate Change. *Nat. Energy* **2017**, *2*, 17140. [CrossRef]
16. Vlados, C.; Chatzinikolaou, D. The Emergence of the New Globalization: The Approach of the Evolutionary Structural Triptych. *JGR* **2025**, *16*, 139–161. [CrossRef]
17. Häckel, B.; Pfosser, S.; Tränkler, T. Explaining the Energy Efficiency Gap—Expected Utility Theory versus Cumulative Prospect Theory. *Energy Policy* **2017**, *111*, 414–426. [CrossRef]
18. Herring, H.; Roy, R. Technological Innovation, Energy Efficient Design and the Rebound Effect. *Technovation* **2007**, *27*, 194–203. [CrossRef]
19. Thaler, R.; Sunstein, C.R. *Nudge: Improving Decisions About Health, Wealth, and Happiness*; Yale University Press: New Haven, CT, USA, 2008; ISBN 978-0-300-12223-7.
20. Smil, V. *Energy and Civilization: A History*; MIT Press: Cambridge, MA, USA, 2018; ISBN 978-0-262-53616-5.
21. Herring, H.; Sorrell, S. (Eds.) *Energy Efficiency and Sustainable Consumption*; Palgrave Macmillan UK: London, UK, 2009; ISBN 978-1-349-35753-6.
22. Jevons, W.S. (Ed.) *The Coal Question: An Inquiry Concerning the Progress of the Nation: And the Probable Exhaustion of Our Coal-Mines*; Macmillan and Co.: London, UK; Cambridge, UK, 1865.
23. Nordhaus, W.D.; Houthakker, H.S.; Sachs, J.D. Oil and Economic Performance in Industrial Countries. *Brook. Pap. Econ. Act.* **1980**, *1980*, 341–399. [CrossRef]
24. Yergin, D. *The Prize: The Epic Quest for Oil, Money, and Power*; Simon & Schuster: New York, NY, USA, 1991; ISBN 978-0-671-50248-5.

25. Gillingham, K.; Newell, R.G.; Palmer, K. Energy Efficiency Economics and Policy. *Annu. Rev. Resour. Econ.* **2009**, *1*, 597–620. [CrossRef]
26. Jaffe, A.B.; Stavins, R.N. The Energy-Efficiency Gap What Does It Mean? *Energy Policy* **1994**, *22*, 804–810. [CrossRef]
27. Sovacool, B.K.; Griffiths, S. The Cultural Barriers to a Low-Carbon Future: A Review of Six Mobility and Energy Transitions across 28 Countries. *Renew. Sustain. Energy Rev.* **2020**, *119*, 109569. [CrossRef]
28. Stern, D.I. The Role of Energy in Economic Growth. *Ann. N. Y. Acad. Sci.* **2011**, *1219*, 26–51. [CrossRef]
29. Hotelling, H. The Economics of Exhaustible Resources. *J. Political Econ.* **1931**, *39*, 137–175. [CrossRef]
30. Bridge, G. Past Peak Oil: Political Economy of Energy Crises. In *Global Political Ecology*; Peet, R., Robbins, P., Watts, M., Eds.; Routledge: London, UK; New York, NY, USA, 2011; pp. 307–324. ISBN 978-0-415-54814-4.
31. Nersesian, R. *Energy Economics: Markets, History and Policy*; Routledge: New York, NY, USA, 2016; ISBN 978-1-317-51186-1.
32. Bridge, G.; Billon, P.L. *Oil*; John Wiley & Sons: Hoboken, NJ, USA, 2013; ISBN 978-0-7456-7595-4.
33. Vernay, A.-L.; Sebi, C. Energy Communities and Their Ecosystems: A Comparison of France and the Netherlands. *Technol. Forecast. Soc. Change* **2020**, *158*, 120123. [CrossRef]
34. Vlados, C.; Chatzinikolaou, D.; Kapaltzoglou, F. Energy Market Liberalisation in Greece: Structures, Policy and Prospects. *Int. J. Energy Econ. Policy* **2021**, *11*, 115–126. [CrossRef]
35. Stern, N. *The Economics of Climate Change: The Stern Review*; Cambridge University Press: Cambridge, UK, 2007; ISBN 978-0-521-70080-1.
36. Dopfer, K.; Foster, J.; Potts, J. Micro-Meso-Macro. *J. Evol. Econ.* **2004**, *14*, 263–279. [CrossRef]
37. Vlados, C.; Chatzinikolaou, D. Macro, Meso, and Micro Policies for Strengthening Entrepreneurship: Towards an Integrated Competitiveness Policy. *J. Bus. Econ. Policy* **2020**, *7*, 1–12. [CrossRef]
38. Chatzinikolaou, D.H. The Energy Transition in Europe from the Perspective of International Political Economy, Business Ecosystems and Entrepreneurship. Ph.D. Thesis, Department of Economics, Democritus University of Thrace, Komotini, Greece, 2024.
39. Chatzinikolaou, D.; Vlados, C. International Political Economy, Business Ecosystems, Entrepreneurship, and Sustainability: A Synthesis on the Case of the Energy Sector. *Sustainability* **2024**, *16*, 10092. [CrossRef]
40. Chatzinikolaou, D.; Vlados, C.; Kokkinaki, A. The Ecosystems Perspective in Energy Research: A New Field Is Born? *Int. J. Energy Econ. Policy* **2024**, *15*, 249–266. [CrossRef]
41. Wittmayer, J.M.; Campos, I.; Avelino, F.; Brown, D.; Doračić, B.; Fraaije, M.; Gährs, S.; Hinsch, A.; Assalini, S.; Becker, T.; et al. Thinking, Doing, Organising: Prefiguring Just and Sustainable Energy Systems via Collective Prosumer Ecosystems in Europe. *Energy Res. Soc. Sci.* **2022**, *86*, 102425. [CrossRef]
42. Hellström, M.; Tsvetkova, A.; Gustafsson, M.; Wikström, K. Collaboration Mechanisms for Business Models in Distributed Energy Ecosystems. *J. Clean. Prod.* **2015**, *102*, 226–236. [CrossRef]
43. Hou, R.; Li, S.; Chen, H.; Ren, G.; Gao, W.; Liu, L. Coupling Mechanism and Development Prospect of Innovative Ecosystem of Clean Energy in Smart Agriculture Based on Blockchain. *J. Clean. Prod.* **2021**, *319*, 128466. [CrossRef]
44. Kim, E.; Ha, Y. Vitalization Strategies for the Building Energy Management System (BEMS) Industry Ecosystem Based on AHP Analysis. *Energies* **2021**, *14*, 2559. [CrossRef]
45. Adu-Kankam, K.O.; Camarinha-Matos, L.M. Emerging Community Energy Ecosystems: Analysis of Organizational and Governance Structures of Selected Representative Cases. In *Technological Innovation for Industry and Service Systems. DoCEIS 2019. IFIP Advances in Information and Communication Technology*; Camarinha-Matos, L.M., Almeida, R., Oliveira, J., Eds.; Springer: Cham, Switzerland, 2019; Volume 553, pp. 24–40. ISBN 978-3-030-17771-3.
46. Küfeoğlu, S.; Açıkgöz, E.; Taşcı, Y.E.; Arslan, T.Y.; Priesmann, J.; Praktiknjo, A. Designing the Business Ecosystem of a Decentralised Energy Datahub. *Energies* **2022**, *15*, 650. [CrossRef]
47. Ben Letaifa, S. How to Strategize Smart Cities: Revealing the SMART Model. *J. Bus. Res.* **2015**, *68*, 1414–1419. [CrossRef]
48. Peneder, M. Competitiveness and Industrial Policy: From Rationalities of Failure towards the Ability to Evolve. *Camb. J. Econ.* **2017**, *41*, 829–858. [CrossRef]
49. De Geus, A. *The Living Company*; Harvard Business School Press: Boston, MA, US, 2002; ISBN 978-1-57851-820-3.
50. Vlados, C. Change Management and Innovation in the "Living Organization": The Stra.Tech.Man Approach. *Manag. Dyn. Knowl. Econ.* **2019**, *7*, 229–256. [CrossRef]
51. Burns, T.; Stalker, G.M. Mechanistic and Organic Systems of Management. In *Sociology of Organizations: Structures and Relationships*; SAGE Publications: Los Angeles, CA, USA; London, UK; New Delhi, India; Singapore; Washington, DC, USA, 2011; pp. 14–18. ISBN 978-1-4129-9196-4.
52. Nelson, R.R. Economics from an Evolutionary Perspective. In *Modern Evolutionary Economics: An Overview*; Nelson, R.R., Ed.; Cambridge University Press: Cambridge, UK, 2018; pp. 1–34. ISBN 978-1-108-42743-2.
53. Ruggiero, S.; Varho, V.; Rikkonen, P. Transition to Distributed Energy Generation in Finland: Prospects and Barriers. *Energy Policy* **2015**, *86*, 433–443. [CrossRef]

4. Vlados, C. La Dynamique du Triangle Stratégie, Technologie et Management: L'insertion des Entreprises Grecques Dans la Globalisation [The Dynamics of the Triangle of Strategy, Technology and Management: The Insertion of Greek Enterprises into Globalization]. Ph.D. Thesis, Université de Paris X-Nanterre, Nanterre, France, 2004.
5. Chatzinikolaou, D.; Vlados, C. Evolution of Business Physiology in the Wine Industry: Insights from the Stra.Tech.Man Scorecard in the Cephalonian Robola Sector. *J. Wine Res.* **2023**, *34*, 210–231. [CrossRef]
6. Vlados, C.; Koronis, E.; Chatzinikolaou, D. Entrepreneurship and Crisis in Greece from a Neo-Schumpeterian Perspective: A Suggestion to Stimulate the Development Process at the Local Level. *Res. World Econ.* **2021**, *12*, 1–16. [CrossRef]
7. Chatzinikolaou, D.; Vlados, C. Weak Sustainable Development Trajectories and Evolving Organisational Physiologies: Empirical Evidence from Greece. In *Business for Sustainability, Volume I*; Vrontis, D., Thrassou, A., Efthymiou, L., Weber, Y., Shams, S.M.R., Tsoukatos, E., Eds.; Palgrave Studies in Cross-disciplinary Business Research, In Association with EuroMed Academy of Business; Springer International Publishing: Cham, Switzerland, 2023; pp. 201–226. ISBN 978-3-031-37360-2.
8. Vlados, C.; Chatzinikolaou, D. Navigating the Crisis toward the New Globalization through Innovation and Change Management: An Examination of the STRA.TECH.MAN–R.A.S.I. Framework. In *Crisis, Creativity, and Innovation*; Reiter-Palmon, R., Ivcevic, Z., Grohman, M., Tang, M., Eds.; Palgrave: Cham, Switzerland, 2023.
9. Chatzinikolaou, D.; Vlados, C. Green Organisational Reorientations for the New Globalisation. *J. Sustain. Sci. Manag.* **2024**, *19*, 43–68. [CrossRef]
0. Hammad, M.A.; Elgazzar, S.; Obrecht, M.; Sternad, M. Compatibility about the Concept of Energy Hub: A Strict and Visual Review. *Int. J. Energy Sect. Manag.* **2021**, *16*, 1–20. [CrossRef]
1. İpin, S.; Ercan, T. Financing Hydroelectric Power Plants: A Review and Evaluation of Financing Channels. *Int. J. Energy Sect. Manag.* **2020**, *15*, 58–80. [CrossRef]
2. Singh, K.; Vashishtha, S. Relationship Trend of Energy Consumption and Economic Growth Studies: A Global Examination Based on Bibliometric and Visualization Analysis. *Int. J. Energy Sect. Manag.* **2022**, *17*, 310–332. [CrossRef]
3. Snyder, H. Literature Review as a Research Methodology: An Overview and Guidelines. *J. Bus. Res.* **2019**, *104*, 333–339. [CrossRef]
4. Tranfield, D.; Denyer, D.; Smart, P. Towards a Methodology for Developing Evidence-Informed Management Knowledge by Means of Systematic Review. *Br. J. Manag.* **2003**, *14*, 207–222. [CrossRef]
5. Vlados, C.; Chatzinikolaou, D.; Demertzis, M. Development, Competitiveness and Institutional Modernisation: Towards a New Approach to the Greek Crisis. *Int. J. Compet.* **2019**, *1*, 293–318. [CrossRef]
6. Region of Eastern Macedonia–Thrace. *Το Επιχειρησιακό Πρόγραμμα «Ανατολική Μακεδονία και Θράκη» 2014–2020 [The Operational Programme "Eastern Macedonia and Thrace" 2014–2020]*; Region of Eastern Macedonia and Thrace: Komotini, Greece, 2014.
7. Creswell, J.W.; Creswell, J.D. *Research Design: Qualitative, Quantitative, and Mixed Methods Approaches*, 5th ed.; SAGE: Los Angeles, CA, USA, 2018; ISBN 978-1-5063-8670-6.
8. Denzin, N.K. *Interpretive Interactionism*; Sage: Newbury Park, CA, USA, 1989.
9. Benavides-Velasco, C.A.; Quintana-García, C.; Guzmán-Parra, V.F. Trends in Family Business Research. *Small Bus. Econ.* **2013**, *40*, 41–57. [CrossRef]
0. Supino, P.G. The Research Hypothesis: Role and Construction. In *Principles of Research Methodology: A Guide for Clinical Investigators*; Supino, P.G., Borer, J.S., Eds.; Springer: New York, NY, USA, 2012; pp. 31–53. ISBN 978-1-4614-3360-6.
1. Mortimore, M. Corporate Strategies for FDI in the Context of Latin America's New Economic Model. *World Dev.* **2000**, *28*, 1611–1626. [CrossRef]
2. Adamkaite, J.; Streimikiene, D.; Rudzioniene, K. The Impact of Social Responsibility on Corporate Financial Performance in the Energy Sector: Evidence from Lithuania. *Corp. Soc. Responsib. Environ. Manag.* **2023**, *30*, 91–104. [CrossRef]
3. Sadorsky, P. Some Future Scenarios for Renewable Energy. *Futures* **2011**, *43*, 1091–1104. [CrossRef]
4. Jamasb, T.; Llorca, M.; Meeus, L.; Schittekatte, T. Energy Network Innovation for Green Transition: Economic Issues and Regulatory Options. *Econ. Energy Environ. Policy* **2023**, *12*, 81–96. [CrossRef]
5. Lund, P. Effectiveness of Policy Measures in Transforming the Energy System. *Energy Policy* **2007**, *35*, 627–639. [CrossRef]
6. Bryant, S.; Straker, K.; Wrigley, C. Designing Our Sustainable Energy Future: A Shock Doctrine for Energy. *Energy Policy* **2020**, *147*, 111914. [CrossRef]
7. Darby, S. Demand Response and Smart Technology in Theory and Practice: Customer Experiences and System Actors. *Energy Policy* **2020**, *143*, 111573. [CrossRef]
8. Erlinghagen, S.; Markard, J. Smart Grids and the Transformation of the Electricity Sector: ICT Firms as Potential Catalysts for Sectoral Change. *Energy Policy* **2012**, *51*, 895–906. [CrossRef]
9. Ford, R.; Walton, S.; Stephenson, J.; Rees, D.; Scott, M.; King, G.; Williams, J.; Wooliscroft, B. Emerging Energy Transitions: PV Uptake beyond Subsidies. *Technol. Forecast. Soc. Change* **2017**, *117*, 138–150. [CrossRef]
0. Dranev, Y.; Izosimova, A.; Meissner, D. Organizational Ambidexterity and Performance: Assessment Approaches and Empirical Evidence. *J. Knowl. Econ.* **2020**, *11*, 676–691. [CrossRef]

81. Dodd, T.; Nelson, T. Trials and Tribulations of Market Responses to Climate Change: Insight through the Transformation of the Australian Electricity Market. *Aust. J. Manag.* **2019**, *44*, 614–631. [CrossRef]
82. Musiolik, J.; Markard, J. Creating and Shaping Innovation Systems: Formal Networks in the Innovation System for Stationary Fuel Cells in Germany. *Energy Policy* **2011**, *39*, 1909–1922. [CrossRef]
83. Jamasb, T.; Thakur, T.; Bag, B. Smart Electricity Distribution Networks, Business Models, and Application for Developing Countries. *Energy Policy* **2018**, *114*, 22–29. [CrossRef]
84. Kealy, T. Does an Embedded Wind Turbine Reduce a Company's Electricity Bill? Case Study of a 300 kW Wind Turbine in Ireland. *J. Bus. Ethics* **2017**, *145*, 417–428. [CrossRef]
85. Franzò, S.; Natalicchio, A.; Frattini, F.; Magliocca, P. How Digital Technologies Enable Business Model Innovation in the Energy Sector: An Empirical Study of Italian Energy Service Companies. *IEEE Trans. Eng. Manag.* **2023**, *71*, 13646–13659. [CrossRef]
86. Palmié, M.; Boehm, J.; Friedrich, J.; Parida, V.; Wincent, J.; Kahlert, J.; Gassmann, O.; Sjödin, D. Startups versus Incumbents in "green" Industry Transformations: A Comparative Study of Business Model Archetypes in the Electrical Power Sector. *Ind. Mark. Manag.* **2021**, *96*, 35–49. [CrossRef]
87. Del Vecchio, P.; Urbinati, A.; Kirchherr, J. Enablers of Managerial Practices for Circular Business Model Design: An Empirical Investigation of an Agro-Energy Company in a Rural Area. *IEEE Trans. Eng. Manag.* **2022**, *71*, 873–887. [CrossRef]
88. Bolton, R.; Hannon, M. Governing Sustainability Transitions through Business Model Innovation: Towards a Systems Understanding. *Res. Policy* **2016**, *45*, 1731–1742. [CrossRef]
89. Richter, M. Business Model Innovation for Sustainable Energy: How German Municipal Utilities Invest in Offshore Wind Energy. *Int. J. Technol. Manag.* **2013**, *63*, 24–50. [CrossRef]
90. Provance, M.; Donnelly, R.; Carayannis, E. Institutional Influences on Business Model Choice by New Ventures in the Microgenerated Energy Industry. *Energy Policy* **2011**, *39*, 5630–5637. [CrossRef]
91. Pereira, G.; Niesten, E.; Pinkse, J. Sustainable Energy Systems in the Making: A Study on Business Model Adaptation in Incumbent Utilities. *Technol. Forecast. Soc. Change* **2022**, *174*, 121207. [CrossRef]
92. Nair, S.; Paulose, H. Emergence of Green Business Models: The Case of Algae Biofuel for Aviation. *Energy Policy* **2014**, *65*, 175–184. [CrossRef]
93. Zhu, Q.; Jia, R.; Lin, X. Building Sustainable Circular Agriculture in China: Economic Viability and Entrepreneurship. *Manag. Decis.* **2019**, *57*, 1108–1122. [CrossRef]
94. Specht, J.; Madlener, R. Energy Supplier 2.0: A Conceptual Business Model for Energy Suppliers Aggregating Flexible Distributed Assets and Policy Issues Raised. *Energy Policy* **2019**, *135*, 110911. [CrossRef]
95. Chaiyapa, W.; Esteban, M.; Kameyama, Y. Why Go Green? Discourse Analysis of Motivations for Thailand's Oil and Gas Companies to Invest in Renewable Energy. *Energy Policy* **2018**, *120*, 448–459. [CrossRef]
96. Gupta, D.; Ghersi, F.; Vishwanathan, S.; Garg, A. Achieving Sustainable Development in India along Low Carbon Pathways: Macroeconomic Assessment. *World Dev.* **2019**, *123*, 104623. [CrossRef]
97. Korhonen, J.; Snäkin, J. Quantifying the Relationship of Resilience and Eco-Efficiency in Complex Adaptive Energy Systems. *Ecol. Econ.* **2015**, *120*, 83–92. [CrossRef]
98. Drago, C.; Gatto, A. Policy, Regulation Effectiveness, and Sustainability in the Energy Sector: A Worldwide Interval-Based Composite Indicator. *Energy Policy* **2022**, *167*, 112889. [CrossRef]
99. Pätäri, S.; Jantunen, A.; Kyläheiko, K.; Sandström, J. Does Sustainable Development Foster Value Creation? Empirical Evidence from the Global Energy Industry. *Corp. Soc. Responsib. Environ. Manag.* **2012**, *19*, 317–326. [CrossRef]
100. McGovern, G. Capturing Community Value in Civic Energy Business Model Design. *Energy Policy* **2021**, *156*, 112468. [CrossRef]
101. Likert, R. A Technique for the Measurement of Attitudes. *Arch. Psychol.* **1932**, *22*, 5–55.
102. Donaldson, G. A New Tool for Boards: The Strategic Audit. *Harv. Bus. Rev.* **1995**, *73*, 99–108.
103. Gordon, P.J. Ten Strategic Audit Questions. *Bus. Horiz.* **1997**, *40*, 7–14. [CrossRef]
104. Grundy, T. Strategic Audit. *Strateg. Dir.* **2008**, *24*, 44–45. [CrossRef]
105. Kaplan, R.S.; Norton, D.P. *The Balanced Scorecard: Measures That Drive Performance*; Harvard Business Review: Boston, MA, USA, 2005; Volume 70, pp. 71–75.
106. Lewin, K. Action Research and Minority Problems. *J. Soc. Issues* **1946**, *2*, 34–46. [CrossRef]
107. Chatzinikolaou, D.; Vlados, C. New Globalization and Energy Transition: Insights from Recent Global Developments. *Societies* **2024**, *14*, 166. [CrossRef]
108. Newell, P. Trasformismo or Transformation? The Global Political Economy of Energy Transitions. *Rev. Int. Political Econ.* **2019**, *26*, 25–48. [CrossRef]
109. Newell, P. *Power Shift: The Global Political Economy of Energy Transitions*; Cambridge University Press: Cambridge, UK; New York, NY, USA, 2021; ISBN 978-1-108-83285-4.

0. Chatzinikolaou, D.; Vlados, C. On a New Sustainable Energy Policy: Exploring a Macro-Meso-Micro Synthesis. *Energies* **2025**, *18*, 260. [CrossRef]
1. Vlados, C. The Stra.Tech.Man Scorecard. *Int. J. Bus. Adm.* **2021**, *12*, 36–63. [CrossRef]

energies

MDPI

Article

From Organizational Readiness to Industry 5.0: An EFQM Model Pathway to Net Zero

Joanna Martusewicz [1,*], Kamil Suchorski [2], Iwona Chomiak-Orsa [1], Joanna Usyk [2], Łukasz Bednarowicz [2] and Marcin Łukaszewicz [3,*]

1 Departament of Management, Wroclaw University of Economics and Business, Komandorska 118/120, 53-345 Wroclaw, Poland; iwona.chomiak-orsa@ue.wroc.pl
2 Toyota Boshoku Poland Sp. z o.o., ul. Wyzwolenia 56, 59-730 Wykroty, Poland; kamil.suchorski@toyota-boshoku.com (K.S.); joanna.usyk@toyota-boshoku.com (J.U.); lukasz.bednarowicz@toyota-boshoku.com (Ł.B.)
3 Department of Biotransformation, University of Wroclaw, F. Joliot-Curie 14a, 50-383 Wroclaw, Poland
* Correspondence: joanna.martusewicz@ue.wroc.pl (J.M.); marcin.lukaszewicz@uwr.edu.pl (M.Ł.)

Abstract: The automotive industry faces urgent pressures to transition to carbon-neutral operations amid evolving policies, shifting consumer demands, and stringent environmental regulations. This study examines how implementing the EFQM Model 2020 can drive sustainability-oriented transformation in a leading European automotive plant. Over a two-year period (November 2021–December 2023), the company reduced CO_2 emissions by 17%, decreased water usage by 9.3%, and elevated recycling rates from 93.3% in FY19 to 98.1% in FY23. Although these improvements demonstrate the EFQM Model's effectiveness in integrating economic, social, and environmental objectives, further progress toward net zero remains challenging due to diminishing returns on efficiency. Sustaining momentum will require continuous innovation such as passive building designs and on-site renewable energy generation supported by robust stakeholder engagement and compliance with evolving ESG reporting standards. These findings affirm the value of the holistic management framework for operational excellence and environmental stewardship, providing a replicable pathway toward carbon neutrality in resource-intensive industries.

Keywords: EFQM Model; SDG; Corporate Sustainability Reporting Directive (CSRD); sustainability; energy efficiency; Industry 5.0

1. Introduction

The automotive industry faces significant challenges in its transition toward hybrid, electric, and alternative fuel vehicles like hydrogen. While initial market growth was buoyed by consumer interest and government subsidies, the surge in electric vehicle (EV) adoption has recently stalled due to consumer hesitancy and shifting subsidy policies. At the same time, the industry's traditionally stable market dominated by a few large manufacturers is now disrupted by new brands, increasing market competition and reshaping consumer preferences [1–6].

In today's global economy, the pursuit of carbon neutrality, frequently termed "net zero" or "climate neutrality", has evolved from a voluntary sustainability goal to a critical business imperative. Heightened concerns over climate change, fueled by rising greenhouse gas (GHG) emissions, have compelled industries and governments worldwide to adopt aggressive carbon reduction targets. In the European Union, for instance, the Corporate Sustainability Reporting Directive (CSRD), which took effect in January 2023, significantly expands corporate reporting obligations regarding environmental, social, and governance (ESG) issues under the European Sustainability Reporting Standards (ESRS) [7]. Given its considerable contributions to global GHG emissions, the automotive sector finds itself at the forefront of this transformation [8]. As mobility demands continue to escalate, automakers face the dual challenge of increasing production while significantly lowering their carbon footprint [9,10]. This tension is magnified by stricter global emissions regulations, advancements in fuel economy standards, and rising consumer demand for more sustainable products.

In response, leading automotive companies have pledged to align with the global Sustainable Development Goals (SDGs), adopting comprehensive strategies for achieving carbon neutrality [11]. This transition spans all operational facets, from manufacturing processes and materials management to logistics and after-sales services [12]. Notably, these changes are resource-intensive and require modernizing production facilities, upskilling the workforce, and investing in advanced technologies [13]. Consequently, manufacturers are seeking innovative management frameworks that integrate environmental, social, and economic objectives [14].

One prominent example is the EFQM Model, which provides a holistic strategic perspective, focusing simultaneously on operational improvements and triple-bottom-line outcomes (economic, social, and environmental). Yet several other well-established excellence and quality management models exist, including ISO 9001 [15], the Baldrige Excellence Framework (BEF), and the Shingo Model, each with distinct emphases on continuous improvement, leadership, and stakeholder engagement [16–20]. ISO 9001 is renowned worldwide for its foundational quality management principles based on the Plan–Do–Check–Act cycle, although its most recent significant update in 2015 did not substantially address emerging issues like sustainability or digital transformation [21,22]. The Baldrige Excellence Framework, structured around seven categories (Leadership, Strategy, Customers, Measurement, Workforce, Operations, and Results), underscores innovation and organizational learning but has not been extensively refreshed to encompass Industry 5.0 and advanced technologies [19]. The Shingo Model prioritizes operational excellence through cultural transformation and respect for people but does not explicitly integrate sustainability or the broader ESG considerations that are becoming pivotal in today's business environment [19]. In comparing the EFQM with ISO 9001, the Baldrige Excellence Framework, and the Shingo Model, the choice of the EFQM emerges as particularly advantageous because it holistically addresses sustainability, digital transformation, and the human-centric thrust of Industry 5.0 within a single, integrated framework [16–18,23]. Moreover, the RADAR assessment tool facilitates quantitative benchmarking, offering a tangible pathway for continuous improvement [17,18]. Whereas ISO 9001 focuses on process standardization and compliance, Baldrige excels in innovation and knowledge management, and Shingo emphasizes cultural transformation, the EFQM uniquely foregrounds modern megatrends such as ESG reporting, AI integration, and carbon neutrality across diverse organizational contexts [17–20,24]. Accordingly, the EFQM's comprehensive approach aligns with the SDGs, supports a robust transition to carbon neutrality, and fosters enduring organizational resilience, core imperatives for today's automotive sector and beyond. Its latest iteration, the EFQM Model 2025, was introduced at the EFQM Forum in Istanbul in June 2024 and offers updated guidance to tackle contemporary challenges and leverage emerging opportunities [17,18,23,25]. The EFQM fosters a rigorous yet flexible structure that integrates economic, social, and environmental goals [26].

Originally established in 1991 to enhance quality in European organizations, the EFQM has evolved into a globally relevant framework that adapts to diverse cultural contexts and rapidly shifting business landscapes [27]. The model's iterative refinements increasingly foreground sustainability, echoing the global emphasis on responsible resource management, climate action, and stakeholder well-being [19]. In particular, the EFQM Model 2020 expanded on sustainable development themes, and the new 2025 iteration sharpens its focus on Industry 5.0 paradigms [28–31]. While Industry 4.0 primarily concentrated on digitization, automation, and the internet of things (IoT) for productivity gains, Industry 5.0 goes further by underscoring human-centric approaches, resilient value networks, and environmental stewardship [32]. The EFQM closely aligns with this shift by helping organizations strategically integrate emerging technologies, such as artificial intelligence and big data analytics, with socially responsible practices, thereby enhancing competitiveness while reducing carbon footprints [33,34].

In tandem with these technological and societal transformations, regulatory pressures, exemplified by the Corporate Sustainability Reporting Directive (CSRD), are accelerating the need for transparent and standardized ESG reporting through frameworks like the European Sustainability Reporting Standards (ESRS) [30]. The EFQM Model can play a pivotal role here, enabling companies to embed carbon reduction strategies into operational processes while documenting these efforts in ways that satisfy or surpass evolving auditing and disclosure mandates [18].

Despite the EFQM Model's strong emphasis on sustainability, empirical evidence of its implementation in real-world industrial contexts, particularly under the 2020 and 2025 iterations, remains scarce. New versions of the EFQM generally outpace the availability of systematic case studies; it often takes years for organizations to fully implement the frameworks and generate measurable results. Hence, most current research still relies on earlier versions of EFQM, creating a temporal gap between theoretical development and validated practice. Real-world industrial contexts are limited, primarily because practical implementation lags theoretical model updates [18,24]. In addition, a significant portion of empirical studies related to EFQM implementation rely primarily on subjective interviews, which, while providing valuable qualitative insights, often lack the robust quantitative data necessary to comprehensively validate the model's impact on organizational performance and sustainability.

Within this context, the present study investigates whether implementing the EFQM Model 2020 in a high-performing automotive plant can reinforce effective sustainable management by embedding key sustainability principles, such as renewable energy and carbon neutrality, alongside improvements in economic performance. This inquiry is timely amid ongoing geopolitical tensions affecting energy markets, persistent supply chain disruptions, and the residual effects of the COVID-19 pandemic, all of which underscore the importance of agility and resilience. Focusing on an industry-leading automotive plant, we examine carbon-neutral policy initiatives, wastewater and waste reduction efforts, CO_2 emissions tracking, and resource consumption optimization. Applying the EFQM as a structured methodology for self-assessment and continuous improvement, we highlight its adaptability within Industry 5.0 contexts [35], which emphasize the convergence of digital innovation, ethical considerations, and sustainability [36]. Our study provides empirical insights into how EFQM-based frameworks can help companies satisfy stringent environmental targets, comply with fast-evolving ESG reporting requirements, and achieve durable value creation.

By centering on the automotive industry, this study thus connects theoretical models with on-the-ground applications, offering both empirical validation of the EFQM Model and a roadmap for entities seeking to align organizational readiness with Industry 5.0 philosophies [31,37]. Ultimately, our findings aim to contribute to global efforts toward net-zero emissions and sustainable industrial transformation.

2. Materials and Methods

This study was conducted in collaboration with one large enterprise from November 2021 to December 2023. The company's primary objective was to achieve a net-zero environmental impact, guided by the implementation of the EFQM Model 2020. The participating plant is part of a larger international corporation operating in the automotive sector, with a staff of over 1000 employees.

To facilitate organizational transformation toward net zero goals, the research team adopted the EFQM Model as a comprehensive framework [21,38]. Implementation followed a structured, multi-step procedure to align leadership, strategy, and continuous improvement activities with the company's economic, social, and environmental objectives. A schematic overview of the study approach is presented in Figure 1.

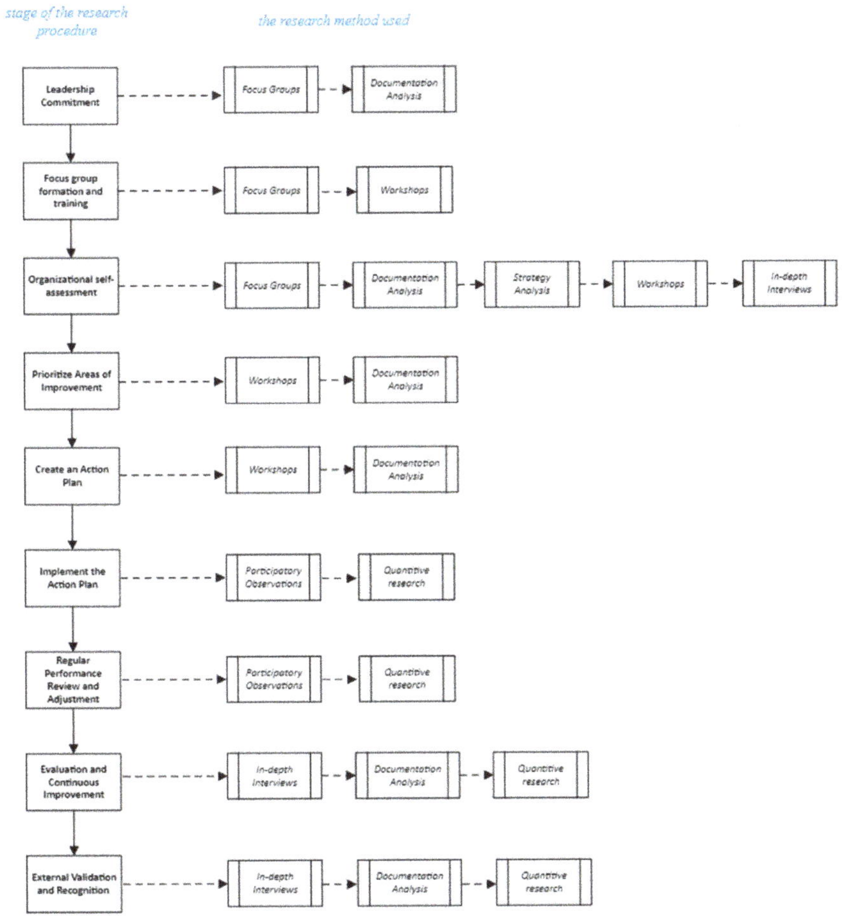

Figure 1. Schematic presentation of the study approach in accordance with EFQM Model implementation.

Figure 1 presents the nine stages of research:

1. Leadership Commitment

Securing top management's commitment to implement EFQM principles was the first step. The company's purpose, mission, and values were clearly defined and aligned with strategic objectives, including net zero targets. Long-term sustainability and performance goals were set to integrate economic, social, and environmental dimensions.

2. Focus Group Formation and Training

Key personnel from diverse functional areas were selected to form the EFQM implementation team. These individuals received in-depth training on EFQM methodology, ensuring a consistent understanding of its framework and criteria. The implementation team consisted of 14 people from different management levels and functional divisions. They included production managers; finance managers; middle managers from procurement, logistics, HR, IT, quality, production, and design; and senior specialists in quality, human resources, CSR, sales, kaizen, logistics. In preparation for the implementation of the EFQM Model, four training courses were held in accordance with EFQM standards:

- Introduction to the EFQM Model (5 h)—intended for the company's management board and all directors.
- EFQM Foundation Training (2 days)—training for the whole team.
- Self-Assessment and Use of the RADAR tool (1 day)—for the whole team.
- EFQM Certified Assessor (3 days plus 2 days of self-study)—for selected team members indicated by the project manager.

Each training course was preceded and ended with a survey and tests to check the level of knowledge.

3. Organizational Self-Assessment

A comprehensive self-assessment, based on EFQM criteria, was carried out to identify strengths and gaps in current practices. Benchmarking was performed against internal corporate standards, including established practices, methods, solutions, and productivity parameters. The benchmarking process involved comparing the studied organization's performance with that of 11 top-performing factories located across Europe for a given criterium. These factories were selected based on their exemplary performance in areas such as productivity, efficiency, and sustainability, as measured by corporate metrics.

The self-assessment and benchmarking revealed critical areas for improvement, which were prioritized based on their alignment with the corporation's strategic goals and expected KPIs for the coming years. Key areas included energy efficiency, waste reduction, water usage, and CO_2 emissions, reflecting the organization's commitment to sustainability and operational excellence. This process provided actionable insights to guide the implementation of targeted improvements in line with the EFQM Model's holistic framework.

4. Prioritize Areas of Improvement

The organization's target was net-zero CO_2 per vehicle produced. The improvement areas identified through self-assessment were prioritized according to their relevance for net zero ambitions and EFQM criteria, for example, renewable energy. Specific, measurable objectives were set (for example, energy economies, reduction in waste and recycling, self-generation of energy, green certificates), with resources and responsibilities clearly assigned within kaizens.

5. Create an Action Plan

A detailed action plan was developed to map out each priority area, associated tasks, deadlines, and roles. Key performance indicators (KPIs) and milestones were established to track progress toward the net zero goal.

6. Implement the Action Plan

Implementation activities (kaizens) were communicated to relevant stakeholders (internal teams, including production, sustainability, and management; suppliers; and local regulators), and necessary training was provided by corporate sustainability officers and external experts, covering EFQM methodology, carbon accounting, and ISO 14001 compliance [39]. Progress was monitored (also by the implementation of new online sensors—IoT) and evaluated against KPIs; adjustments were made as required.

7. Regular Performance Review and Adjustment

Periodic performance reviews were conducted to assess the effectiveness of improvement actions. Results were used to refine the action plan, responding dynamically to in-ternal progress and external factors such as policy changes. Reviewed actions encompassed all company activities and kaizen initiatives. In the present analysis, we focus exclusively on sustainability measures, including energy efficiency upgrades, waste segregation protocols, recycling, employee engagement, and heat recovery systems. External factors considered include the EU Corporate Sustainability Reporting Directive (CSRD), supply chain disruptions due to COVID-19, and energy price fluctuations, particularly those driven by the war in Ukraine. Organizations aim to gather and monitor data in real time using IoT technologies. Regular meetings are held at intervals tailored to specific subjects, while reviews related to EFQM projects and sustainability are scheduled according to each project's timeline.

8. Evaluation and Continuous Improvement

A select group of employees received training as EFQM assessors to ensure ongoing internal evaluations. Comprehensive assessments were carried out regularly to compare performance with organizational and EFQM objectives. Insights from these assessments informed updates to improvement practices, fostering an internal culture of continuous improvement. Eight employees were trained as EFQM assessors: a production director, a finance director, middle-level managers (logistics, HR, IT, production), and senior specialists (quality, kaizen). The selection of assessors was intentionally designed to cover the broadest areas of the organization's activities and its hierarchical structure, ensuring that multiple perspectives and levels of operation were considered. The evaluation aimed to enhance overall organizational performance in line with EFQM guidelines. The EFQM Model offers a universal, quantitative assessment tool applicable to all organizations. The organization's ambitious goal was attaining five-star EFQM recognition, the highest accolade for EU corporations. The results presented here focus specifically on net zero, energy efficiency, and broader sustainability initiatives.

9. External Validation and Recognition

In addition to ongoing internal assessments, the organization underwent an external evaluation by independent EFQM assessors. As a result of this evaluation, the organization was awarded a five-star EFQM recognition in 2023, demonstrating alignment with best practices in quality management and sustainability. The recognition by EFQM assessment was conducted by the Foundation for the Development of Wroclaw University of Economics and Business, a Polish partner of the EFQM. Four certified EFQM assessors with many years of international experience in assessing organizations were invited to the assessment. Publicizing this achievement serves not only as a benchmark for performance but also provides additional motivation for all stakeholders to maintain and enhance continual improvement efforts.

A multi-method research design was used to capture both qualitative and quantitative dimensions of EFQM-driven improvements, especially those targeting sustainable development and specific Sustainable Development Goals (SDGs). Integrating multiple methods allowed for the robust triangulation of data, enhancing the validity of the findings [22]. An overview of this research design is shown in Table 1. Publicly available documents can be accessed at toyota-boshoku.com. Focus groups were composed of 14 trained employees and consultants, with participant selection tailored to topics related to the EFQM criteria.

The number of sessions was determined by the complexity of each topic. Evaluation was grounded in internal strategic documents, annual reports, and established strategic objectives. Researchers participated in all meetings related to EFQM model implementation spanning several sessions between 2021 and 2023. All 14 EFQM project participants engaged in workshops designed to identify the organization's strengths and improvement areas, which subsequently formed the basis for defining enhancement projects. In-depth interviews were conducted with the board and directors nominated by the board. Data collection methods were similarly aligned, acknowledging that the specifics are inherently unique to each organization due to variations in machinery, systems, sensors, software, and other technological components. New servers were implemented (increased efficiency lowered energy consumption related to data collection, analysis, and storage). The large IT infrastructure is beyond scope of this study.

Table 1. Research methods and approaches used in the study.

Approach	Description
Focus groups	Focus groups comprising organizational stakeholders convened to identify the primary needs and improvement areas. Discussions centered on internal perceptions of sustainability, readiness for EFQM implementation, and alignment with the SDGs [40].
Documentation analysis	Policy papers, annual reports, and key performance parameters were systematically reviewed for the period of 2019–2023. This analysis provided a baseline of historical performance and strategic intent, informing targets and KPIs.
Participatory observations	Researchers observed steering committee meetings to gain real-time insights into decision-making, implementation practices, and stakeholder engagement [41]. Observations provided a contextual understanding of how EFQM principles were operationalized.
Strategy analysis	Both current and planned development strategies were scrutinized to assess the organization's alignment with EFQM Model principles and SDG-related objectives. The review included mapping corporate goals to relevant EFQM criteria and sustainability metrics.
Workshops	Interactive workshops facilitated collaborative brainstorming and strategy validation. Participants co-developed action plans for sustainability and EFQM alignment, thereby refining the organization's approach to net zero targets.
In-depth interviews	Seven top-level managers were interviewed to elicit executive-level perspectives on implementing EFQM and sustainability practices [42]. Interviews covered the challenges, opportunities, and strategic rationale behind net zero initiatives.
IoT-based data collection	Quantitative data were captured via sensors and detection systems, integrated into an internet of things (IoT) platform. Real-time measurements of energy, water, and other resource usage were aggregated to facilitate detailed performance evaluation against set EFQM and SDG metrics.

Data from the above methods were cross-verified through methodological triangulation. For instance, quantitative metrics on resource efficiency and emissions were compared with qualitative feedback from focus groups, interviews, and steering committee observations. This process ensured that any reported improvements or barriers to EFQM implementation were validated from multiple perspectives. By blending empirical measurements (energy consumption, emissions reduction, etc.) with stakeholder perceptions, the study captured a holistic view of the EFQM-driven transformation toward net zero.

3. Results and Discussion

3.1. Overview of the EFQM Model's Role in Sustainable Development

The pursuit of strategic solutions to achieve the Sustainable Development Goals (SDGs) has become a primary mission for many global corporations, particularly those operating in resource-intensive sectors such as the automotive industry. One such holistic framework, the EFQM Model 2020, has gained prominence as a promising approach to integrating economic, social, and environmental objectives into core corporate operations. This study builds on a multi-year, multi-phase collaboration with an international automotive corporation renowned for its innovative and sustainability-focused practices.

Over an extensive observation period (November 2021–December 2023), the research team gathered multifaceted data on how EFQM-driven processes support the adoption of carbon-neutral practices, as well as broader environmental, social, and governance (ESG) targets. The overarching aim was to assess whether EFQM implementation could help align strategic and operational activities with the SDGs while maintaining robust management efficiency.

Within the European Union (EU), organizations are increasingly guided by evolving regulations that mandate transparent reporting on sustainability actions and SDG contributions. The European Sustainability Reporting Standards (ESRS), published in 2023 and 2024, comprise two cross-cutting standards (ESRS 1 and ESRS 2) and ten thematic standards addressing environmental (ESRS E1–E5), social (ESRS S1–S4), and governance (ESRS G1) dimensions. These standards share a consistent four-pillar framework—Governance, Strategy, Impact and Risk Management, and Metrics and Targets—designed to facilitate a coherent reporting structure across different industries.

To comply with these directives, organizations must identify the most relevant ESG issues for stakeholders and the most impactful concerns for business performance. This "double materiality" principle obliges companies to consider both financial implications and wider environmental and social factors when reporting sustainability outcomes. Such a reporting landscape underscores the importance of adopting comprehensive management systems capable of gathering, analyzing, and communicating ESG data effectively.

Given the ESRS mandates, the EFQM Model offers an integrated approach that unifies operational excellence with sustainability targets, thus helping organizations both implement and report on their SDG progress. Through its structured criteria and continuous improvement philosophy, the EFQM facilitates the monitoring of economic, social, and environmental metrics, key components of modern sustainability frameworks [43]. An overview of how the EFQM aligns with the ESRS is presented in Table 2, illustrating the synergistic relationship between a proven quality management methodology and emerging regulatory standards.

By capturing the dual priorities of performance excellence and environmental responsibility, the EFQM positions organizations to meet increasingly stringent sustainability requirements. This alignment also enhances stakeholder trust, as the transparent reporting of ESG metrics is quickly becoming a competitive imperative. The subsequent sections delve deeper into the implementation of the EFQM Model within an automotive production plant, highlighting specific outcomes in terms of carbon reduction, resource efficiency, and stakeholder engagement.

Table 2. Alignment of the EFQM Model with the ESRS.

Area	Criteria	Criterion Parts	ESG Factor
Direction	1. Purpose, vision, and strategy	1.1 Define purpose and vision	ESRS 1; ESRS 2
		1.2 Identify and understand stakeholders' needs	ESRS S3
		1.3 Understand the ecosystem, own capabilities, and major challenges	ESRS E1; ESRS E2; ESRS E3; ESRS E4
		1.5 Design and implement a governance and performance management system	ESRS G1; ESRS 1; ESRS 2
	2. Organizational culture and leadership	2.1 Steer the organization's culture and nurture values	ESRS G1
Execution	3. Engaging stakeholders	3.1 Customers: build sustainable relationships	ESRS S4
		3.2 People: attract, engage, develop, and retain	ESRS S1
		3.4 Society: contribute to development, well-being, and prosperity	ESRS S3
		3.5 Partners and suppliers: build relationships and ensure support for creating sustainable value	ESRS G1; ESRS S2
	4. Creating sustainable value	4.1 Define the value and how it is created	ESRS S4
		4.2 Communicate and sell the value	
		4.3 Deliver the value	ESRS S4
		4.4 Define and implement the overall experience	
	5. Driving performance and transformation	5.1 Drive performance and manage risk	ESRS G1; ESRS1; ESRS 2
		5.2 Transform the organization for the future	ESRS E1; ESRS E2; ESRS E3; ESRS E4
		5.3 Drive innovation and utilize technology	ESRS E5
		5.4 Leverage data, information, knowledge	ESRS 1; ESRS 2
		5.5 Manage assets and resources	ESRS E1; ESRS E2; ESRS E3; ESRS E4; ESRS E5
Results	6. Stakeholder perceptions		ESRS 1; ESRS 2
	7. Strategic and operational performance		ESRS 1; ESRS 2; ESRS E1; ESRS E2; ESRS E3; ESRS E4; ESRS E5

The EFQM Model aligns well with the European Sustainability Reporting Standards (ESRS) through a structured approach that incorporates both performance management and sustainability reporting criteria across all areas:

Within the direction area, purpose, vision, and strategy (Criterion 1) in the EFQM Model aligns with ESRS 1 and ESRS 2 by guiding organizations to define their long-term mission and strategic vision, integrating sustainability objectives such as those outlined in the SDGs. The stakeholder needs component under Criterion 1.2 aligns with ESRS S3, emphasizing the importance of understanding stakeholder expectations in sustainability and governance. The EFQM Model's approach to ecosystem understanding (Criterion 1.3) mirrors ESRS E1–E4 by requiring companies to assess environmental challenges and their own sustainability capabilities. Criterion 2 (organizational culture and leadership) in the EFQM Model is directly supported by ESRS G1, which emphasizes governance structures that cultivate a sustainability-oriented culture and reinforce core values across the organization.

Within the execution area, including engaging stakeholders and creating sustainable value (Criterion 3), the EFQM Model's focus on stakeholder engagement aligns with multiple ESRS standards, where customer relationships are linked to ESRS S4, people management to ESRS S1, societal impact to ESRS S3, and supplier partnerships to ESRS G1 and ESRS S2. Creating sustainable value (Criterion 4), which involves defining, communicating, and delivering value, is directly aligned with ESRS S4 in the ESRS framework. This ensures that organizations communicate sustainability value effectively to stakeholders. Performance and risk management (Criterion 5.1) aligns with ESRS G1, ESRS 1, and ESRS 2, focusing on robust governance and risk management in sustainability initiatives. Transformation initiatives (Criterion 5.2) and innovation (Criterion 5.3) align with ESRS E1–E5, requiring organizations to innovate in environmentally responsible ways and leverage technology (particularly ESRS E5). Resource and asset management (Criterion 5.5) further emphasizes sustainable operations, aligning with ESRS E1–E5 to promote effective resource use and mitigate environmental impacts.

Within the results area, stakeholder perceptions (Criterion 6) in the EFQM Model ties into ESRS 1 and ESRS 2, highlighting the importance of transparent stakeholder feedback in sustainability performance. Strategic and operational performance (Criterion 7) is aligned with a broad range of ESRS (ESRS E1–E5), reinforcing comprehensive performance tracking and reporting on environmental impact and sustainability outcomes. Analysis of the documentation [28,44] showed the compatibility of the adopted SDGs with the EFQM Model 2020 and the European Sustainability Reporting Standards (ESRS) in the environmental area (Table 3).

Table 3. Alignment of the EFQM Model with SDGs implemented within the organization and related ESRS standards.

<table>
<tr><th rowspan="2">SDG</th><th rowspan="2">ESRS</th><th colspan="3">Model EFQM</th></tr>
<tr><th>Direction</th><th>Execution</th><th>Results</th></tr>
<tr><td>Goal 6—Clean water and sanitation: ensure access to water and sanitation for all through sustainable water resource management</td><td>ESRS E3—Water and marine resources</td><td rowspan="5">1.3 Understand the ecosystem, own capabilities, and major challenges</td><td rowspan="5">5.2 Transform the organization for the future
5.5 Manage assets and resources
5.3 Drive innovation and utilize technology</td><td rowspan="5">7. Strategic and operational performance</td></tr>
<tr><td>Goal 7—Clean and accessible energy: ensure affordable access to stable, sustainable, and modern energy for all</td><td>ESRS E2—Pollution</td></tr>
<tr><td>Goal 9—Innovation, industry, infrastructure: build resilient infrastructure, promote sustainable industrialization, and foster innovation</td><td>ESRS E2—Pollution</td></tr>
<tr><td>Goal 12—Responsible consumption and production: ensure sustainable consumption and production patterns</td><td>ESRS E5—Resource utilization and the circular economy</td></tr>
<tr><td>Goal 13—Climate action: take urgent action to combat climate change and its impacts</td><td>ESRS E1—Climate change</td></tr>
</table>

3.2. Driving Sustainable Development with the EFQM Model

This study focused on a leading international automotive factory that began implementing the EFQM Model 2020 between November 2021 and December 2023. Situated in Poland within the European Union (EU), the plant employs over 1000 staff and manufactures components both for its own brand and for external automotive partners. Against a backdrop of increasingly stringent environmental legislation and rising stakeholder expectations, the company sought to leverage EFQM principles to accelerate progress toward carbon neutrality and other environmental targets.

The primary research objective was to determine whether the EFQM Model could effectively guide the factory in aligning management and operational practices with the Sustainable Development Goals (SDGs). Drawing on the established EFQM assessment criteria, the project aimed to achieve the following:

1. Evaluate how the factory's existing culture, leadership commitment, and resources measured up against EFQM guidelines.
2. Pinpoint specific areas needing improvement to meet regulatory and societal expectations for low-emission, resource-efficient operations.
3. Use the EFQM's structured approach to devise, implement, and track systematic sustainability measures at both the strategic and shop-floor levels.
4. Translate the EFQM framework into measurable outcomes in energy savings, waste reduction, and greenhouse gas (GHG) emissions abatement.

The quantitative evaluation of the criteria was performed through structured self-assessment workshops and surveys, where employees provided quantitative ratings based on the EFQM's RADAR logic. These scores facilitated a measurable understanding of strengths and areas for improvement within the analyzed criteria.

Resource allocation was examined through both tangible and intangible assets (e.g., technology, human capital, and financial support for sustainability initiatives) using internal performance metrics aligned with the EFQM criteria.

This iterative self-assessment process, similar to the PDCA cycle, supports continuous improvement and is further validated by external EFQM assessors when the organization opts for official recognition.

Evolving EU directives, particularly the European Sustainability Reporting Standards (ESRS), demand greater transparency and accountability in ESG performance. In conjunction with these requirements, the EFQM Model serves as a complementary system for organizations to embed sustainable practices, track progress, and report on outcomes. The following points highlight the synergy between the EFQM and ESRS mandates:

- The EFQM prompts leaders to integrate sustainability considerations into their organizational vision, mission, and strategic plans, which aligns with ESRS pillars on governance and strategy.
- The EFQM's continuous improvement cycle encourages ongoing risk assessments and stakeholder engagement, mirroring the ESRS emphasis on identifying and managing material, environmental, and social impacts.
- The EFQM's results-oriented criteria facilitate structured data collection critical for meeting ESRS metrics and targets, particularly around energy use, carbon footprint, and water management.

Building on recent research [45,46], this study hypothesized that EFQM implementation enables systematic sustainable development while preserving robust management performance. Automotive manufacturing, characterized by high energy consumption and substantial global supply chains, presents a stringent test case for EFQM-driven improvements. Moreover, the mandatory shift toward lower-emission operations, exacerbated by regulatory frameworks such as the Corporate Sustainability Reporting Directive (CSRD), offers fertile ground to assess the EFQM's suitability for navigating complex environmental targets.

A pivotal element of contemporary sustainability efforts is the concept of double materiality, which compels organizations to evaluate both financially material factors (implications for profitability, risk exposure, etc.) and environmental–social materiality (broader impacts on ecosystems and communities). The EFQM's stakeholder-focused structure facilitated the plant's capacity to achieve the following:

- Map relevant stakeholders (internal teams, suppliers, local communities, customers), assessing how each group perceives or is affected by the firm's operations.
- Prioritize environmental and social issues in a way that resonates with stakeholder concerns, thus channeling finite resources toward the most pressing sustainability challenges.
- Integrate continuous feedback through regular reviews and surveys, ensuring that the firm remains responsive to shifting stakeholder priorities and market conditions.

The EFQM Model provided a robust framework within which the automotive factory could synchronize its strategic goals, operational activities, and ESG reporting obligations.

3.3. *Implementation of the Net Zero Strategy in the Automotive Factory*

In tandem with the EFQM-based improvements, the organization enacted a comprehensive strategy to achieve net-zero environmental impact. The initiative centered on five UN SDGs, Goals 6, 7, 9, 12, and 13, deemed critical for addressing water usage, clean energy, responsible consumption, industrial innovation, and urgent climate action (Table 3). By targeting these SDGs, the company sought to translate high-level sustainability commitments into operational realities on the production floor.

Figure 2 illustrates the four interlocking pillars that constitute the company's net zero roadmap: (1) team awareness, (2) data collection and analysis, (3) kaizen-driven process improvements, and (4) on-site green energy generation. Each pillar is grounded in EFQM Model principles, ensuring that every step toward decarbonization aligns with broader organizational excellence and stakeholder engagement.

1. Team Awareness

Intensive training programs and workshops increased employees' understanding of the net zero vision. Cross-functional teams were empowered to design improvement projects, linking the EFQM focus on leadership and culture (Criteria 1 and 2) with practical sustainability actions [8].

2. Data Collection and Analysis

Advanced metering devices and IoT platforms were installed to capture real-time energy consumption and emissions data. Regular performance reviews and analytics provided an evidence-based foundation for decision-making, fulfilling EFQM's emphasis on data-driven management (Criterion 5).

3. Kaizens for Continuous Improvement

A series of iterative, small-scale improvements targeted reductions in energy usage, water consumption, and waste. Quality control circles (QCCs) and Kaizen activities helped operationalize the EFQM's "improvement" ethos, fostering employee ownership of sustainability initiatives.

4. Green Energy Self-Generation

Initially, the company prioritized reducing consumption through efficiency gains. Subsequent plans called for solar photovoltaics and other renewable technologies to cover remaining energy needs. This sequential approach, with efficiency first and renewables second, aligns with the EFQM's lifecycle thinking, minimizing both resource use and the carbon footprint of new infrastructure.

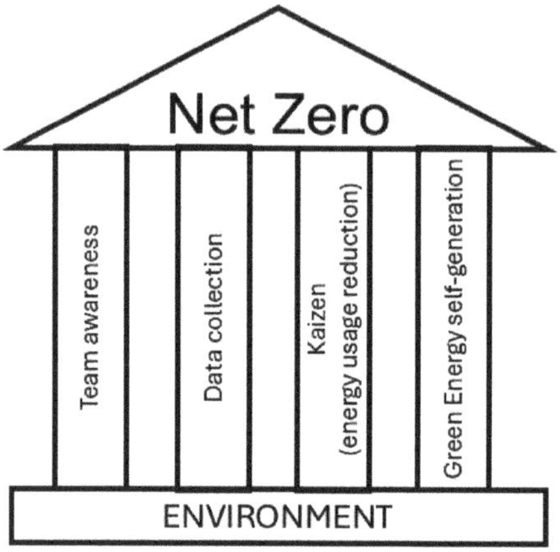

Figure 2. Four fundamental pillars leading to a net-zero environmental impact.

Tables 4–6 link each EFQM criterion to specific environmental initiatives. This integrated framework facilitated a seamless progression from strategic intent (direction) to execution (execution) and outcome measurement (results).

Table 4. Actions taken as part of the implementation of the EFQM Model 2020 across the area of direction.

Direction			
Criteria	**Guidelines**	**Description**	**Company Actions/Kaizen**
1. Purpose, vision, and strategy	1.3 Understand the ecosystem, own capabilities, and major challenges	Explores and understands the ecosystem, including megatrends and the implications of the UN Sustainable Development Goals	The company's strategy up to 2030 includes 10 UN goals, of which 5 are environmental. The company using the strategy decided to achieve following environmental goals: 1. Net-zero GHG emissions by 2035 2. Net-zero GHG emissions by 2050 3. Minimizing natural resource usage 5. Minimizing waste production 6. Participating in the whole-group goal: planting 1.32 million trees by 2050
		Assesses and values knowledge, data, and information collected from across the ecosystem to understand the main current and future challenges	Expands the data collection and monitoring system, the analysis of which enables the implementation of key actions, such as within kaizens
2. Organizational culture and leadership	2.1 Steer the organization's culture and nurture values	Expresses and promotes concern for the natural environment and the scarcity of resources, raising awareness of the importance of a responsible approach to the environment	Builds employee engagement in quality control circles (QCCs) by monitoring satisfaction levels and involvement in kaizen activities

Table 5. Actions taken as part of the implementation of the EFQM Model 2020 across the area of execution.

Execution			
Criteria	**Guidelines**	**Description**	**Company Actions/Kaizen**
3. Stakeholder engagement	3.4 Society: contribute to development, well-being, and prosperity	People or groups outside the organization who represent the immediate community or wider society, including, for example, interest groups focusing on issues such as the environment	Planting 1.32 million trees: As part of reforestation efforts, the company engages in protecting habitats of endemic species in various regions and countries, conserving forests and restoring abundant habitats.
4. Creating sustainable value	4.1 Design the value and how it is created	Designs value and value creation approaches to reflect the lifecycle of products, services, and solutions in a responsible manner, considering the impact on public health, safety, and the environment	Using recyclable or biodegradable materials in production
	4.3 Deliver the value	Provides its products, services and solutions in a way that minimizes the negative impact on society and the natural environment	Automatic management of lighting and ventilation
5. Driving performance and transformation	5.2 Transform the organization for the future	Transforms the value creation process and other organizational processes to adapt them to future needs	Cooling system energy reduction during non-production periods; solar farm installation; heat recovery from cutting process; A/C in transformer station instead of ventilation
	5.5 Manage assets and resources	Identifies assets and resources that are no longer needed and disposes of them responsibly based on the principles of the circular economy	External lighting management Recycling

Table 6. Actions taken as part of the implementation of the EFQM Model 2020 across the area of results.

Results			
Criteria	**Guidelines**	**Description**	**Company Actions/Kaizen**
6. Stakeholder perceptions	6.1 Customer perception results	The brand and reputation of the organization, including its impact on society and the environment	Production of low-carbon-intensity vehicles
	6.3 Business and governing stakeholder perception results	Perception of the organization's social and environmental responsibility	Good relations with local authorities
	6.4 Society perception results	Sustainability of the effects of the organization's activities on the community in economic, social, and environmental terms	Restoring natural forest environments
7. Strategic and operational performance	Considers the current and future needs and expectations of key stakeholders when selecting the most appropriate performance indicators aligned with strategic and operational goals		Recycling

The table is divided into three areas of the EFQM Model: direction, execution, and results. The columns include, from the left, model criteria within the area, guidelines, a description, and, in the last column, examples of implementation by the company.

The initiatives included the following:

- Resource monitoring: deploying an extensive sensor network to measure electricity, water, and natural gas consumption in real time.
- Exceeding a 98% recycling rate by FY23 (Table 7), reflecting a circular economy focus and the EFQM's results-driven perspective.
- Replacing traditional lighting systems with automated, energy-saving solutions, alongside improved insulation and heat recovery measures to curtail fixed energy consumption.

Table 7. Improvement of the recycling ratio over five last fiscal years.

Fiscal Year	Recycling Ratio
FY19	93.3%
FY20	96.9%
FY21	97.2%
FY22	98.0%
FY23	98.1%

These efforts collectively reduced CO_2 emissions by 17% over the study period, underscoring the pragmatic benefits of EFQM-based continuous improvement.

To evaluate the net zero trajectory, the company tracked CO_2 emissions per vehicle produced (Figures 3 and 4). Figure 3 illustrates the roadmap toward net-zero emissions for the EU region, showing the percentage reduction in CO_2 emissions per vehicle relative to the FY19 baseline. In this figure, upward arrows denote periods where significant improvements were achieved, reflecting the effective implementation of energy efficiency measures and operational optimizations. In contrast, downward arrows indicate intervals during which progress plateaued or experienced minor regressions, likely due to external challenges or transitional phases in the implementation process. This visual representation provides readers with a clear understanding of the dynamic trajectory of emissions reduction over time. Over a four-year span (FY19–FY23), total CO_2 output per vehicle decreased to 65.19% of the initial baseline in the broader EU region, whereas the specific plant achieved a reduction to 54.43% of its baseline, indicating a more pronounced improvement at the plant. It is important to note that these percentages are relative and depend on each facility's baseline and operational characteristics, such as energy intensity and the degree of baseline optimization. While these figures highlight significant progress, further gains are anticipated through additional Kaizen projects and the planned expansion of on-site renewables [47].

Approximately half of total energy use remains fixed, underscoring ongoing challenges in scaling down baseline consumption. To close the gap in on-site generation capacity, the company supplemented its carbon reduction goals by purchasing green energy certificates, reflecting the EFQM's risk management approach to achieving environmental targets.

By systematically mapping its actions to the five targeted SDGs and adhering to emerging ESRS guidelines, the factory demonstrated a robust alignment between operational excellence and regulatory compliance. This alignment extends beyond internal efficiency gains to broader ESG reporting. The EFQM's stakeholder-centric structure enabled the company to address both financial materiality (e.g., cost savings from energy efficiency) and environmental–social materiality (e.g., local community benefits, greenhouse gas reduction). The enterprise documented its progress in annual sustainability reports, showcasing verified metrics that support ESRS obligations. By reinforcing EFQM principles, particularly Criterion 5 (driving performance and transformation), the company built adaptability into its net zero strategy, anticipating evolving regulations and stakeholder expectations.

These outcomes highlight that EFQM Model implementation can serve as a powerful catalyst for achieving ambitious sustainability objectives in the automotive sector. The synergy of structured management practices, employee-driven improvement, and technology-enabled monitoring underscores the feasibility of transitioning to carbon-neutral operations while maintaining high standards of product quality and commercial viability.

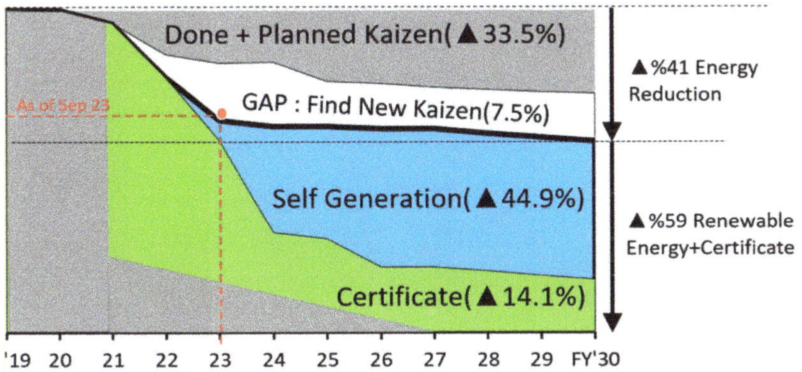

Figure 3. Roadmap of the EU region to net-zero environmental impact shown as the progression in CO_2 emissions reduction per vehicle as a percentage of the initial emissions in FY19.

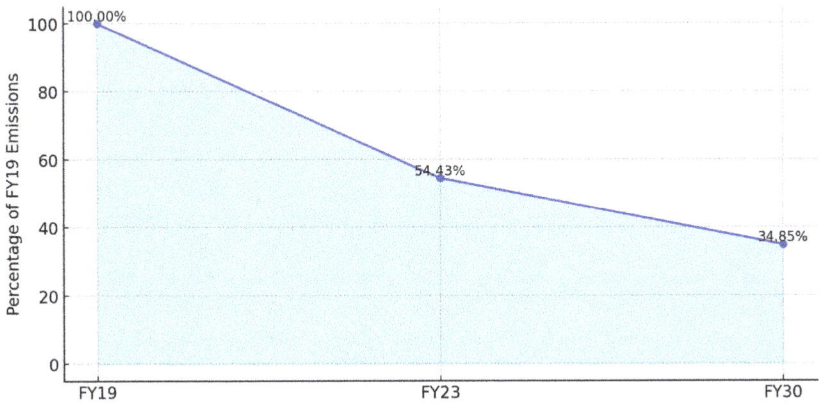

Figure 4. Roadmap of the studied plant to net-zero environmental impact shown as the progression in CO_2 emissions reduction per vehicle as a percentage of the initial emissions in FY19.

3.4. Strengthening Organizational Engagement: Kaizens, Team Awareness, and Data-Driven Insights

A core principle of the EFQM Model is fostering a shared vision and culture of continuous improvement, which this automotive factory operationalized through employee awareness initiatives, kaizen activities, and rigorous data monitoring. This integrated approach not only supported carbon reduction targets but also enhanced staff commitment and operational resilience.

From the outset, leadership placed significant emphasis on team awareness of sustainability goals. The organization's central aim to achieve net-zero emissions was reinforced at every level through training sessions, workshops, and regular communications. Figure 5 illustrates an upward trend in quality control circle (QCC) participation from 2019 to 2022, climbing from approximately 20% to nearly 60%. Although this still fell short of the 70% benchmark, it indicates substantial progress in both engagement and collective problem-solving. The 70% benchmark referenced in our study is derived from an internal corporate target, which is based on the average participation rate of 67% observed across the corporate group. This benchmark reflects the level of employee engagement in continuous improvement activities, a key indicator of the organizational culture and the values embraced by our workforce. Each idea approved by the review committee contributes to enhancing operational efficiency [8]. Achieving or approaching this 70% participation rate is indicative of effective collective problem-solving and is linked to optimal quality improvements. It is worth noting that kaizen personal engagement tripled over the same period, demonstrating the positive impact of EFQM-driven leadership and culture criteria (Criteria 1 and 2).

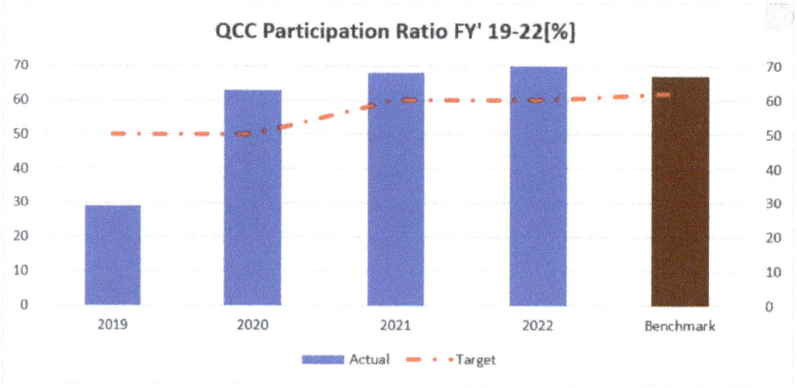

Figure 5. QCC participation ratio FY19–FY22.

Such an improvement in employee involvement proved instrumental in reducing turnover rates, which had been a significant concern given tight labor markets and low unemployment rates (2.9%) at the end of January 2023 [48]. Externally, the company's commitment to continuous improvement and sustainability has been validated through prestigious awards. Notably, the plant was recognized in the Global QCC Competition, and in 2022, it was the only plant in Europe that received the Gold Award at an international competition held in Japan. These accolades underscore the effectiveness of stakeholder-driven initiatives and further enhance the company's reputation among local communities and business partners, in alignment with ESRS requirements for transparent ESG reporting. By cultivating a sense of shared ownership over sustainability outcomes, the factory not only minimized recruitment and training costs but also bolstered morale and retained valuable expertise.

Kaizen initiatives formed the operational backbone of the company's sustainability strategy. These small-scale, iterative improvements aligned with the EFQM's execution dimension (Criteria 3–5) and directly targeted reductions in energy consumption and resource waste. Examples of kaizen-driven results include the following:

1. Installing inverters for pumps and advanced sensors for lighting and air circulation systems, thereby allowing for on-demand rather than continuous operation.
2. Capturing excess heat from certain production processes (e.g., cutting, press operations) to warm adjacent areas of the plant, ultimately lowering total energy demand.

3. Segmenting production lines and auxiliary systems (e.g., compressed air, ventilation) so that individual sections could be shut down during downtime, tackling previously "fixed" energy loads.

Collectively, these kaizen activities accounted for a measurable portion of the 17% carbon dioxide emissions reduction observed during the study period (Figure 6). Incremental gains in each micro-process, when scaled across the facility, resulted in significant energy savings and cost reductions.

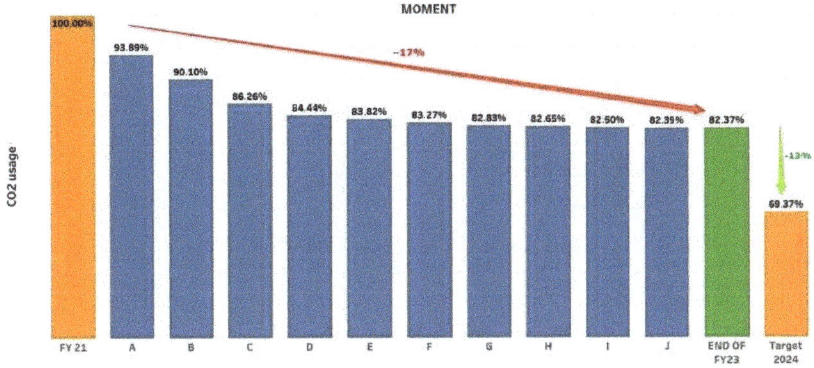

Figure 6. Implementation of 10 kaizen initiatives (A–J) leading to a 17% reduction in carbon dioxide emissions.

Data collection and analysis form the second pillar of the organization's net zero strategy, providing an objective basis for identifying high-impact kaizen initiatives. Advanced monitoring tools, such as metering devices and IoT platforms, continuously capture real-time data on electricity, water, and natural gas consumption (Criterion 5.5 in the EFQM Model). These data are then aggregated and reviewed during weekly steering committee meetings, informing both immediate operational improvements and long-term sustainability planning. For example, analysis revealed that approximately 50% of the plant's energy consumption was fixed (Figure 7), prompting targeted initiatives such as improved insulation and reduced equipment idling. Likewise, a review of documentation showed an increase in the recycling rate from 93.3% in FY19 to 98.1% in FY23 (Table 5), illustrating the continuous improvement ethos in waste management and material recovery. By linking specific Kaizen projects to key performance indicators (KPIs), such as energy use per unit of production, managers can benchmark progress and effectively communicate improvements to both internal stakeholders and external audiences.

This robust, data-driven approach not only facilitates the rapid iteration of environmental solutions but also underpins the organization's annual sustainability disclosures. Although these annual reports summarize retrospectively aggregated data, the underlying continuous monitoring ensures that the most accurate and timely information informs these reports, thereby enhancing transparency and credibility in ESG reporting. This integration of real-time operational data with periodic reporting strengthens internal stakeholder engagement and bolsters the company's reputation among local communities and suppliers. These efforts, aligned with the European Sustainability Reporting Standards (ESRS), demonstrate the synergy between EFQM-facilitated operational excellence and a proactive, stakeholder-centric sustainability strategy.

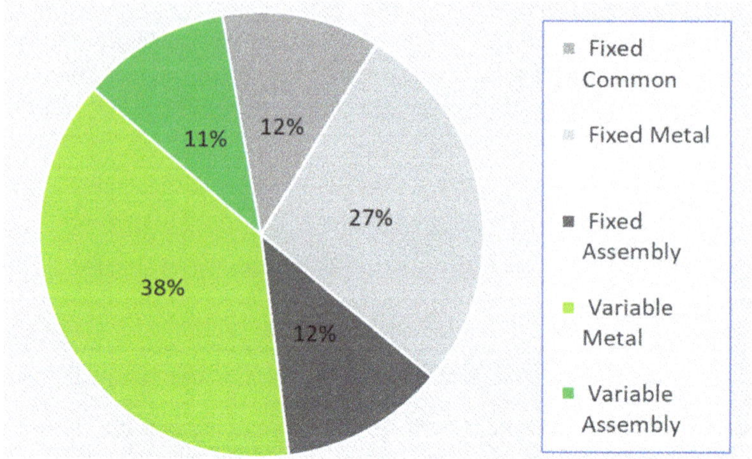

Figure 7. Breakdown of energy consumption to fixed and variable in FY23.

3.5. Green Energy Self-Generation and the Path to Net Zero

While the factory's incremental improvements in energy efficiency and resource optimization significantly contributed to its carbon reduction targets, the final stages of achieving net zero depend on the self-generation of green energy. This approach is driven by the EFQM Model's emphasis on long-term strategic outcomes (Criterion 7—strategic and operational performance) and the need for both environmental and economic sustainability.

Initially, the organization focused on reducing overall energy consumption to maximize the eventual impact of new renewable energy installations. By identifying and eliminating inefficiencies first, the factory minimized the capacity requirements for solar photovoltaics (PV) or other renewable infrastructure, thus reducing both cost and embodied emissions. Early pilot installations of PV arrays tested the feasibility of larger-scale deployment. However, the main production facility occupies a relatively confined area due to its specific location and building design, which limits the total capacity for rooftop or ground-mounted solar installations. To address these space constraints, the organization purchased additional land and developed a dedicated photovoltaic farm, thereby supplementing on-site generation and overcoming limitations imposed by the factory's physical footprint. Planners also evaluated alternate sites or partnerships (e.g., community solar) as supplementary measures.

This strategy, combined with staggered photovoltaic adoption that benefits from ongoing improvements in solar panel efficiency and battery storage, aligns with the EFQM's continual transformation perspective (Criterion 5.2). Although on-site generation improvements are significant, purchasing green certificates remains a key bridging strategy (Figures 4 and 5), representing the external renewable energy required to offset any shortfall in internal production.

Although certificates do not directly reduce on-site emissions, they signal a market demand for greener power sources. From an EFQM standpoint, this practice reflects a pragmatic approach to risk management (Criterion 5.1) and resource acquisition (Criterion 5.5) while the company scales its renewable capabilities.

As the factory continued lowering its carbon footprint, each percentage increment of reduction became more difficult to achieve (Figures 6 and 7). This phenomenon underscores the concept of diminishing returns, wherein easily attainable efficiency improvements have already been captured, and further emissions cuts require more advanced or more expensive technological interventions. To overcome these hurdles, the plant has explored the following:

- Passive building approaches, high-efficiency HVAC systems, and integrated energy storage solutions that reduce reliance on external power grids.
- Collaborations with local utilities to synchronize on-site generation with broader network demands, potentially offering load-balancing services for additional revenue or credits.

Throughout this phase, alignment with the EFQM Model continued to guide strategic decisions. The organization systematically assessed whether each new technology or pilot project advanced its long-term sustainability goals (Criterion 1.3—understanding major challenges; Criterion 2.1—steering culture), while maintaining economic viability (Criterion 7—strategic and operational results). Furthermore, these renewable energy initiatives directly supported multiple SDGs, particularly Goal 7 (affordable and clean energy) and Goal 13 (climate action).

From a reporting and compliance perspective, the systematic documentation of renewable energy installations and certificate purchases contributed to meeting the European Sustainability Reporting Standards (ESRS), ensuring that both direct and indirect emissions reductions were transparently communicated to stakeholders and regulators.

3.6. Challenges, Lessons Learned, and Future Outlook

Despite marked successes in reducing resource consumption and laying the groundwork for net-zero emissions, the transition guided by the EFQM Model was not without its obstacles. This section synthesizes the key lessons learned during its implementation, highlighting both the benefits of an EFQM-driven approach and the areas requiring sustained effort and innovation.

One of the most prominent issues was staff fatigue, fueled by the breadth of improvement projects and new tasks introduced alongside regular operational responsibilities. While kaizens and quality control circles significantly enhanced employee engagement and skill development, the rapid pace of change risked overwhelming frontline workers and middle managers. This phenomenon underscores the importance of the following:

- Staggering major projects and kaizen rollouts to avoid overwhelming personnel.
- Providing regular updates, success stories, and peer-to-peer learning sessions to maintain morale and a sense of shared purpose.

By acknowledging these human factors, the company strategically leveraged the EFQM's focus on culture and leadership (Criteria 1–2) to foster resilience and maintain overall organizational cohesion.

As the factory's resource use and emissions approached lower thresholds, further improvements became increasingly challenging. Much of the "low-hanging fruit" had been addressed by the following:

1. Upgrading lighting and insulation systems.
2. Optimizing fixed loads through advanced monitoring.
3. Implementing continuous improvement cycles.

Additional reductions, particularly in water usage and Scope 2 emissions, now necessitate technological breakthroughs (e.g., advanced building envelopes, new production methods) or significant capital expenditure (e.g., large-scale photovoltaic arrays, battery storage). This plateau effect aligns with other industrial decarbonization experiences and reinforces the EFQM Model's guidance on proactively transforming for the future (Criterion 5.2).

While the EFQM Model provides a robust framework for aligning strategy and operations, regulatory requirements such as the European Sustainability Reporting Standards (ESRS) add a layer of complexity. Organizations must address multiple stakeholder demands, each having different reporting expectations. At the same time, they must maintain high-quality, real-time data capture, ensuring the accurate monitoring of all relevant ESG indicators, a demanding practice that can strain existing IT and administrative systems.

Nevertheless, the plant's structured EFQM approach with clear KPIs and regular self-assessment proved advantageous in fulfilling ESRS obligations. Documenting and communicating improvements validated the company's progress and built trust among internal and external stakeholders (Criterion 6—stakeholder perceptions; Criterion 7—results).

Ultimately, achieving net zero will require a continued commitment to process innovation and advanced technologies, complementing incremental kaizen efforts. Potential avenues under exploration include leveraging naturally regulated indoor climates, reducing HVAC loads, and integrating construction materials that help lower embodied emissions. Very promising to achieve Sustainable Development Goals such as reducing carbon emissions are new digital technologies such as IoT, especially online sensors such as temperature and vibroacoustic sensors, energy consumption, digital twins, new efficient databases, big data, AI-driven predictive maintenance, and process optimization, critical for detecting inefficiencies that are no longer visible to conventional monitoring systems. Digital innovations align directly with the Industry 5.0 paradigm, where human-centric, data-driven approaches converge to meet ambitious environmental targets without compromising economic viability.

The enterprise's five-star EFQM recognition demonstrates that its experience can serve as a benchmark for other automotive and manufacturing facilities. Critical success factors include the following:

- Holistic strategy.
- Incremental, data-driven improvements.
- Balancing near-term efficiency measures with strategic investments in future-ready technologies.

By exemplifying these best practices, the company sets a precedent for others pursuing carbon neutrality in a manner consistent with both corporate objectives and evolving global standards.

4. Conclusions

The automotive industry stands at the forefront of a global transition toward sustainability, driven by tightening environmental regulations, evolving market expectations, and heightened awareness of climate change. This study examined the application of the EFQM Model 2020 in a large automotive production plant, with a particular focus on achieving a net-zero environmental impact and contributing to the United Nations Sustainable Development Goals (SDGs). Across an extended implementation window (November 2021 to December 2023), the following key conclusions were drawn:

1. EFQM as a Catalyst for Sustainability

The EFQM Model's holistic criteria provided an effective roadmap for embedding sustainability objectives into both strategic and operational levels. Compared to other quality and excellence frameworks such as ISO 9001, the Baldrige Excellence Framework, and the Shingo Model, the EFQM uniquely integrates continuous improvement with sustainability, particularly through its alignment with emerging regulatory frameworks like the European Sustainability Reporting Standards (ESRS). This integration enabled the organization to systematically address carbon reduction, water conservation, waste minimization, and energy efficiency.

2. Measurable Progress Toward Net Zero

Through iterative kaizen projects, advanced data analytics, and strategic investments in green energy, the plant achieved a 17% reduction in carbon dioxide emissions, significant improvements in water usage efficiency, and a recycling rate exceeding 98%. The staged approach prioritizing cost-effective efficiency gains before scaling on-site renewable installations ensured the optimal use of resources and capital, highlighting how a data-driven and structured application of EFQM principles can yield tangible environmental benefits.

3. Enhancing Employee Engagement and Organizational Culture

Sustained leadership commitment and widespread employee involvement were pivotal to the success of the initiative. Quality control circles (QCCs) and Kaizen activities not only increased staff participation and reduced turnover but also fostered a strong culture of continuous improvement. This internal engagement, reinforced by a shared sustainability mission, illustrates how effective change management can support operational excellence in energy-intensive industries.

4. Challenges and Limitations

As efficiency levels improved, the plant encountered diminishing returns, with incremental gains requiring more advanced or capital-intensive interventions. Concurrent project rollouts posed risks of overwhelming employees, underscoring the need for phased implementation and robust change management. Additionally, while on-site renewable installations were constrained by the plant's limited physical footprint, a challenge mitigated by purchasing green certificates and developing additional photovoltaic capacity, the experience underscores the importance of contextual factors such as location and facility design in shaping sustainability strategies.

5. Future Directions

To bridge the gap between current performance and genuine net zero status, the plant must continue to adopt advanced technologies (e.g., AI-driven predictive maintenance, passive building designs) and expand collaborations (e.g., microgrids, waste-to-energy consortia). Future research should explore the EFQM Model 2025 and its added value in achieving sustainability goals, with particular attention to practical applications in Industry 5.0. Moreover, differentiated application suggestions tailored to automotive companies of varying sizes and developmental stages are essential. Given the automotive sector's strategic importance and its influence on broader manufacturing industries, our study provides a replicable blueprint for sustainable transformation that can serve as a starting point for wider industry adoption.

Overall, our study confirms that the EFQM Model is a powerful enabler of sustainability-driven transformation. By combining data-driven decision-making, robust leadership engagement, and a culture of continuous improvement, automotive companies can not only meet regulatory and societal expectations but also set new benchmarks for environmental performance. This work contributes both to management practice and academic discourse, illustrating how holistic quality management frameworks can be effectively adapted to address emergent climate and resource challenges.

Author Contributions: Conceptualization: J.M. and M.Ł.; methodology: J.M., M.Ł. and I.C.-O.; formal analysis and investigation: J.M., K.S., ŁB., M.Ł. and J.U.; resources: K.S., Ł.B., J.U. and J.M.; writing—original draft preparation: J.M. and M.Ł.; writing—review and editing: J.M. and M.Ł., visualization: J.M., M.Ł., I.C.-O. and Ł.B.; Supervision: J.M., M.Ł. and K.S. All authors have read and agreed to the published version of the manuscript.

Funding: This research received no external funding. The Open Access publication of this article was partially funded by the Institutional Open Access Program (IOAP).

Informed Consent Statement: Informed consent was obtained from all subjects involved in the study.

Data Availability Statement: The original contributions presented in the study are included in the article, further inquiries can be directed to the corresponding author.

Conflicts of Interest: Authors Kamil Suchorski, Joanna Usyk and Łukasz Bednarowicz were employed by the company Toyota Boshoku Poland Sp. z o.o. The remaining authors declare that the research was conducted in the absence of any commercial or financial relationships that could be construed as a potential conflict of interest.

References

1. Lee, J.; Koo, Y. A General Equilibrium Analysis of Individual Choice Behavior on Alternative Fuel Vehicles. *Ecol. Econ.* **2023**, *204*, 107685. [CrossRef]
2. Singh, S.; Rao, D.G.; Dixit, M. Future of Transportation-A Comparison between Internal Combustion Engine, Electric Vehicles and Fuel Cell Vehicles. *Univ. J. Mech. Eng.* **2023**, *11*, 13–23. [CrossRef]
3. Jia, W.; Jiang, Z.; Wang, Q.; Xu, B.; Xiao, M. Preferences for Zero-Emission Vehicle Attributes: Comparing Early Adopters with Mainstream Consumers in California. *Transp. Policy* **2023**, *135*, 21–32. [CrossRef]
4. Hassan, Q.; Azzawi, I.D.J.; Sameen, A.Z.; Salman, H.M. Hydrogen Fuel Cell Vehicles: Opportunities and Challenges. *Sustainability* **2023**, *15*, 11501. [CrossRef]
5. Aiello, G.; Quaranta, S.; Inguanta, R.; Certa, A.; Venticinque, M. A Multi-Criteria Decision-Making Framework for Zero Emission Vehicle Fleet Renewal Considering Lifecycle and Scenario Uncertainty. *Energies* **2024**, *17*, 1371. [CrossRef]
6. Kokkinos, K.; Nathanail, E.; Gerogiannis, V.; Moustakas, K.; Karayannis, V. Hydrogen Storage Station Location Selection in Sustainable Freight Transportation via Intuitionistic Hesitant Decision Support System. *Energy* **2022**, *260*, 125008. [CrossRef]
7. Allgeier, S.; Feldmann, R. CSRD Sustainability Reporting For Non-Listed SMEs: European Regulators Remain Challenged. *Eur. Co. Financ. Law Rev.* **2023**, *20*, 438–446. [CrossRef]
8. Palhau, M.; Sá, J.C.; Ávila, P.; Dinis-Carvalho, J.; Rodrigues, C.; Santos, G. Toyota Way–the Heart of Toyota Production System and Its Impact on Sustainable Company Growth. *Qual. Innov. Prosper.* **2024**, *28*, 23–45. [CrossRef]
9. Giampieri, A.; Ling-Chin, J.; Ma, Z.; Smallbone, A.; Roskilly, A.P. A Review of the Current Automotive Manufacturing Practice from an Energy Perspective. *Appl. Energy* **2020**, *261*, 114074. [CrossRef]
10. Gebler, M.; Cerdas, J.F.; Thiede, S.; Herrmann, C. Life Cycle Assessment of an Automotive Factory: Identifying Challenges for the Decarbonization of Automotive Production–A Case Study. *J. Clean. Prod.* **2020**, *270*, 122330. [CrossRef]
11. Lukin, E.; Krajnović, A.; Bosna, J. Sustainability Strategies and Achieving SDGs: A Comparative Analysis of Leading Companies in the Automotive Industry. *Sustainability* **2022**, *14*, 4000. [CrossRef]
12. Silva, F.J.G.; Pereira, M.T.; Sá, J.C.; Ferreira, L.P.; Sá, J.C.; Ferreira, L.P. *Innovation, Digital Transformation and Process Improvement Towards a Better Efficiency on Industrial and Management Systems*; Silva, F.J.G., Pereira, M.T., Sá, J.C., Ferreira, L.P., Eds.; MDPI: Basel, Switzerland, 2024; ISBN 978-3-7258-0781-9.
13. Wei, X.; Qiu, R.; Liang, Y.; Liao, Q.; Klemeš, J.J.; Xue, J.; Zhang, H. Roadmap to Carbon Emissions Neutral Industrial Parks: Energy, Economic and Environmental Analysis. *Energy* **2022**, *238*, 121732. [CrossRef]
14. Fonseca, L.; Amaral, A.; Oliveira, J. Quality 4.0: The EFQM 2020 Model and Industry 4.0 Relationships and Implications. *Sustainability* **2021**, *13*, 3107. [CrossRef]
15. Saizarbitoria, I.H.; Viadiu, F.M.; Fa, M.C. ISO 9001 Vs. EFQM: A comparative analisys based on external independent data. In Proceedings of the 4th International Conference on Industrial Engineering and Industrial Management, Donostia-San Sebastián, Spain, 8–10 September 2010; pp. 1575–1582.
16. Susanto, D.A.; Suef, M.; Karningsih, P.D.; Prasetya, B. ISO 9001 Implementation Model: A Review and Future Research Agenda. *TQM J.* **2024**, *37*, 1342–1363. [CrossRef]
17. Andraws, R.S.; Bellaaj, M. Organizational Performance Evaluation Using European Foundation for Quality Management (EFQM) Model: Case Study of the Central Bank of Iraq. *J. Ecohuman.* **2025**, *4*, 1. [CrossRef]
18. Martusewicz, J.; Wierzbic, A.; Łukaszewicz, M. Strategic Transformation and Sustainability: Unveiling the EFQM Model 2025. *Sustainability* **2024**, *16*, 9106. [CrossRef]
19. Sá, J.C.; Oliveira, A.R.; Hines, P.; Nicola, S.; Mourão, F.; Noites, H.; Silva, F. Study of the Impact of the Shingo Model and the EFQM Model on Occupational Safety in Portuguese Organizations. In *Lecture Notes in Mechanical Engineering*; Springer: Berlin/Heidelberg, Germany, 2024; pp. 100–114.

Wiśniewska, M.Z. Cele Zrównoważonego Rozwoju i Ich Realizacja w Świetle Norm Dotyczących Systemów Zarządzania i Ich Integracji. *Probl. Jakości* **2022**, *54*, 16–26. [CrossRef]

Martusewicz, J.; Szewczyk, K.; Wierzbic, A. The Environmental Protection and Effective Energy Consumption in the Light of the EFQM Model 2020—Case Study. *Energies* **2022**, *15*, 7260. [CrossRef]

Kaplan, B.; Maxwell, J.A. Qualitative Research Methods for Evaluating Computer Information Systems. In *Evaluating the Organizational Impact of Healthcare Information Systems*; Springer: New York, NY, USA, 2005; pp. 30–55.

Fonseca, L. The EFQM 2020 Model. A Theoretical and Critical Review. *Total Qual. Manag. Bus. Excell.* **2022**, *33*, 1011–1038. [CrossRef]

Sá, J.C.; Oliveira, A.R.; Dinis-Carvalho, J.; Santos, G.; Silva, F.J.G. A New Conceptual Model for Excellence in Business Towards Sustainable Development. *Qual. Innov. Prosper.* **2023**, *27*, 33–60. [CrossRef]

Wierzbic, A.; Martusewicz, J. The Evolution of the EFQM Model in the Context of Contemporary Challenges for Organizations. In Proceedings of the International Business Information Management Associacion, Proceedings of the Vision 2025: Education Excellence and Management of Innovations Through Sustainable Economic Competitive Advantage, Madrid, Spain, 13–14 November 2019; pp. 11200–11209.

Martusewicz, J.; Wierzbic, A. The Level of Maturity and the Use of Management Methods in Business Excellence Models. *Przedsiębiorczość I Zarządzanie* **2018**, *19*, 273–285.

Manresa, A.; Escobar Rivera, D. Excellence in Sustainable Management in a Changing Environment. *Sustainability* **2021**, *13*, 2296. [CrossRef]

EFQM. The EFQM Model. Available online: https://efqm.org/the-efqm-model/ (accessed on 30 July 2024).

Jankalová, M.; Jankal, R. How to Characterize Business Excellence and Determine the Relation between Business Excellence and Sustainability. *Sustainability* **2020**, *12*, 6198. [CrossRef]

de Menezes, L.M.; Escrig-Tena, A.B.; Bou-Llusar, J.C. Sustainability and Quality Management: Has EFQM Fostered a Sustainability Orientation That Delivers to Stakeholders? *Int. J. Oper. Prod. Manag.* **2022**, *42*, 155–184. [CrossRef]

Garrido, S.; Muniz, J.; Batista Ribeiro, V. Operations Management, Sustainability & Industry 5.0: A Critical Analysis and Future Agenda. *Clean. Logist. Supply Chain* **2024**, *10*, 100141. [CrossRef]

Cunha, F.; Dinis-Carvalho, J.; Sousa, R.M. Assessment of Performance Measurement Systems' Ability to Mitigate or Eliminate Typical Barriers Compromising Organisational Sustainability. *Sustainability* **2024**, *16*, 2173. [CrossRef]

Carayannis, E.G.; Grigoroudis, E.; Campbell, D.F.J.; Katsikas, S.K. (Eds.) *The Elgar Companion to Digital Transformation, Artificial Intelligence and Innovation in the Economy, Society and Democracy*; Edward Elgar Publishing: Cheltenham, UK, 2023; ISBN 9781839109362.

Bresciani, S.; Huarng, K.-H.; Malhotra, A.; Ferraris, A. Digital Transformation as a Springboard for Product, Process and Business Model Innovation. *J. Bus. Res.* **2021**, *128*, 204–210. [CrossRef]

Leng, J.; Sha, W.; Wang, B.; Zheng, P.; Zhuang, C.; Liu, Q.; Wuest, T.; Mourtzis, D.; Wang, L. Industry 5.0: Prospect and Retrospect. *J. Manuf. Syst.* **2022**, *65*, 279–295. [CrossRef]

Akter, S.; Michael, K.; Uddin, M.R.; McCarthy, G.; Rahman, M. Transforming Business Using Digital Innovations: The Application of AI, Blockchain, Cloud and Data Analytics. *Ann. Oper. Res.* **2022**, *308*, 7–39. [CrossRef]

European Commission: Directorate-General for Research and Innovation. *Industry 5.0, a Transformative Vision for Europe: Governing Systemic Transformations Towards a Sustainable Industry*; Publications Office of the European Union: Luxembourg, 2022.

Martusewicz, J.; Szewczyk, K.; Wierzbic, A. EFQM RADAR-Based Assessment of RFID System as Part of Industry 4.0 Implementation-A Case Study of a Production Plant. In *Industry 4.0: A Glocal Perspective*; Routledge: London, UK, 2021; pp. 95–108. [CrossRef]

ISO 14001:2015; Environmental Management Systems—Requirements with Guidance for Use. International Organization for Standardization: Geneva, Switzerland, 2015.

Acocella, I. The Focus Groups in Social Research: Advantages and Disadvantages. *Qual. Quant.* **2012**, *46*, 1125–1136. [CrossRef]

Busetto, L.; Wick, W.; Gumbinger, C. How to Use and Assess Qualitative Research Methods. *Neurol. Res. Pract.* **2020**, *2*, 1–10. [CrossRef] [PubMed]

Garcia, D.; Gluesing, J.C. Qualitative Research Methods in International Organizational Change Research. *J. Organ. Change Manag.* **2013**, *26*, 423–444. [CrossRef]

Asif, M.; Searcy, C.; Garvare, R.; Ahmad, N. Including Sustainability in Business Excellence Models. *Total Qual. Manag. Bus. Excell.* **2011**, *22*, 773–786. [CrossRef]

Dathe, T.; Helmold, M.; Dathe, R.; Dathe, I. *ESG Reporting Rules*; Springer: Berlin/Heidelberg, Germany, 2024; pp. 147–158.

Politis, Y.; Grigoroudis, E. Incorporating the Sustainability Concept in the Major Business Excellence Models. *Sustainability* **2022**, *14*, 8175. [CrossRef]

Chomiak-Orsa, I.; Martusewicz, J. Creating Good Practice in Effective Sustainability Management by Implementing the EFQM Model. *Procedia Comput. Sci.* **2023**, *225*, 3517–3526. [CrossRef]

47. Prayuda, R.Z. Continuous Improvement Through Kaizen In An Automotive Industry. *J. Ind. Eng. Manag. Res.* **2020**, *1*, 37–42. [CrossRef]
48. Eurostat. Euro Area Unemployment at 6.3%—Euro Indicators. Available online: https://ec.europa.eu/eurostat/web/products-euro-indicators/w/3-31102024-bp (accessed on 15 February 2025).

energies

MDPI

rticle

)evelopment of Recurrent Neural Networks for Thermal/ :lectrical Analysis of Non-Residential Buildings Based on :nergy Consumptions Data

lisa Belloni [1,*], Flavia Forconi [2], Gabriele Maria Lozito [3], Martina Palermo [2], Michele Quercio [2] ıd Francesco Riganti Fulginei [2]

1 Department of Engineering, University of Perugia, 06125 Perugia, Italy
2 Department of Industrial, Electronic and Mechanical Engineering, Roma Tre University, 00146 Rome, Italy; flavia.forconi@uniroma3.it (F.F.); martina.palermo@uniroma3.it (M.P.); michele.quercio@uniroma3.it (M.Q.); francesco.rigantifulginei@uniroma3.it (F.R.F.)
3 Department of Information Engineering, University of Firenze, 50139 Firenze, Italy; gabrielemaria.lozito@unifi.it
* Correspondence: elisa.belloni@unipg.it; Tel.: +39-075-5853562

Abstract: Extensive research has focused on optimizing energy consumption in residential buildings based on indoor thermal conditions. However, modeling the energy and thermal behavior of non-residential buildings presents greater challenges due to their complex geometries and the high computational cost of detailed simulations. Simplifying input variables can enhance the applicability of artificial intelligence techniques in predicting energy and thermal performance. This study proposes a neural network-based approach to characterize the thermal–energy relationship in commercial buildings, aiming to provide an efficient and scalable solution for performance prediction. Consumptions trends for a building are generated using the EnergyPlus™ dynamic simulation software over a timespan of a year in different locations, and the data are then used to train neural network models. Uncertainty analyses are carried out to evaluate the behavior effectiveness of the artificial neural networks (ANNs) in different weather conditions, and the root mean square error (RMSE) is calculated in terms of mean air temperatures. The results show that this approach can reproduce the functional relationship between input and output data. Three different ANNs are trained for the northern, central, and southern climatic zones of Italy. The southern region's models achieved the highest accuracy, with an RMSE below 0.5 °C; whereas the model for the northern cities was less accurate, since no specific trend in plant management was present, but it still achieved an acceptable accuracy of 1.0 °C. This approach is computationally lightweight; inference time is below 5 ms, and can be easily embedded in optimization algorithms for load dispatch or in microcontroller applications for building automation systems.

Keywords: mathematical modeling; artificial neural network; non-residential buildings; energy consumptions; optimization function; thermal conditions

1. Introduction

Buildings represent a significant opportunity for cost-effective demand-side management, as they are the primary consumers of electricity in both Europe and the United States. In the U.S., approximately 75% of total electricity consumption is attributed to buildings. In this context, the U.S. Department of Energy (DOE), through the Building Technologies Office (BTO), published a technical report on the opportunities for grid-interactive efficient buildings [1]. The scenario in the USA shows that today primary energy use in buildings is about 41% (22% for residential and 19% for commercial type). In Europe, primary energy for buildings reaches 65% of the total amount, split into 43% for households and 22% for commercial buildings. At the EU level, two-thirds of the building's consumption is therefore residential [2,3]. Nevertheless, for the residential sector in Europe only 33% of the buildings consumption is covered by electricity, natural gas is still the most widely used energy source by households (35%). Different scenarios are observed for commercial and non-residential buildings. The increasing energy demand in the commercial sector is due to the many air-conditioned commercial buildings, mostly in metropolitan areas [4–7].

Many studies have been related to the importance of forecasting the energy consumption for these kinds of buildings. The relevant studies have shown the possibility of determining the thermal and electrical loads of commercial [8–24] or residential buildings [25–27] through algorithmic approaches, such as artificial neural networks (ANNs). Robinson et al. [8] presented a novel technique for the estimation of commercial buildings loads: several models were used for the training (energy consumption of different case studies in New York city) and were then applied in Atlanta cases to obtain an aggregate energy consumption dataset. The models could be also applied in other U.S. areas exploiting this method. In [10], the authors developed a short-term electrical load estimation algorithm that considers the variation in building operation. In this case non-residential buildings were chosen and, specifically, five commercial buildings of five different types. The case studies were also positioned in different countries all over the world. The most important focus is short-term load forecasting procedures (STLF) (from 1 h to 1 week). The deep learning algorithms proposed in this work allow for the load forecasting framework that is more accurate than similar models in the literature (with an improvement of 20–45%). In [13], a predictive control strategy in a commercial building energy management system (BEMS) was proposed for the management of boilers, where the neural network also takes into account the environmental conditions and the maintenance of the control levels. The BEMS uses real-time automatic control techniques, and it make buildings more intelligent.

Due to the interaction between occupants and buildings, occupancy behavior is irregular and difficult to model, which indirectly introduces uncertainties in the energy use of the building systems and plants [28–31]. Therefore, by predicting how many people will be in a building in a specific period, systems such as heating, ventilation, and lighting can be regulated more efficiently. An interesting study [30] analyzed the relationship between occupancy, air conditioning usage, and tolerance temperatures in office buildings in China, proposing a control strategy that takes multiple behavioral patterns into account. The consideration of the multiple impacts of the users' behavior on building energy efficiency allows for accurate energy performance predictions and the adoption of optimization strategies. Forecasting building energy consumption for enhancing flexibility services, controlling building energy systems, and reducing the carbon footprint of the building stock is widely discussed in some other interesting review papers [32–34]. A review of the role of artificial intelligence in home energy management systems (HEMS) was presented in [32]. HEMS is fundamental for the optimization of energy consumption and the reduction of electricity costs in smart grids (SGs). For the correct operation of these applications, smart meters are important to track the bidirectional flows of energy, and the integration of renewables (RES), energy storage systems (ESS), electric vehicles (EVs), and other electricity devices are essential.

Another interesting case study is described in [17], the object of which was the energy analysis of commercial buildings. The daily electricity demand was evaluated by means of the EnergyPlus program [35], and an ANN model was developed in MATLAB r2024. This ANN can predict energy consumption for lighting, cooling, and other equipment with mean errors of 13%, 8%, and 4%, respectively. In a paper written by Afzal et al. [14], three different artificial neural networks and regression models were used to predict cooling and heating loads, and the results were compared. A hybrid model was proposed (biogeography-based optimization (BBO) algorithm together with an extreme learning machine (ELM)), and the results demonstrated it was the best one in terms of correlation factor for the estimation of the consumptions. In these wide scenarios, one of the main purposes is the optimization of the building performance that can be obtained by the implementation of techniques that are complex and computationally expensive, above all for the description of the shape and form of the buildings, their thermal behavior, and their insulation features [36–39]. This is the reason why ANNs are able to replace detailed building models whose simulations are time-consuming. In [37], it was shown that neural networks with multi-objective optimization algorithms represent a good solution to improve energy efficiency and indoor thermal comfort. Also, in [40], a genetic algorithm (GA) with multi-objective functions (cooling energy consumption, discomfort hours, equivalent carbon emissions to be minimized) was proposed, and the authors obtained excellent optimization results with a reduction of electricity consumption of about 40%, with a decrease of discomfort hours equal to 63%. To overcome the problems related to the low accuracy of surrogate models, ANNs are used as an interaction between input and output data for many types of building stock [41,42].

Another important application of ANNs were carried out by Wong et al. [39]. The authors applied ANNs to predict the electricity consumed by office buildings for heating, cooling, and lighting. Also, in this case, simulations were carried out in EnergyPlus [35] to obtain a daily building energy load database for the training and testing of the ANNs. Excellent predictive ANNs were obtained, above all for the forecasting of the electrical demand for lighting. A modeling method for cooling and heating loads prediction was proposed by Chou and Bui [43] with an approach that considered both support vector regression (SVR) and ANNs; an absolute percentage error lower than 4% was obtained with effective and accurate results.

Furthermore, the electricity use in buildings is a key energy factor when also considering the implementation of renewables for the distributed generation of photovoltaic (PV) technologies [44–47]. The implementation of smart grids and smart buildings to cover this electricity demand in a cost-effective way is fundamental for minimizing greenhouse gas emissions [6]. The integration of intermittent renewable sources implies the dynamic management of smart buildings, and accurate forecasting of electricity consumption is therefore necessary. This is the final aim of a research paper carried out by Belloni et al. [47–50]: in the first work [47], a computationally lean neural network was developed based on a dataset obtained from a model of the E+ program [35] allowing for the prediction of the thermal–electrical behavior of a building. The added value of these works is the use of these models among a renewable energy community (REC) in which the PV plant generation is also considered for the balance of the loads [50]. In general, machine learning tools are valuable solutions for a variety of applications at the building level or within the context of an energy community (EC), particularly for load forecasting and EC optimization [51–54].

The work presented in this study investigates using recurrent neural networks (RNNs) as a black-box model for estimating a building's thermal behavior. The primary objective is to create a model that generalizes well to different climatic conditions and has low computational overhead, enabling its use in building management systems, optimization algorithms for RECs, and embedded applications. In particular, a case study of thermal and energy demand in HVAC systems for a commercial building in different Italian locations, is considered. The selected building is modeled in terms of geometry, insulation features, and HVAC system, and the simulations are carried out in many Italian cities characterized by different weather conditions (global solar radiation, mean air temperatures, relative humidity). Combining the use of a dynamic simulation software with an ANN is another innovative aspect of this approach; the building simulation program can characterize buildings and model their features (envelopes, exposures, geometry, HVAC systems and other electrical devices). Once properly trained with data obtained from the simulations, the ANN will be able to recreate the behavior of buildings in many different locations and environmental conditions. Three different Italian climatic categories are considered, and four cities for each of these categories (for a total number of simulating scenarios equal to twelve). The proposed approach can also be extended to other countries providing the same input data for the ANNs (mean air temperatures, global solar radiation, relative humidity, and wind speed), as well as electrical consumption for heating or cooling purposes. This data can be collected, measured, or derived from meteorological files in other countries.

The remainder of this paper is structured as follows: in Section 2 the methodology and the overall approach to the problem is described. The case study analyzed in this work is described along with the modelling of the office/commercial building in the EnergyPlus program, specifying the assumptions and the boundary conditions. Sections 2.3 and 2.4 show the general features of the ANN models, with details on the architecture, the input and outputs, and the training hyperparameters. In Section 3.1, the results are analyzed in terms of thermal/electric model output (electricity consumptions for heating/cooling and internal thermal behavior) obtained with the E+ models. Sections 3 and 4 describe the validation of the ANNs and the critical discussion of the results, respectively. Future works, limitations, and potential are discussed in Section 5. Conclusions and final remarks close the manuscript.

2. Materials and Methods

2.1. Approach

This study aims at developing an RNN model for a commercial/office building located in central Italy by applying a novel method able to correlate the mean inside air temperatures with the hourly energy consumptions of the building itself. The first stage of this approach is the creation of a dataset for the RNN training, for which the methodology is shown in Figure 1.

The EnergyPlus™ (E+) software (version 23.2.0) [35,55] is used to simulate the case study and model the building's thermal behavior. This program is particularly suited for analyzing heat transmission through opaque and transparent envelopes, making it ideal for complicated geometry and for highly glazed buildings [56]. The program is used to evaluate the electrical energy demand for different profiles of HVAC heating and cooling, for zone programs and loads, for lighting systems, and for people's occupancy throughout the year. A large dataset is obtained for the training of the RNNs, configured with inputs and outputs with the aim to estimate the internal temperature profile knowing the electrical loads obtained from the E+ simulations.

Twelve datasets of simulation cases are created; each one composed of 8760 vector samples related to hourly energy consumption and mean air temperatures inside the building. Each dataset considers a geographical location for one of twelve Italian cities, grouped by climate

- Brindisi, Trapani, Messina, and Catania—southern regions with a warm climate.
- Bolzano, Trieste, Aviano, Udine—northern regions with a typical cold mountain climate.
- Milano, Firenze, Piacenza, and Perugia—central regions with an intermediate climate.

The thermostat of each zone in the E+ setting is programmed to allow for the mean air temperatures to be valued inside a range that maintains comfort conditions both in the winter and in the summer.

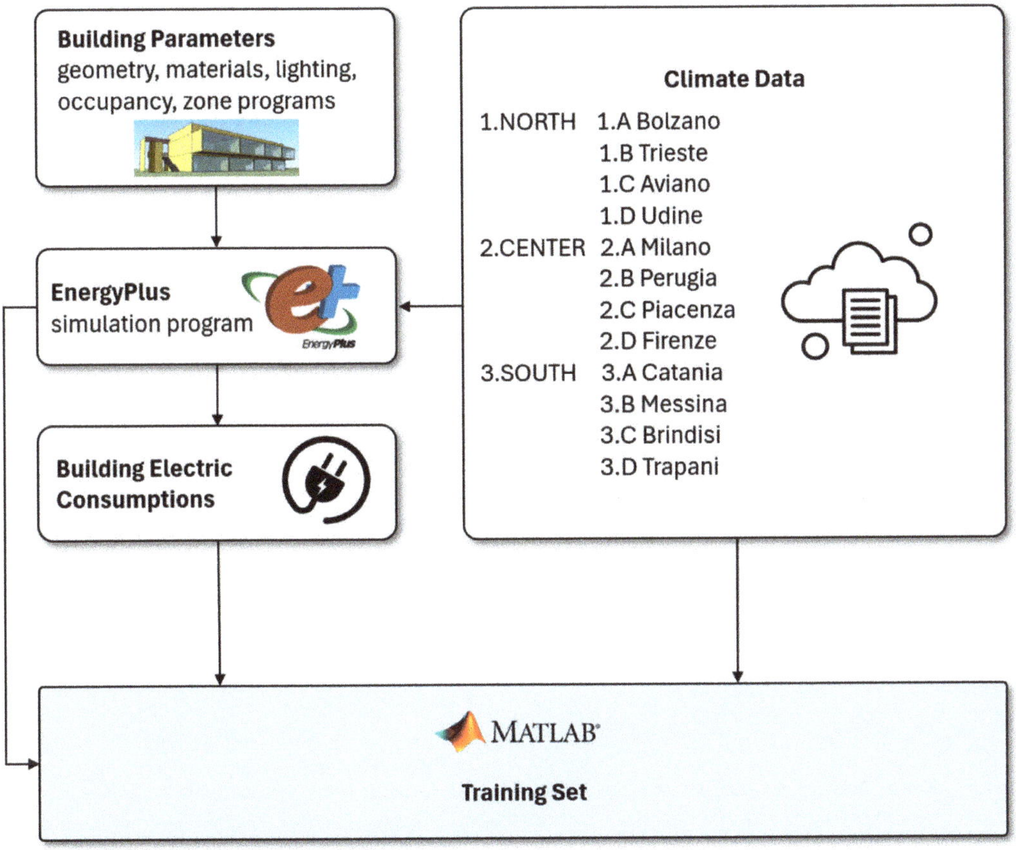

Figure 1. Flowchart of the training set creation approach.

The dataset was then used to create a standard training dataset for a neural network, considering a standard 70%/30% division for training, validation, and testing. The datasets were used to train three different RNNs (one for each climate type) and used for the estimation of the thermal behavior of the building when the electricity consumption is given. The neural networks were designed with a gradual increase in weight and connection complexity to preserve their ability to generalize. To prevent overfitting, early stopping techniques were also applied during the training process. Additionally, the training and inference times of the networks were profiled to verify the methodology's low computational demands.

2.2. Case Study Description and the EnergyPlus Model

The commercial building chosen as the case study was built in 2005 in a suburb zone in central Italy (Figure 2). It consists of two floors, each one including five zones. The total height of the building is more than 10 m. The east and west facades are small and without windows; very large glass walls (with a total area of about 25 m^2 for every zone) are in the south facades: they are provided with sunshades that can be used at peak hours. The facades of the ground floor are partially shaded by the upper floor, and an arcade is present in front of the commercial activities. The thermal transmittance of the opaque envelope is very low (0.25 W/(m^2K)), and the total area of the activities (shops, coffee, restaurants, etc.) is about 600 m^2. The first floor of the building presents 5 offices with an L-shaped plant (total floor area of 80 m^2). Also, these rooms have large facades with southern exposure: the thermal transmittance of the double-glazed wall is 1.0 W/m^2K. The mean window-to-wall (WWR) ratio of the rooms is 75%. The total inside volume of the building is about 12,000 m^3 (considering both the offices and the shops), and the occupancy period chosen for all the final use is 8 a.m.–7 p.m. (except for the restaurant). Internal load contribution is about 6 W/m^2 for offices and a mean value of 8 W/m^2 for the shops/coffee bar.

Figure 2. Front and back side of the commercial/office building.

The geometry of this building was implemented by means of the Google SketchUp plug-in, and the energy performances were evaluated by means of the E+ simulation code (Figure 3). EnergyPlus is able to model the building's thermal performance, and the energy/electricity demands related to the HVAC system, equipment/plug loads, and lighting system. The inside vertical surfaces and the horizontal floor between the different zones allow for the transfer of heat, but they are at the same temperature as they are all air-conditioned. The external walls are considered to be heat exchanging surfaces with the outside boundary air conditions; also the ground floor can exchange heat with the soil. The room's air set point temperature is fixed at 20 °C in the winter, and 26 °C in the summer periods.

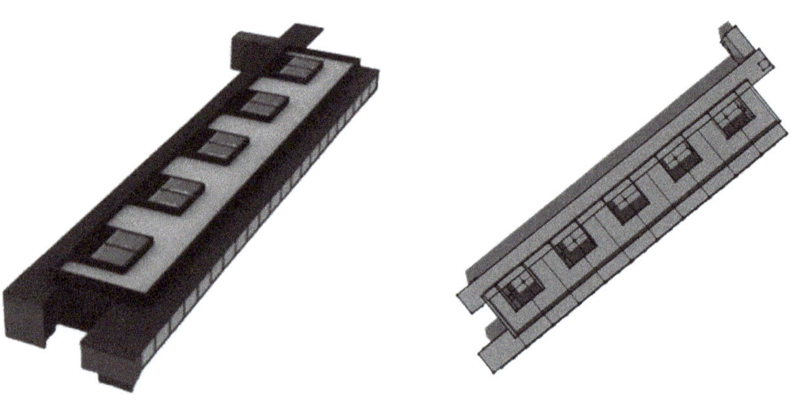

Figure 3. 3D model of the building chosen as the case study.

Both opaque and transparent building envelope features are implemented in the E+ interface, where it is possible to specify the different layers that make up walls, ceilings, roofs, and floors. The thermal conductance, density, and heat capacity of each component can be included. The HVAC system features are also included in E+, where the peak power and efficiency of the heat pumps can be set for heating in terms of COP (coefficient of performance) and for cooling in terms of EER (energy efficiency ratio). The model also considers the lighting equipment; the number of LEDs and power are chosen to achieve 500 lumens per square meter inside the offices, and 300 lumens per square meter in the shop/coffee bar. The on/off switching periods have been set based on occupancy periods.

The building features (envelopes and plants) and occupancy schedules are the same in all the EnergyPlus simulations because the reference building is unique, and it should be consistent regardless of its location.

The energy demand for the maintenance of these temperatures inside the zones is calculated, and the corresponding electricity consumptions associated to a heat pump are evaluated in post-processing phases. The ON period in the winter season is not the same for all the selected Italian cities; the national limits are fixed based on the climatic classification of each location, and they are shown in Table 1.

Table 1. Simulated case studies considered for the analysis and the ON-period of the HVAC.

	NORTH Climatic Zone 1		CENTER Climatic Zone 2		SOUTH Climatic Zone 3	
	Heating Period	**Cooling Period**	**Heating Period**	**Cooling Period**	**Heating Period**	**Cooling Period**
Annual ON period	01/10–30/04	01/07–15/09	15/10–15/04	01/06–30/09	15/11–15/03	01/05–15/10
Design Temperatures	20 ± 1 °C	26 ± 1 °C	20 ± 1 °C	26 ± 1 °C	20 ± 1 °C	26 ± 1 °C
Daily ON period	8 a.m.–7 p.m.					

2.3. RNN Model

The nonlinear functional relationship described by the thermoelectrical model of the building located was implemented through an RNN. The network architecture proposed is a network with recurrent hidden layers. This network implements internal feedback loops with variable delays, avoiding direct feedback between input and output. The architecture is defined by the number of hidden layers, the number of neurons for each layer, and the number of delays (i.e., states) for each layer.

With the goal of containing the computational costs associated with the neural estimator, a network with a single hidden layer was used, leaving the number of neurons and the number of delays as variable parameters. The same architecture for each climatic zone was used. For the training, 12 datasets, containing at each timestep the outside temperature (T_{out}), the relative humidity (RH), direct radiation (R_{direct}), diffuse radiation ($R_{diffuse}$), wind speed (WS), the mean inside air temperature at the beginning of the hour (T_i), the energy demand for heating (E_h), and the energy demand for cooling (E_c) are used in the implemented artificial neural network.

Each timestep corresponds to one hour of data. This temporal scale has been chosen because the start/stop schedules for the plants are generally programmed hour-by-hour. Furthermore, this timestep is commonly employed for monitoring the mean air temperatures and energy consumption of HVAC systems via smart meters. This could be useful for future applications of the model when real data is available.

The only data taken from the E+ program as input are related to the energy consumption for heating or cooling (E_h and E_c, respectively). The other input data are the environmental conditions, and they are taken from the EnergyPlus Weather Files (.epw) [55]. The output of the RNN is the hourly mean air temperatures of each building zone (T_i), for which the targets are obtained by the building simulation in E+. The input/output representation of the RNN is shown in Figure 4.

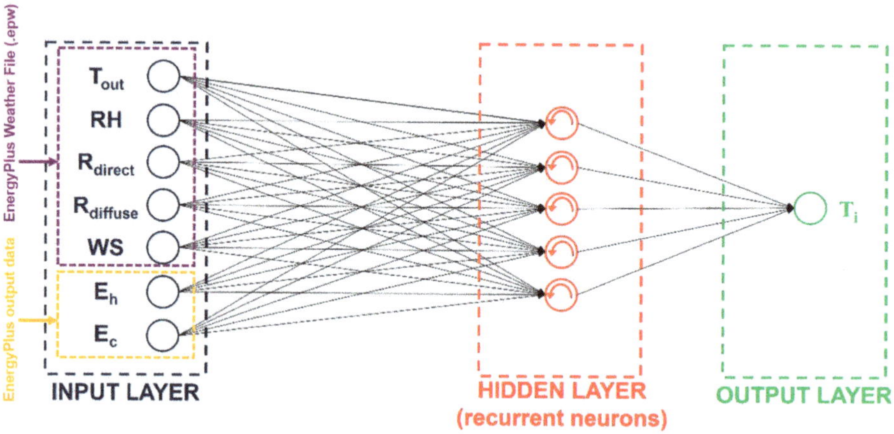

Figure 4. Architecture of the RNN used for the thermo-electrical model of the commercial/office building.

To determine the optimal size of the RNN in terms of hidden neurons and delay depth, an exhaustive test was performed by increasing progressively the hidden layer size and the delay depth, following the approach described in [57]. The optimal configuration of the RNN that yielded the best results in terms of fitting and generalization capabilities was represented by two delays and five neurons for the hidden layer. The RNN architecture and the most important training hyperparameters are reported in Table 2.

Table 2. Summary of the RNN architecture and training hyperparameters.

Parameter	Value or Description
Architecture	Layer Recurrent Neural Network (layrecnet)
Number of Hidden Layers	1
Activation Function	Tangent Sigmoid (tansig)
Feedback Delays	2
Sample Time	1
Total Number of Weights	91
Training Algorithm	Levenberg–Marquardt (trainlm)
Performance Metric	Mean Squared Error (MSE)
Data Division	Random by sample (dividerand)
Training/Validation Split	70%/30%
Maximum Training Epochs	10.000
Minimum Gradient	1×10^{-7}
Maximum Validation Failures	250
Average Training Time	93 s
Average Inference Time	4.4 ms
Average Epochs for Training	1500–2000

It is worth noting that each dataset, containing one year of hourly data, is divided monthly into twelve datasets, thus including several values equal to 672 (28 days for February), 720 (30 days for April, June, September, and November), and 744 (31 days for January, March, July, August, October, and December). Indeed, each neural network is trained and tested monthly to avoid temporal correlation between the data. Since each climatic zone is composed of four different cities and, therefore, by four databases, for each zone, one of them was used for testing the RNN, while the remaining three were used for the training. This test is provided to understand the RNN's capabilities in a real scenario where training and deployment could happen for different climatic conditions. The performance metric is the RMSE calculated over each month. The algorithms of training and testing are implemented in the MATLAB r2024 environment, running on a 13th Gen Intel Core i9 CPU, with 32 GB of RAM [58].

2.4. Model Inversion for REC Applications

The implemented neural model can be used to assess the thermal effects in the building for a known energetic input. However, to design HVAC controls, it is also useful to have an instrument able to estimate the energetic input necessary to achieve a specific thermal response. It is possible to define the direct problem as the determination of the thermal response from an energetic input, for a known "state vector" describing the environmental quantities of interest. The inverse problem then is the determination of the energetic demand for a known state vector, if a specific thermal response is required.

Such an inverse problem could be solved, in theory, simply by exchanging the inputs and outputs of the dataset, and thus training an RNN to solve the inverse problem. However, this approach required a large RNN, and proved very poor in terms of generalization capabilities. A different approach consists of posing the problem of determining the energetic demand as an optimization problem, as shown in Figure 5. In this case the RNN model is implemented inside an inverse iterative loop featuring an optimization algorithm. The algorithm optimizes the energetic demand sent to the RNN direct model until the difference between the model temperature and the demanded temperature is minimal. In this context, the inverted model features as input the demanded temperature, and as output the required energetic demand. Previous research works [44–46] have proposed the application of both direct and inverse models of RNNs, with the ultimate goal of predicting energy/electrical consumption and using the resulting data to optimize the operation of renewable energy communities (RECs) incorporating those investigated buildings.

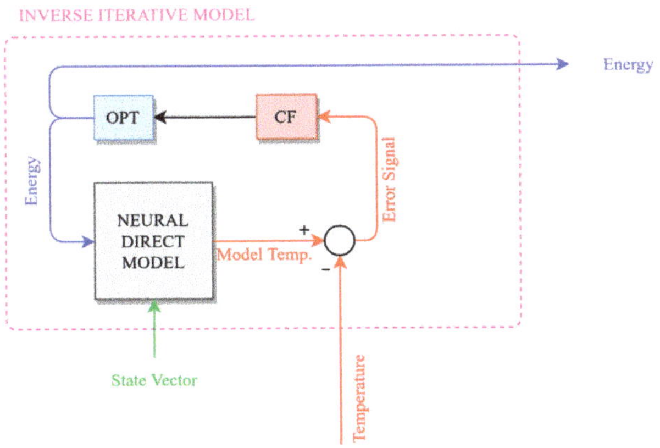

Figure 5. Block diagram of the direct and inverse modes of operation of the algorithms.

3. Results

3.1. EnergyPlus Simulations Results

To find the Italian cities characterized from similar climatic conditions behavior, a preliminary analysis on their weather files was carried out in terms of monthly mean air temperatures and global solar radiation. As previously explained, three climatic zones were selected, typical of the north, central, and south regions of Italy. For each zone four cities were chosen, for a total number of twelve case studies.

The weather files used for the EnergyPlus simulation activities (EnergyPlus Weather data file [55]) used the mean data characteristics of every city that were averaged for about ten years, and they included hourly data for a total number of 8760 data (in particular, for mean air temperatures, for direct, diffuse, and global solar radiation, for relative humidity, and for mean air speed and direction). Figure 6a,b shows the comparison between the mean monthly data in terms of global solar radiation on horizontal surface and outdoor temperatures; it is possible to observe that the three climatic zones are clearly identifiable in the graph because the maximum solar radiation is always higher than 6000 Wh/m^2 for the southern region, between 5000 and 6000 Wh/m^2 for the central region, and about 5000 for the northern cities. For Bolzano, Trieste, Aviano, and Udine values lower than 1000 Wh/m^2 are measured in the winter. For temperatures, Messina, Catania, Trapani, and Brindisi have values always higher than the other locations, and very similar to each other. Milano, Perugia, and Piacenza (central zone) also have similar trends, whereas Firenze has a behavior that can be compared with the northern locations and, in particular, it is very similar to Udine. More differences were observed for the northern cities above all in the winter months: Trieste has higher values comparable to central Italy, and both Aviano and Bolzano present air temperature values typical of mountain villages.

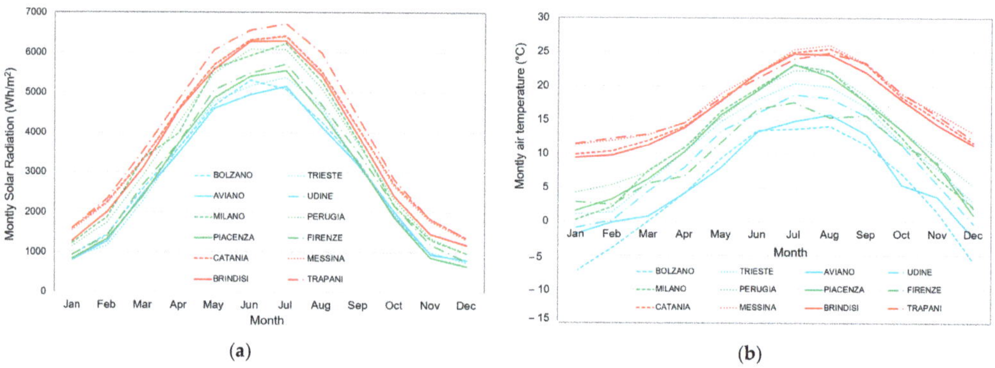

Figure 6. Comparison between the monthly mean values in terms of solar radiation (**a**) and mean air temperature (**b**) obtained from the EnergyPlus weather files.

On the basis of these input data, yearly simulations were carried out considering the same case study situated in each of these cities with the on/off profiles of the plants specified in Table 1. A typical weekly behavior in December and in August is analyzed in Figure 7a,b for the three Italian zones in terms of indoor air temperatures and hourly electricity consumptions (kWh) required for the maintenance of the design air temperature values in winter (20 °C) and summer (26 °C). It is possible to observe that, in winter, to maintain the temperature between 19 °C and 21 °C, the heating consumptions in Udine are approximately twice the ones of Firenze (Figure 7a). Very low energy consumptions for heating are obtained in Trapani, mostly in the last two days when the outside air temperatures are high in southern Italy. It is important to observe two peak values of electricity consumption at the beginning and at the end of the heating period due to the solar radiation contribution during the central hours of the day: the large glazing façade with southern exposure allows the entry of heat and natural light. In the summer (Figure 7b), the hourly cooling consumptions are very low during the winter: maximum peak values of 8–10 kWh were obtained for Trapani, and for Firenze the maintenance of a temperature lower than 26 °C can be obtained with very low consumption (maximum values of 2–5 kWh). Finally, in Udine, the mean zone air temperatures are never higher than 26 °C, and for the whole week it is not necessary to turn on the plant (except for the first day).

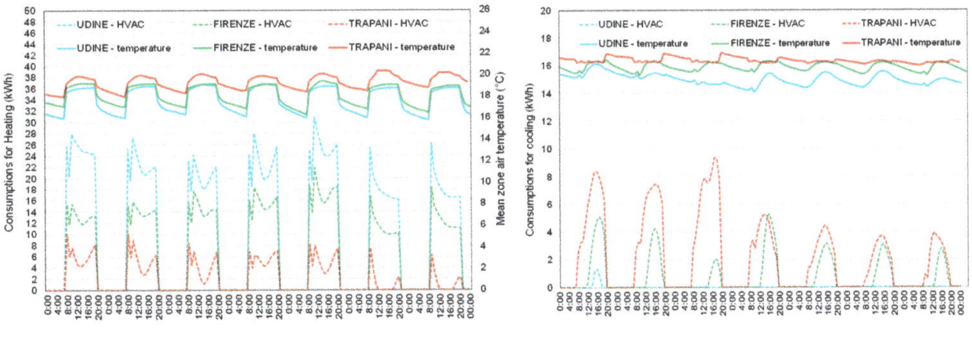

(**a**) (**b**)

Figure 7. EnergyPlus simulation results in terms of indoor air temperature and electricity consumptions: (**a**) winter week; (**b**) summer week.

3.2. RNN Model Validation and Results

As mentioned above, for each one of the three climatic zones, four different RNN training sessions were performed using a dataset consisting of three cities. Every training session was tested using the fourth remaining city. Figures 8–10 show the results of the predicted mean inside air temperature, Ti, for the three cities (Udine, Firenze, and Catania) that best represent the thermal behavior of the commercial buildings located in the north, center, and south of Italy, respectively. The mean inside-air temperatures trends are shown for two winter and summer months, i.e., January–February and July–August (the periods when the heating or the cooling systems are fully operational are chosen to avoid the mean-seasons when the building may require both heating and cooling depending on the outdoor weather conditions).

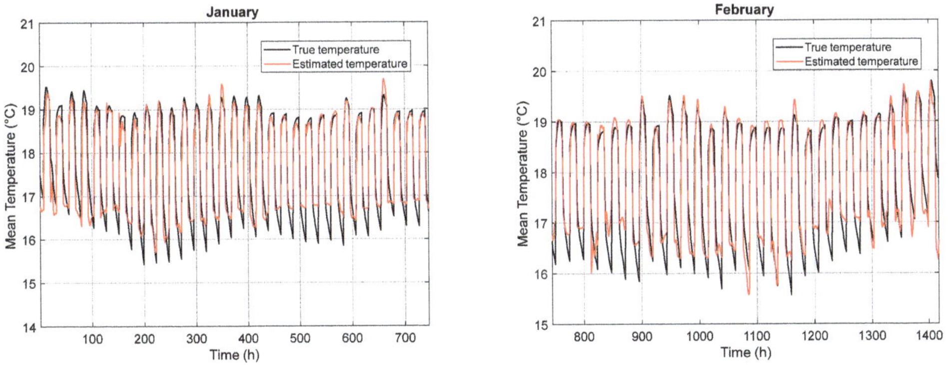

Figure 8. *Cont.*

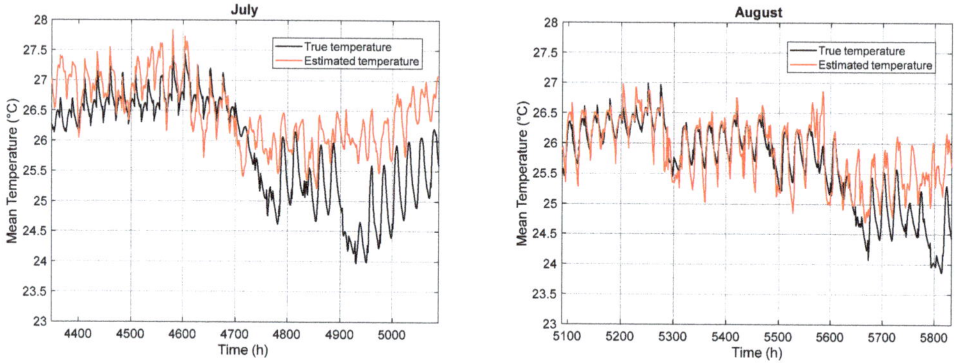

Figure 8. Prediction of the mean inside air temperature (Ti) in January, February, July, and August for Udine (northern Italy). The estimated temperature is represented by the red line, while the true temperature is represented by the black line.

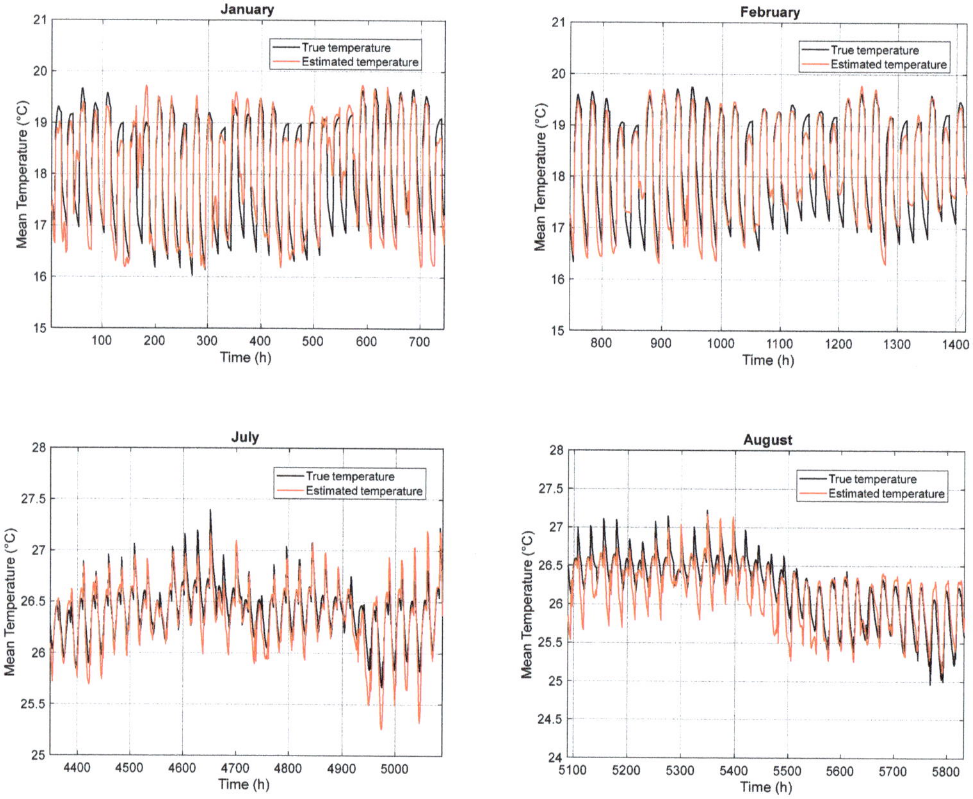

Figure 9. Prediction of the mean inside air temperature (Ti) in January, February, July, and August for Firenze (central Italy). The estimated temperature is represented by the red line, while the true temperature is represented by the black line.

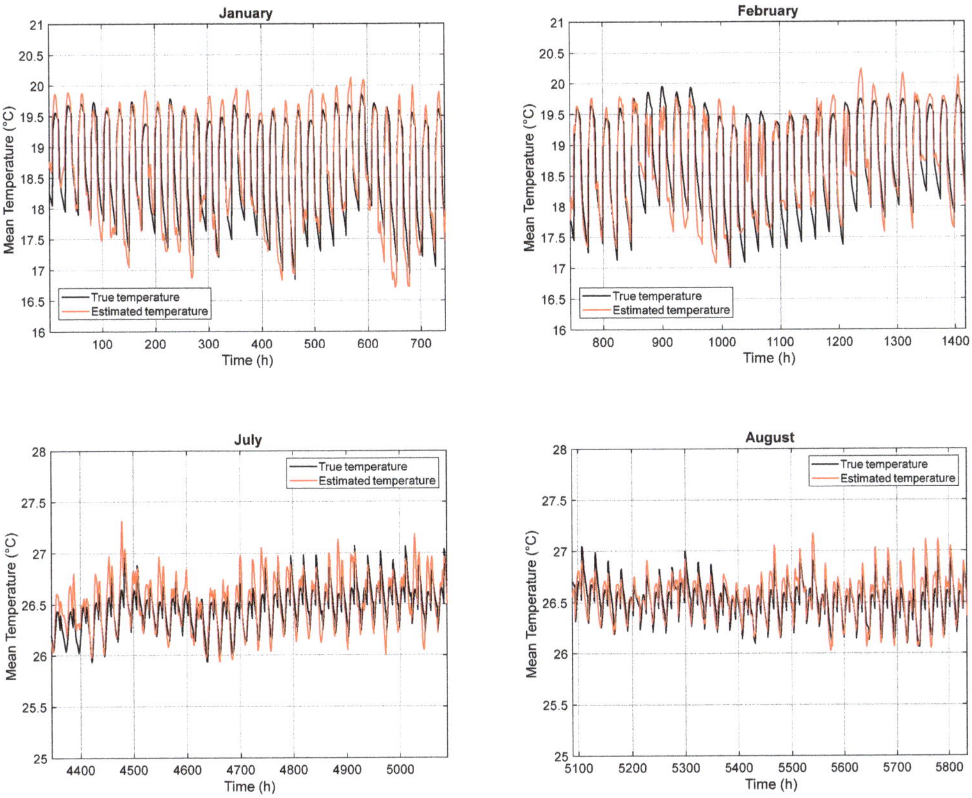

Figure 10. Prediction of the mean inside air temperature (Ti) in January, February, July, and August for Catania (southern Italy). The estimated temperature is represented by the red line, while the true temperature is represented by the black line.

In general, it is possible to note that the trend of the estimated T_i is quite similar to the true one, for the three representative cities, showing good performance of the network in terms of being able to generalize towards new city climate conditions. However, a greater deviation is observed between the estimated and true temperatures in Udine during the summer months, because the training cities exhibited higher temperatures during the same months. Specifically, the maximum difference between the estimated and the true temperatures is less than 2 °C. This behavior is clearly an outlier, probably due to the lack of data on mean air temperature for this location in the north. This discrepancy highlights the limitations of the RNN during certain periods of the year.

For the remaining cities of each climatic zone (Aviano–Bolzano–Trieste, Milano–Perugia–Piacenza, Messina–Brindisi–Trapani), the average monthly RMSE was calculated, and the results are reported in Figures 11–13. The tested RNN can predict the thermo-electrical behavior of the commercial buildings, exhibiting better results for the winter and summer months, i.e., when the heating and cooling systems were activated at full load. This result is pivotal in the smart management of energy aimed at reducing as much energy wastage as possible. For the mid-season, the figures show that, in the spring and autumn months, the average RMSE is higher than in the other months. In detail, for the northern climatic zone, it is lower than 6 °C, while for the central zone it is lower than 4 °C. Finally, for the southern climatic zone, it is lower than 2 °C. In these periods, the climatic conditions are highly variable in the Italian territory, and neighboring cities can also present different behaviors, especially in recent years where the climate changes have been most frequent. To obtain a better RNN performance during the spring and autumn seasons, it would be useful to consider a higher number of cities for each climatic zone.

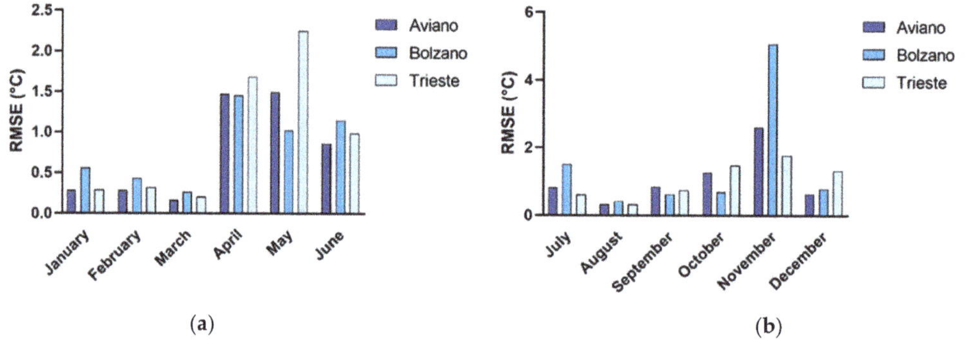

Figure 11. Monthly average RMSE from January to June (**a**) and from July to December (**b**) for Aviano, Bolzano, and Trieste, representatives of the northern climatic zone.

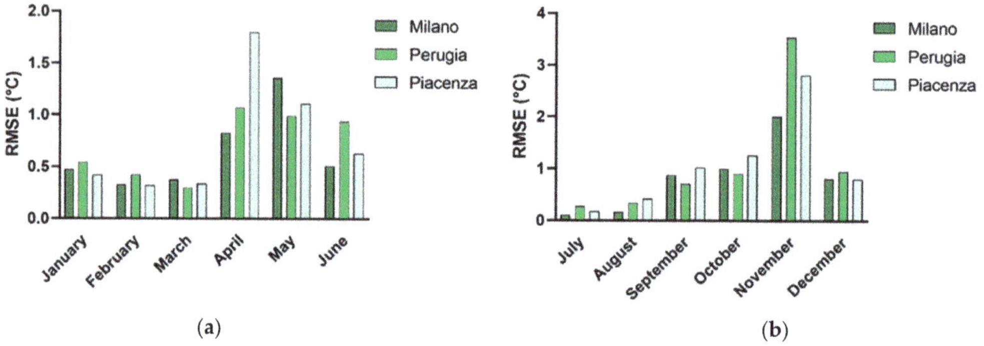

Figure 12. Monthly average RMSE from January to June (**a**) and from July to December (**b**) for Milano, Perugia, and Piacenza, representatives of the central climatic zone.

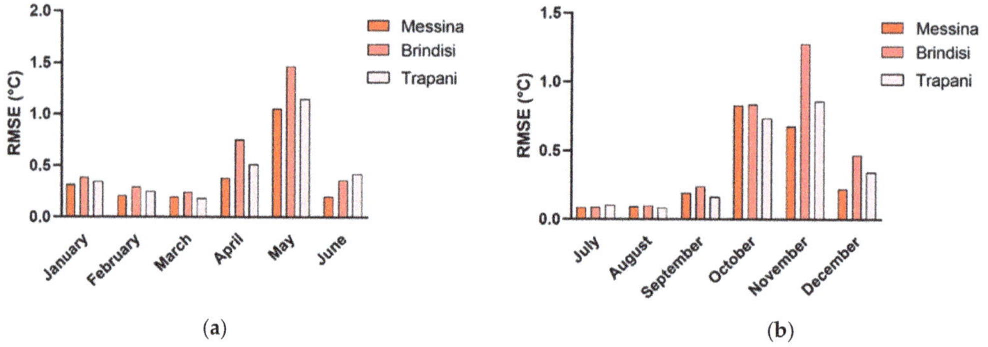

Figure 13. Monthly average RMSE from January to June (**a**) and from July to December (**b**) for Messina, Brindisi, and Trapani, representatives of the southern climatic zone.

Finally, the model inversion was tested on a small subset of the dataset to show the accuracy of the reconstructed energy profile using, as an optimization algorithm, the standard non-linear least square solver implemented in MATLAB. Over a period of 8 winter days, the inverted model reconstructed the energy profile with an RMSE of 0.31 Wh. The reconstructed profiles are shown in Figure 14. As can be seen, the model can be inverted without problems through a suitable optimization algorithm.

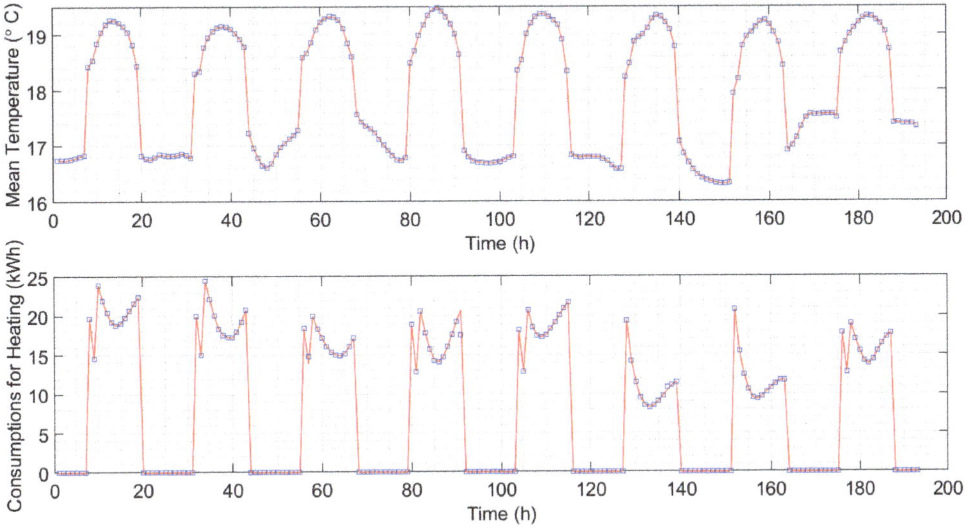

Figure 14. Results of the model inversion approach. (**Bottom**), the true energy demand (red curve) and the reconstructed energy demand (blue square). (**Top**), the reference temperature (red curve) and the temperature resulting from the reconstructed energy demand (blue square).

4. Critical Discussion

To comprehensively evaluate the neural network's ability to predict internal temperatures that ensure thermal comfort, especially during the winter and summer, final performance metrics were calculated on a seasonal and annual basis. Specifically, Figure 15 reports the average root mean square error (RMSE) across the four seasons for each climatic zone.

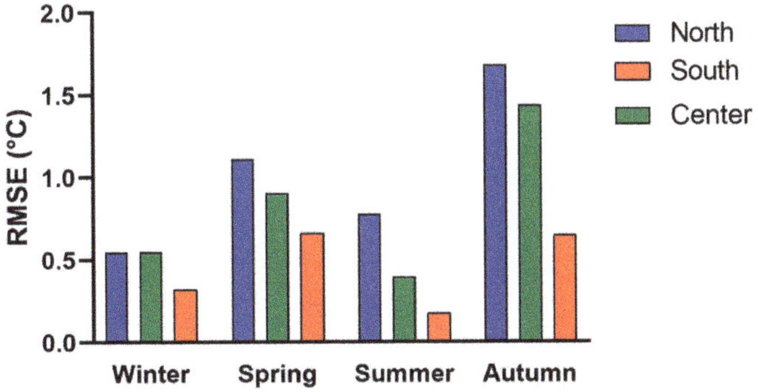

Figure 15. Seasonal average RMSE for the three climatic zones.

As expected, the winter and summer seasons exhibit the best performance with an overall RMSE lower than 1 °C. This is predictable, as heating and cooling systems are primarily used during the winter and summer months, respectively. During the mid-seasons, the need for heating or cooling varies from day to day. Consequently, the neural network performs better in identifying the internal thermo-electric relationship precisely in those periods in which it is possible to control the internal temperature of the rooms by consciously switching the heating and cooling systems on and off. During the transitional seasons, the contribution of the energy demand of thermal systems is less significant. Therefore, the internal temperature of the site is primarily dependent on weather conditions and location. It is important to note that, in the southern regions, the RMSE is less than 0.7–0.8 °C, even in the summer and spring, due to the stable climatic conditions. The climate in the winter is also very mild and consistent throughout the year. The maximum error value is approximately 1.65 °C for the northern cities during the autumn. There are significant temperature fluctuations throughout the day, and users tend to turn off their plants for extended periods.

The average annual RMSE was then computed for each climate zone, as depicted in Figure 16. This allows for a direct comparison between the southern and northern climatic zones, indicating superior prediction accuracy in the former and less accurate predictions in the latter. This discrepancy could be attributed to more uniform occupant behavior in southern urban areas and a more diverse pattern of user activity in the northern regions, potentially influenced by greater climatic variability. Notably, the overall prediction error does not exceed 1 °C.

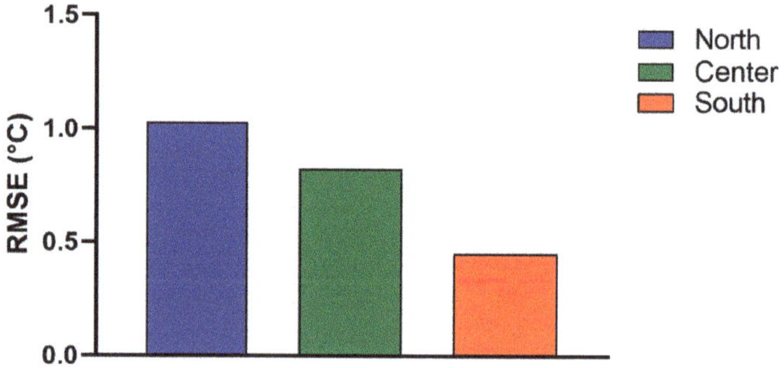

Figure 16. Annual average RMSE for the three climatic zones.

5. Limitations, Potential, and Future Works

Possible extensions of this work include exploring alternative sources of reference data or adopting different algorithmic implementations. In the first case, real-world experimental data could be used to train the artificial neural network (ANN). This approach has been adopted in previous studies, such as [50], where experimental measurements were collected from an educational building to train both direct and inverse ANN models. These models were then employed to simulate and optimize the operation of a renewable energy community (REC). In that study, environmental data from weather stations and energy consumption data from smart meters were used as ANN inputs, enabling the estimation of the mean indoor air temperature. For the inverse model, accurate hourly values of indoor temperature are required to estimate the corresponding energy consumption. This methodology, however, presents certain limitations. Simulation tools, such as EnergyPlus, allow for the same case study to be analyzed under different geographic and environmental conditions (a flexibility not afforded by the monitoring of a single real-world case). Therefore, a combined approach using both simulation data and ANN models provides enhanced versatility, as it requires only the outdoor environmental boundary conditions available in standard weather files for different locations. In future work, the authors intend to incorporate real data from non-residential buildings (including commercial, public, and industrial facilities), considering various boundary conditions (such as occupancy schedules and HVAC system operation). The analysis will also be extended to include additional geographic locations.

In the case of Italy, the national classification of climatic zones (six zones, from A to F) has been considered [59]. This classification defines the permitted heating and cooling periods for buildings during summer and winter, depending on the specific thermal needs of each zone. Twelve cities were selected, with four from each zone, to train three recurrent neural networks (RNNs), one for each major climatic area in Italy. This methodology can also be extended to other countries, as the input data required for the ANNs consists of outdoor environmental conditions (including mean air temperature, global solar radiation, relative humidity, and wind speed) and electrical consumption data related to heating or cooling. These data can be measured, collected, or derived from meteorological datasets available for other countries. In such cases, it is also possible to group cities from different countries that exhibit similar climatic characteristics. Regarding future development and application, the proposed framework could be scaled up for use across a national building stock or integrated into energy planning at the policy level. Since each building is characterized by specific thermal, electrical, and energy-related properties, the method can only be generalized by applying it to a sufficiently large number of buildings with similar characteristics (such as year of construction, thermal transmittance of opaque and transparent envelope components, and HVAC system specifications). Buildings with comparable features could be grouped together to represent segments of the building stock, and RNNs could be trained and implemented to analyze the behavior of each group. This approach could prove useful for forecasting the energy and electrical performance of representative buildings in each territory, as well as in the context of renewable energy community (REC) scenarios.

An additional consideration for future development concerns the structure and implementation potential of the proposed RNN models. Like neural models, RNNs are inherently scalable and can be extended to include additional input variables if required by specific applications or scenarios. Although the current implementation does not incorporate long-term memory, this limitation can be easily addressed by adopting more advanced architecture, such as long short-term memory (LSTM) or gated recurrent unit (GRU) networks, which are well-suited for capturing temporal dependencies over longer horizons [60,61].

Furthermore, due to their relatively low computational complexity and lightweight design, the proposed RNN models are well-suited for deployment on microcontroller-based systems that support TinyML frameworks, such as STMicroelectronics platforms [62,63]. This feature will be explored in future work with the goal of embedding models in edge devices for building automation. Such integration is particularly relevant in the context of smart buildings and intelligent energy management systems, where on-device inference enables real-time, energy-efficient operation without reliance on cloud-based infrastructure.

6. Conclusions

In Italy, non-residential buildings consume a significant amount of energy for heating/cooling, lighting, and other appliances due to their inadequate construction characteristics, which do not meet current legislative requirements. This is particularly true for buildings that are at least 20 years old. The present paper presents an interesting approach to model the relationship between thermal and energy demand in HVAC systems in different Italian locations. A commercial/office building was chosen as a case study, every feature of the building envelopes and plants were known, and it was modelled in the EnergyPlus program. Yearly simulations were carried out for twelve Italian cities, distributed evenly across the north, center, and south. Three artificial recurrent neural networks were developed based on the E+ output data: the first one was trained with the simulation results of the northern cities in Italy; the second was related to the behavior of the same building in central Italy; the third was suitable for the southern regions. The hourly electricity consumptions obtained for the occupancy periods were used as input in the RNN, and the temperature profiles of each zone are obtained as output. The monthly evaluation of the root mean square error reveals consistently low values, particularly during the winter and summer when the heating and cooling systems are activated in a distinct on–off manner. In this case the values are always lower than 1 °C. The RNN for the south also performs well in intermediate seasons when it is often necessary for the activation of the heat pumps in cooling mode or when the air conditioning is nowhere required. For the northern regions, the values of the seasonal RMSE are high above all for autumn (higher than 1.5 °C) when no specific trend for managing the plant has been identified. Future work could enhance the RNN model by incorporating data from additional Italian cities that represent the country's diverse climatic zones. Furthermore, the inversion of the RNNs is being explored to predict electricity consumption (output) based on user temperature profiles (input). This inversion would enable the minimization of electricity loads by adjusting indoor thermal conditions (e.g., different target temperatures and operational schedules). This approach has promising applications, such as load dispatch optimization and building automation systems within smart grids and renewable energy communities.

Author Contributions: Conceptualization, G.M.L. and F.R.F.; methodology, G.M.L. and E.B.; software, M.P., M.Q. and F.F.; validation, E.B., M.P., M.Q. and F.F.; formal analysis, E.B. and G.M.L.; investigation, E.B. and G.M.L.; resources, F.R.F.; data curation, E.B., M.P. and G.M.L.; writing—original draft preparation, E.B., M.P., F.F. and G.M.L.; writing—review and editing, F.R.F. and M.Q.; visualization, E.B. and G.M.L.; supervision, F.R.F. and G.M.L.; project administration, F.R.F.; funding acquisition, F.R.F. All authors have read and agreed to the published version of the manuscript.

Funding: This study was carried out within the "Innovative Solutions for Renewables in Energy Communities (ISoREC)" project n. 202054TZLF—funded by European Union within the PRIN 2020 program (D.D. 1628—16/10/2020 Ministero dell'Università e della Ricerca—CUP: J57G20000100008). This manuscript reflects only the authors' views and opinions, and the Ministry cannot be considered responsible for them.

Data Availability Statement: Data will be made available on request.

Conflicts of Interest: The authors declare no conflicts of interest.

References

1. U.S. Department of Energy. Grid-Interactive Efficient Buildings Technical Report Series Overview of Research Challenges and Gaps. December 2019. Available online: https://www.energy.gov/eere/buildings/articles/grid-interactive-efficient-buildings-technical-report-series-overview (accessed on 3 April 2025).
2. Martirano, L.; Rotondo, S. Modello di Microgrid per "Smart Building" Come Energy Community Con Gestione Ottimizzata Delle Risorse Energetiche Parte 1—Analisi di Modelli di Reti Energetiche per Smart Building e NZEB. Ricerca del Sistema Elettrico (RSE) e ENEA. Available online: https://www.ricercasistemaelettrico.enea.it/archivio-documenti/category/490-report-2019-progetto-1-5.html (accessed on 14 April 2025). (In Italian)
3. Household Energy Consumption by Energy in the EU. Available online: https://www.odyssee-mure.eu/publications/efficiency-by-sector/households/energy-consumption-eu.html (accessed on 3 April 2025).
4. Ding, C.; Ke, J.; Levine, M.; Zhou, N. Potential of artificial intelligence in reducing energy and carbon emissions of commercial buildings at scale. *Nat. Commun.* **2024**, *15*, 5916. [CrossRef] [PubMed]
5. Brhane, G.Y.; Oh, E.; Son, S.Y. Virtual Energy Storage System Scheduling for Commercial Buildings with Fixed and Dynamic Energy Storage. *Energies* **2024**, *17*, 3292. [CrossRef]
6. Rahman, A.; Srikumar, V.; Smith, A.D. Predicting electricity consumption for commercial and residential buildings using deep recurrent neural networks. *Appl. Energy* **2018**, *212*, 372–385. [CrossRef]
7. Mocanu, E.; Nguyen, P.H.; Gibescu, M.; Kling, W.L. Deep learning for estimating building energy consumption. *Sustain. Energy Grids Netw.* **2016**, *6*, 91–99. [CrossRef]
8. Robinson, C.; Dilkina, B.; Hubbs, J.; Zhang, W.; Guhathakurta, S.; Brown, M.A.; Pendyala, R.M. Machine learning approaches for estimating commercial building energy consumption. *Appl. Energy* **2017**, *208*, 889–904. [CrossRef]

Michailidis, P.; Michailidis, I.; Gkelios, S.; Kosmatopoulos, E. Artificial Neural Network Applications for Energy Management in Buildings: Current Trends and Future Directions. *Energies* **2024**, *17*, 570. [CrossRef]

Chitalia, G.; Pipattanasomporn, M.; Garg, V.; Rahman, S. Robust short-term electrical load forecasting framework for commercial buildings using deep recurrent neural networks. *Appl. Energy* **2020**, *278*, 115410. [CrossRef]

Huang, H.; Chen, L.; Hu, E. A neural network-based multi-zone modelling approach for predictive control system design in commercial buildings. *Energy Build.* **2015**, *97*, 86–97. [CrossRef]

Fu, H.; Baltazar, J.C.; Claridge, D.E. Review of developments in whole-building statistical energy consumption models for commercial buildings. *Renew. Sustain. Energy Rev.* **2021**, *147*, 111248. [CrossRef]

Macarulla, M.; Casals, M.; Forcada, N.; Gangolells, M. Implementation of predictive control in a commercial building energy management system using neural networks. *Energy Build.* **2017**, *151*, 511–519. [CrossRef]

Afzal, S.; Shokri, A.; Ziapour, B.M.; Shakibi, H.; Sobhani, B. Building energy consumption prediction and optimization using different neural network-assisted models; comparison of different networks and optimization algorithms. *Energy* **2023**, *282*, 128446. [CrossRef]

Han, Y.; Fan, C.; Geng, Z.; Ma, B.; Cong, D.; Chen, K.; Yu, B. Energy efficient building envelope using novel RBF neural network integrated affinity propagation. *Energy* **2020**, *209*, 118414. [CrossRef]

Severiche-Maury, Z.; Uc-Rios, C.E.; Arrubla-Hoyos, W.; Cama-Pinto, D.; Holgado-Terriza, J.A.; Damas-Hermoso, M.; Cama-Pinto, A. Forecasting Residential Energy Consumption with the Use of Long Short-Term Memory Recurrent Neural Networks. *Energies* **2025**, *18*, 1247. [CrossRef]

Uba, F.; Apevienyeku, H.K.; Nsiah, F.D.; Akorli, A.; Adjignon, S. Energy Analysis of Commercial Buildings Using Artificial Neural Network. *Model. Simul. Eng.* **2021**, *2021*, 8897443. [CrossRef]

Genkin, M.; McArthur, J.J. B-SMART: A reference architecture for artificially intelligent autonomic smart buildings. *Eng. Appl. Artif. Intell.* **2023**, *121*, 106063. [CrossRef]

Kulathilaka, M.J.S.; Saravanan, S.; Kumarasiri, H.D.H.P.; Logeeshan, V.; Kumarawadu, S.; Wanigasekara, C. NILM for Commercial Buildings: Deep Neural Networks Tackling Nonlinear and Multi-Phase Loads. *Energies* **2024**, *17*, 3802. [CrossRef]

Lu, Y.; Chen, Q.; Yu, M.; Wu, Z.; Huang, C.; Fu, J.; Yu, Z.; Yao, J. Exploring spatial and environmental heterogeneity affecting energy consumption in commercial buildings using machine learning. *Sustain. Cities Soc.* **2023**, *95*, 104586. [CrossRef]

Liang, W.; Li, H.; Zhan, S.; Chong, A.; Hong, T. Energy flexibility quantification of a tropical net-zero office building using physically consistent neural network-based model predictive control. *Adv. Appl. Energy* **2024**, *14*, 100167. [CrossRef]

Han, F.; Du, F.; Jiao, S.; Zou, K. Predictive Analysis of a Building's Power Consumption Based on Digital Twin Platforms. *Energies* **2024**, *17*, 3692. [CrossRef]

Gao, Z.; Yang, S.; Yu, J.; Zhao, A. Hybrid forecasting model of building cooling load based on combined neural network. *Energy* **2024**, *297*, 131317. [CrossRef]

Jiang, B.; Li, Y.; Rezgui, Y.; Zhang, C.; Wang, P.; Zhao, T. Multi-source domain generalization deep neural network model for predicting energy consumption in multiple office buildings. *Energy* **2024**, *299*, 131467. [CrossRef]

Ye, H.; Zhu, Q.; Zhang, X. Short-Term Load Forecasting for Residential Buildings Based on Multivariate Variational Mode Decomposition and Temporal Fusion Transformer. *Energies* **2024**, *17*, 3061. [CrossRef]

Wang, G.; Mukhtar, A.; Moayedi, H.; Khalilpoor, N.; Tt, Q. Application and evaluation of the evolutionary algorithms combined with conventional neural network to determine the building energy consumption of the residential sector. *Energy* **2024**, *298*, 131312. [CrossRef]

Biswas, M.A.R.; Robinson, M.D.; Fumo, N. Prediction of residential building energy consumption: A neural network approach. *Energy* **2016**, *117 Pt 1*, 84–92. [CrossRef]

Li, T.; Liu, X.; Li, G.; Wang, X.; Ma, J.; Xu, C.; Mao, Q. A systematic review and comprehensive analysis of building occupancy prediction. *Renew. Sustain. Energy Rev.* **2024**, *193*, 114284. [CrossRef]

Azar, E.; O'Brien, W.; Carlucci, S.; Hong, T.; Sonta, A.; Kim, J.; Andargie, M.S.; Abuimara, T.; Asmar, M.E.; Jain, R.K.; et al. Simulation-aided occupant-centric building design: A critical review of tools, methods, and applications. *Energy Build.* **2020**, *224*, 110292. [CrossRef]

Li, T.; Liu, X.; Zhou, W.; Ma, J.; Li, Y.; Gao, J.; Chen, M.; Mao, Q. Optimizing building energy consumption through synchronization and asynchronization of occupancy and air-conditioning behavior. *Energy Build.* **2025**, *331*, 115409. [CrossRef]

Wang, X.; Li, T.; Yu, Y.; Liu, Q.; Shi, L.; Xia, J.; Mao, Q. Performance simulation and energy efficiency analysis of multi-energy complementary HVAC system based on TRNSYS. *Appl. Therm. Eng.* **2024**, *257 Pt B*, 124378. [CrossRef]

Nutakki, M.; Mandava, S. Review on optimization techniques and role of Artificial Intelligence in home energy management systems. *Eng. Appl. Artif. Intell.* **2023**, *119*, 105721. [CrossRef]

Khalil, M.; McGough, A.S.; Pourmirza, Z.; Pazhoohesh, M.; Walker, S. Machine Learning, Deep Learning and Statistical Analysis for forecasting building energy consumption—A systematic review. *Eng. Appl. Artif. Intell.* **2022**, *115*, 105287. [CrossRef]

34. Mehmood, M.U.; Chun, D.; Zeeshan; Han, H.; Jeon, G.; Chen, K. A review of the applications of artificial intelligence and big dat to buildings for energy-efficiency and a comfortable indoor living environment. *Energy Build.* **2019**, *202*, 109383. [CrossRef]
35. EnergyPlus Program, 23.2.0 Version. Available online: https://energyplus.net/ (accessed on 14 April 2025).
36. Chung, W.J.; Liu, C. Analysis of input parameters for deep learning-based load prediction for office buildings in different climat zones using eXplainable Artificial Intelligence. *Energy Build.* **2022**, *276*, 112521. [CrossRef]
37. Si, B.; Wang, J.; Yao, X.; Shi, X.; Jin, X.; Zhou, X. Multi-objective optimization design of a complex building based on an artifici neural network and performance evaluation of algorithms. *Adv. Eng. Inform.* **2019**, *40*, 93–109. [CrossRef]
38. Edwards, R.E.; New, J.; Parker, L.E.; Cui, B.; Dong, J. Constructing large scale surrogate models from big data and artifici intelligence. *Appl. Energy* **2017**, *202*, 685–699. [CrossRef]
39. Wong, S.L.; Wan, K.K.W.; Tony, N.T.L. Artificial neural networks for energy analysis of office buildings with daylighting. *App Energy* **2010**, *87*, 551–557. [CrossRef]
40. Saryazdi, S.M.E.; Etemad, A.; Shafaat, A.; Bahman, A.M. Data-driven performance analysis of a residential building applyin artificial neural network (ANN) and multi-objective genetic algorithm (GA). *Build. Environ.* **2022**, *225*, 109633. [CrossRef]
41. Melo, A.P.; Cóstola, D.; Lamberts, R.; Hensen, J.L.M. Development of surrogate models using artificial neural network fo building shell energy labelling. *Energy Policy* **2014**, *69*, 457–466. [CrossRef]
42. Lazzeroni, P.; Mariuzzo, I.; Quercio, M.; Repetto, M. Economic, Energy, and Environmental Analysis of PV with Battery Storag for Italian Households. *Electronics* **2021**, *10*, 146. [CrossRef]
43. Chou, J.S.; Bui, D.K. Modeling heating and cooling loads by artificial intelligence for energy-efficient building design. *Energ Build.* **2014**, *82*, 437–446. [CrossRef]
44. Belloni, E.; Bianchini, G.; Casini, M.; Faba, A.; Intravaia, M.; Laudani, A.; Lozito, G.M. An overview on building-integrate photovoltaics: Technological solutions, modeling, and control. *Energy Build.* **2024**, *324*, 114867. [CrossRef]
45. Lucaferri, V.; Quercio, M.; Laudani, A.; Riganti Fulginei, F. A Review on Battery Model-Based and Data-Driven Methods fo Battery Management Systems. *Energies* **2023**, *16*, 7807. [CrossRef]
46. Risi, B.; Riganti Fulginei, F.; Laudani, A.; Quercio, M. Compensation Admittance Load Flow: A Computational Tool for th Sustainability of the Electrical Grid. *Sustainability* **2023**, *15*, 14427. [CrossRef]
47. Belloni, E.; Lozito, G.M.; Reatti, A. A Python Tool for Simulation and Optimal Sizing of a Storage Equipped Grid Connecte Photovoltaic Power System. In Proceedings of the 2022 IEEE 21st Mediterranean Electrotechnical Conference (MELECON Palermo, Italy, 14–16 June 2022; pp. 884–889. [CrossRef]
48. Belloni, E.; Fulginei, F.R.; Lozito, G.M.; Poli, D. Direct and Inverse Neural Modelling of Buildings HVAC Systems. In Proceeding of the IEEE EUROCON 2023—20th International Conference on Smart Technologies, Torino, Italy, 6–8 July 2023; pp. 269–27 [CrossRef]
49. Palermo, M.; Forconi, F.; Belloni, E.; Quercio, M.; Lozito, G.M.; Fulginei, F.R. Optimization of a feedforward neural network' architecture for an HVAC system problem. In Proceedings of the 2023 3rd International Conference on Electrical, Compute Communications and Mechatronics Engineering (ICECCME), Tenerife, Spain, 19–21 July 2023; pp. 1–6. [CrossRef]
50. Belloni, E.; Grasso, F.; Lozito, G.M.; Poli, D.; Riganti Fulginei, F.; Talluri, G. Neural-assisted HVACs optimal scheduling fo renewable energy communities. *Energy Build.* **2023**, *301*, 113658. [CrossRef]
51. Liu, B.; Vu-Bac, N.; Zhuang, X.; Fu, X.; Rabczuk, T. Stochastic integrated machine learning based multiscale approach for th prediction of the thermal conductivity in carbon nanotube reinforced polymeric composites. *Compos. Sci. Technol.* **2022**, *22* 109425. [CrossRef]
52. Liu, B.; Vu-Bac, N.; Rabczuk, T. A stochastic multiscale method for the prediction of the thermal conductivity of Polyme nanocomposites through hybrid machine learning algorithms. *Compos. Struct.* **2021**, *273*, 114269. [CrossRef]
53. Ji, J.; Yu, H.; Wang, X.; Xu, X. Machine learning application in building energy consumption prediction: A comprehensive review *J. Build. Eng.* **2025**, *104*, 112295. [CrossRef]
54. Liu, B.; Vu-Bac, N.; Zhuang, X.; Liu, W.; Fu, X.; Rabczuk, T. Al-DeMat: A web-based expert system platform for computationall expensive models in materials design. *Adv. Eng. Softw.* **2023**, *176*, 103398. [CrossRef]
55. EnergyPlus Weather Data File. Available online: https://energyplus.net/weather (accessed on 14 April 2025).
56. Buratti, C.; Moretti, E.; Belloni, E. Nanogel Windows for Energy Building Efficiency. In *Nano and Biotech Based Materials for Energ Building Efficiency*; Pacheco Torgal, F., Buratti, C., Kalaiselvam, S., Granqvist, C.G., Ivanov, V., Eds.; Springer: Cham, Switzerlan 2016. [CrossRef]
57. Lozito, G.M.; Salvini, A. Swarm intelligence based approach for efficient training of regressive neural networks. *Neural Compu Appl.* **2020**, *32*, 10693–10704. [CrossRef]
58. MathWorks. MATLAB r2022. 2022. Available online: https://www.mathworks.com/products/new_products/release2022a.htm (accessed on 3 April 2025).

. Regulation Containing Rules for the Design, Installation, Operation and Maintenance of Heating Systems in Buildings for the Purpose of Containing Energy Consumption. D.P.R 412/1993, Updated on 28 April 2022. Available online: https://www.normattiva.it/uri-res/N2Ls?urn:nir:stato:decreto.del.presidente.della.repubblica:1993-08-26;412!vig= (accessed on 14 April 2025).

. Somu, N.; MR, G.R.; Ramamritham, K. A deep learning framework for building energy consumption forecast. *Renew. Sustain. Energy Rev.* **2021**, *137*, 110591. [CrossRef]

. Mtibaa, F.; Nguyen, K.K.; Azam, M.; Papachristou, A.; Venne, J.S.; Cheriet, M. LSTM-based indoor air temperature prediction framework for HVAC systems in smart buildings. *Neural Comput. Appl.* **2020**, *32*, 17569–17585. [CrossRef]

. Osman, A.; Abid, U.; Gemma, L.; Perotto, M.; Brunelli, D. Tinyml platforms benchmarking. In Proceedings of the International Conference on Applications in Electronics Pervading Industry, Environment and Society, Online, 21–22 September 2021; Springer International Publishing: Cham, Switzerland, 2021; pp. 139–148.

. Sudharsan, B.; Salerno, S.; Nguyen, D.D.; Yahya, M.; Wahid, A.; Yadav, P.; Breslin, J.G.; Ali, M.I. TinyML benchmark: Executing fully connected neural networks on commodity microcontrollers. In Proceedings of the 2021 IEEE 7th World Forum on Internet of Things (WF-IoT), New Orleans, LA, USA, 14 June–31 July 2021; IEEE: Piscataway, NJ, USA; pp. 883–884.

energies

MDPI

Article

Agricultural Value Added, Renewable Energy, and the Environmental Kuznets Curve: Evidence from Turkey

Neslihan Koç [1,*], Özgür Emre Koç [1], Florina Oana Virlanuta [2,*], Orhan Orçun Bıtrak [3], Uğur Çiçek [4], Radu Octavian Kovacs [5], Valentina-Alina Vasile (Dobrea) [5] and Tincuta Vrabie [6]

1 Department of Public Finance, Faculty of Economics and Administrative Sciences, Hitit University, 19030 Corum, Turkey; oemrekoc@hitit.edu.tr
2 Department of Economics, Faculty of Economics and Business Administration, "Dunărea de Jos" University of Galați, 800008 Galați, Romania
3 Department of Banking and Insurance, Yalvaç Vocational School, Isparta University of Applied Sciences, 32400 Yalvaç, Turkey; bitrak.orcun@gmail.com
4 Department of Public Finance, Faculty of Economics and Administrative Sciences, Burdur Mehmet Akif University, 15030 Burdur, Turkey; ugurcicek@mehmetakif.edu.tr
5 Doctoral School of Economic Sciences, Faculty of Economics and Business Administration, "Dunărea de Jos" University of Galați, 800008 Galați, Romania; rokwise@gmail.com (R.O.K.); dobrea.alina@gmail.com (V.-A.V.)
6 Department of History, Philosophy and Sociology, Faculty of History, Philosophy and Theology, "Dunărea de Jos" University of Galați, 800008 Galați, Romania; tincuta.vrabie@ugal.ro
* Correspondence: neslihankoc@hitit.edu.tr (N.K.); florina.virlanuta@ugal.ro (F.O.V.)

Abstract

In this study, the relationship between economic growth and carbon emissions for the period 1968–2022 in Turkey was evaluated within the framework of the EKC (Environmental Kuznets Curve) hypothesis. In addition, the impacts of renewable energy consumption and agricultural value added on carbon emissions were analyzed using the ARDL bounds testing approach. The validity of the results was also tested using the FMOLS and DOLS methods. The findings confirmed the existence of a cointegration relationship between carbon emissions and per capita income, renewable energy consumption, and agricultural value added. Long-term analyses indicate that renewable energy consumption reduces carbon emissions, whereas growth in agricultural value added leads to an increase in emissions. In addition, it has been determined that the EKC hypothesis is valid in both the long and short terms and that increases in per capita income raise emissions up to a certain threshold and have a mitigating effect when this threshold is exceeded. The results of the short-term analysis showed that the effects of renewable energy consumption vary across periods, and that agricultural value added increases emissions in the short term. This study provides empirical evidence for Turkey by incorporating sectoral variables within the EKC framework and offers meaningful insights for policymakers regarding the environmental impacts of agricultural value added and renewable energy use in the context of a developing country. Accordingly, fiscal policy instruments such as green taxation, carbon credit trading mechanisms, and financial and agricultural subsidies should be more effectively utilized in Turkey to support structural transformation in agriculture and promote the use of clean energy, in line with the findings that suggest the need for targeted agricultural and energy policies aligned with Turkey's SDG commitments.

Keywords: environmental Kuznets curve; renewable energy; agricultural value added; carbon emissions; ARDL bounds testing; sustainable development

1. Introduction

The Environmental Kuznets Curve (EKC) hypothesis provides a fundamental framework for understanding the relationship between economic growth and the environment, positing that degradation increases during the early stages of growth but declines after a certain income threshold is surpassed [1]. Many studies do not include sectoral factors that cause emissions in their analysis. This study addresses this gap by examining the environmental impacts of renewable energy consumption and agricultural value added within the EKC framework, focusing on Turkey as a developing economy that is heavily dependent on fossil fuels and traditional agricultural practices. These issues are central to current global sustainability debates, particularly those framed by the United Nations Sustainable Development Goals (SDGs), including SDG 7 (Affordable and Clean Energy), SDG 12 (Responsible Consumption and Production), and SDG 13 (Climate Action).

In addition to macro-level sustainability goals, recent approaches underscore the importance of coherent, sector-specific strategies that are both actionable and aligned with long-term environmental objectives. Within this evolving discourse, the concept of pragmatic sustainability has gained prominence by promoting feasible, innovation-driven frameworks that integrate economic and ecological priorities. Second-hand clothing initiatives, the integration of natural capital into circular economy models, the adoption of hybrid renewable energy systems, and the sustainable management of forests all play important roles in reducing environmental impacts, enhancing agricultural and rural sustainability, and advancing sustainable development [2–5]. This underscores the value of sectoral, integrative, and practice-oriented strategies within the pragmatic sustainability paradigm. Given this context, sector-specific and integrative strategies are essential for advancing environmental sustainability in Turkey, where this study contributes by empirically testing the EKC hypothesis through an analysis of long-term dynamics.

According to the EKC hypothesis, at the first stage of economic growth, environmental degradation increases due to an increase in production volume (scale effect). As growth continues, the services sector, where growth effects are less, begins to grow (composition effect). With economic growth, reducing emissions and increasing waste management allows for investment in technologies (technological impact). But this trend should not mean that environmental degradation is a natural consequence of economic growth and will correct itself over time. On the contrary, in the face of the risk of exceeding environmental thresholds and irreversible damage, the necessity of implementing structural policies arises [6]. Emissions changes cannot be explained by carbon intensity or growth rates alone; more factors affecting carbon emissions should be included in the EKC research framework. In particular, sector-based analysis is crucial for effectively assessing policy monitoring and achieving carbon neutrality targets.

In Turkey, energy consumption plays a decisive role in growth dynamics and the link between energy and growth [7,8]. Given Turkey's dependency on imported fossil fuels, renewable energy holds strategic importance, not only environmentally but also economically. For instance, [9] found that a 1% increase in renewable energy consumption leads to a 0.10% rise in the GDP per capita, underscoring the macroeconomic benefits of renewable energy beyond environmental considerations. Moreover, the use of renewable energy is not limited to reducing environmental impacts; it also provides multidimensional contributions such as supporting rural development, reducing income inequality, and modernizing the agricultural sector, thereby playing an important role in achieving sustainable development goals, especially in developing countries.

The agricultural sector is another critical area that significantly affects both economic and environmental outcomes. Although the share of the agricultural sector in Turkey's GDP declined from 25.8% in 1980 to 4.8% in 2022, it remains strategically important, as it still employs 15.8% of the total workforce [10]. Due to increasing food demand, its role as a fundamental driver of rural development, and its high vulnerability to climate change, the agricultural sector can be considered one of the most critical sectors in terms of environmental sustainability.

Most studies linking agriculture and emissions focus on broad indicators such as agricultural activities or the volume of production. However, this study includes the variable of agricultural value added, which accounted for approximately 4.1% of global national income as of 2021. Notably, the top 20 economies—generating 78% of the global income—also produce 56% of the global agricultural output [10]. Agricultural value added represents the net output derived from the difference between agricultural inputs and outputs; in this respect, it is a key indicator that directly reflects productivity growth and economic contribution in the sector while also allowing for the testing of the EKC theory in the context of agriculture.

The increase in agricultural added value can contribute to environmental sustainability with lower emissions, as opposed to energy-intensive production structures. High technology integration in developed countries makes it possible to achieve high agricultural added value at lower environmental costs. In developing countries, the increase in agricultural value added is usually achieved through traditional and resource-intensive forms of production rather than through modern technologies or sustainable practices. Access to clean technologies in these countries is restricted due to high costs and limited financing opportunities. Since economic growth is prioritized, environmental regulations and sanctions are not sufficiently strict. Therefore, in developing countries, the increase in agricultural value added may lead to short-term economic gains but long-term environmental costs. As a middle-income country with limited technology integration, Turkey is likely to face similar environmental risks.

In light of these observations, the aim of this study is to test the validity of the Environmental Kuznets Curve hypothesis in the Turkish economy and to empirically analyze the effects of agricultural added value and renewable energy use on carbon emissions, identified as complementary indicators for evaluating the effectiveness of climate policies. In this study, agricultural value added as a share of the GDP is used instead of absolute agricultural output to capture the structural composition effect of agriculture within the economy, which provides deeper insights into sectoral environmental dynamics. Similarly, renewable energy consumption is included as a key explanatory variable, not only due to its emission-reducing potential but also because of its broader macroeconomic contributions and strategic relevance in fossil fuel-dependent countries like Turkey. Despite a growing body of EKC-related research, sectoral applications—especially those focusing on agricultural value added as a share of GDP—remain underexplored. In the context of Turkey, empirical studies that isolate the environmental effects of agriculture within the EKC framework are particularly scarce. This study addresses this gap by offering a sector-specific analysis using agricultural value added and renewable energy within the EKC framework for the Turkish economy.

The key contributions of the study include the following: First, it is one of the few studies to jointly analyze the effects of renewable energy consumption and agricultural value added on carbon emissions in Turkey, a developing country with a dual dependency on fossil fuels and traditional agricultural methods. Second, the study incorporates agricultural value added as a share of GDP to reflect sector-specific environmental impacts, an approach that provides more insight than absolute production metrics. Third, it applies a comprehensive econometric methodology combining ARDL bounds testing with FMOLS and DOLS estimators, enhancing the robustness and reliability of the long-run findings.

Lastly, it evaluates both short-run and long-run dynamics, offering a more complete view of the EKC hypothesis over time.

2. Theoretical Background

The EKC has emerged as a synthesis of approaches that argue that increasing income levels can raise environmental quality by increasing demand for environmental protection, as opposed to approaches that consider economic growth as the root cause of environmental degradation. Different results are presented depending on the analysis and variables regarding the validity of the EKC hypothesis.

Many studies on the validity of the EKC hypothesis have presented findings with supportive evidence for the EKC hypothesis [11–13]. Conversely, refs. [14–16] reported that the EKC hypothesis is not valid. However, ref. [17] found that the impact of economic growth on environmental degradation varies depending on a certain threshold level in per capita income and that the EKC hypothesis is partially valid. Accordingly, while below-the-threshold GDP growth increases resource depletion and erosion, above the threshold, these effects weaken, and the impact of growth on some environmental indicators becomes limited or meaningless. These findings suggest that the validity of the EKC hypothesis may differ depending on the country group, periodic scope, variables used, and econometric methods.

Although renewable energy consumption in general has been found to have a reducing effect on carbon emissions in the literature [18,19], studies have also found that renewable energy sources have ineffective or limited effects. As shown in [20], the effects of renewable energy sources on carbon emissions vary: hydropower and wind energy contribute to emission reductions, but solar energy shows no meaningful impact. The study in [21] found no statistically significant relationship between renewable energy consumption and environmental indicators such as ecological footprint and load capacity factor. According to [22], renewable energy improves environmental quality, but the EKC hypothesis is not valid. Also, ref. [12] states that the effects of renewable energy consumption and human capital-reducing carbon emissions differ according to the countries' EKC return points.

There is a mutual interaction between the environment and agriculture. Climatic conditions have impacts on the amount and diversity of agricultural production. For example, ref. [23] revealed that the increase in carbon emissions has negative effects on both economic growth and agricultural added value. Similarly, refs. [24–27] found that climatic conditions such as temperature increases and drought lead to decreases in agricultural productivity and income; in contrast, favorable climatic conditions can have positive effects on agricultural production and yields.

On the other hand, agricultural production also has effects on the environment. Agricultural added value can be considered equivalent to production growth and productivity in the agricultural sector. Further production increases environmental damage directly or indirectly by increasing resource use. Previous studies addressing the impact of agricultural activities on carbon emissions differ depending on the period studied, the country's agricultural policies, modes of practice, and use of technology. Findings from previous studies have revealed that added agricultural value can have effects on carbon emissions in different directions.

Reference [28] found that agricultural added value improves environmental quality in Vietnam, whereas [29] noted that the growth of agricultural added value contributes to the increase in ecological footprint. Similarly, [30] in OECD countries, [31] in Pakistan, and [32] in Brazil found that agricultural added value leads to higher CO_2 emissions. In contrast, ref. [33] concluded that agricultural added value in Somalia reduces environmental quality, while [34,35] reported no statistically significant relationship.

In some studies, it is stated that the increase in agricultural added value increases the environmental impact first, but the environmental cost decreases after a certain income or technology threshold is crossed. Similar to the EKC hypothesis, rising agricultural income levels make it possible to invest in cleaner technologies and move towards sustainable practices. According to [36], agricultural growth in developing countries contributes to increased emissions in the short term, while the long-term effects appear to be less significant.

In the literature, studies that directly use the variable 'agricultural value added' are quite limited; in particular, there is a significant gap in analyses examining its relationship with emissions in the context of Turkey. This gap necessitates a more in-depth investigation and the implementation of original empirical studies tailored to the Turkish economy. Including agricultural value added in such analyses is particularly meaningful for assessing environmentally sustainable growth. References [37,38] stated that agricultural value-added growth in Turkey reduces emissions in the long term. This finding suggests that higher added value and lower emissions may be achieved through improved agricultural technologies, renewable energy, and sustainable farming practices. On the other hand, ref. [39] concluded that agricultural value added increases carbon emissions in Turkey. Similarly, refs. [40,41] found that agricultural value added increases CO_2 emissions in both the short and long runs, whereas renewable energy consumption reduces emissions only in the short run; its long-term effect was found to be statistically insignificant. These mixed results indicate that the long-term impact of renewable energy and agricultural practices in Turkey should be examined more thoroughly.

3. Research Methodology

In the studies that measure the validity of the EKC hypothesis, it is seen that different variables such as income level, foreign direct investments, energy consumption, commercial deficit, urbanization, renewable energy, and industrialization level are included in the analysis. In this study, income, agricultural added value, and renewable energy consumption were used as variables affecting carbon emission. Only the share of agricultural value added in the GDP was used to include, in the analysis, the relative weight of the economy, not the absolute level of agricultural production. This variable reflects the structural composition of the economy rather than the direct scale effect and is therefore suitable for analyzing the relative size of environmental impacts. Although alternative proxies such as direct agricultural emissions or input–output-based environmental indicators could be used, these are often subject to limitations in consistency, comparability, and long-term availability, especially in developing countries. In contrast, agricultural value added as a share of the GDP is a widely available and standardized macroeconomic indicator. This approach is particularly appropriate in the context of the EKC framework, which emphasizes relative sectoral dynamics over absolute outputs. Moreover, using AGRI as a share of the GDP aligns with previous studies [28,29] that employ this measure to explore the sector's contribution to environmental degradation in an aggregated economic context.

In this study, the impact of renewable energy consumption and agricultural value added on carbon emissions is examined within the scope of the Environmental Kuznets Curve (EKC) hypothesis. For this purpose, data from the period 1968–2022 pertaining to Turkey were analyzed. A total of 55 observations were considered. Various trials were conducted to determine the most suitable functional form for the model, and it was concluded that the double-logarithmic form is the most appropriate. The empirical model is expressed as follows:

$$\mathrm{lnCO_2} = \theta_0 + \theta_1 \mathrm{lnGDP_t} + \theta_2 \mathrm{lnGDP^2}_t + \theta_3 \mathrm{lnREW_t} + \theta_4 \mathrm{lnAGRI_t} \quad (1)$$

In the equation, CO_2 represents carbon emissions; GDP represents gross domestic product; REW represents renewable energy consumption; and AGRI represents agricultural value added. In addition, t denotes the time period, the parameters from θ_1 to θ_4 represent the long-term elasticity coefficients, and ln indicates the natural logarithmic transformation. As assumed in the EKC hypothesis, an inverted U-shaped relationship is expected between economic growth and carbon emissions in the model. The validity of the EKC hypothesis depends on the coefficient θ_1 being positive and the coefficient θ_2 being negative. This implies that as the gross domestic product increases, carbon emissions also increase, but after a certain turning point, carbon emissions begin to decrease. The relevant turning point is calculated using the formula $GDP^* = -\theta_1/2\theta_2$, and exp(GDP)* provides the monetary value of the turning point [42] (p. 7743). However, in developing countries, it is anticipated that environmental pressures continue to increase alongside economic growth, and the calculated turning point has not yet been reached [43] (p. 446).

The sign of the coefficient θ_3 is expected to be positive, while the sign of θ_4 may be positive or negative depending on agricultural practices in the country. Comprehensive explanations of the variables in the model are presented in Table 1.

Table 1. Data definitions.

The Symbol	Variable Name	Definition	Source
CO_2	Emissions of carbon	Per capita (metric tons)	Global Footprint Network [44]
GDP	Gross domestic product	Per capita (constant 2015 USD)	World Bank (WDI) [45]
REW	Renewable energy consumption	Per capita (kWh)	Our World in Data [46]
AGRI	Agricultural added value	At constant prices (percentage of GDP)	World Bank (WDI) [45]

3.1. Stationary Test

In this study, the ARDL bounds testing approach was used to determine the impact of economic growth, renewable energy consumption, and agricultural value added on carbon emissions. In the ARDL bounds testing approach, it is first necessary to test the stationarity of the variables. In this approach, the dependent variable must become stationary at the first difference, while the independent variables must be stationary at level or at the first difference. Unit root tests are used to test the stationarity of the variables. In this study, the Augmented Dickey–Fuller (ADF) and Phillips–Perron (PP) unit root tests, which are frequently used in the literature, were applied. The ADF approach is conducted by adding a sufficient number of lagged values of the dependent variable to the regression equation in order to eliminate autocorrelation, assuming that the error terms exhibit autocorrelation [47]. In the ADF test, it is assumed that the distribution of the error term is independent and homoscedastic [48] (p. 10). Unlike this approach, the PP test allows for a non-parametric structure with weak dependence and heteroscedastic distribution of the error term [49]. In both approaches, the null hypothesis suggests that there is a unit root (non-stationary) [50] (p. 264).

3.2. ARDL Bounds Testing Approach

In time-series analyses, there are different approaches for examining the cointegration relationship among non-stationary series. The ARDL approach developed by Pesaran and Shin is a method that can be applied to series that become stationary at level and first difference [51]. In this approach, short- and long-run relationships between the dependent and independent variables can be estimated simultaneously, addressing problems due to omitted variables and autocorrelation [52]. Moreover, it enables reliable results in studies with small samples [53]. In this study, the ARDL bounds testing approach, which offers several advantages due to its greater flexibility and practicality compared to other methods, has been employed. The ARDL bounds testing approach consists of three stages. First, an unrestricted error correction model (UECM) is established to estimate short- and long-run parameters. The model in which the dependent variable used in the analysis is included is presented in the following equation [54,55]:

$$\begin{aligned}\triangle \mathrm{lnCO}_{2t} = \alpha_0 + & \sum_{i=1}^{n} \alpha_{i1} \triangle \mathrm{lnCO}_{2t-i} + \sum_{k=0}^{n} \alpha_{1k} \triangle \mathrm{lnGDP}_{t-k} \\ & + \sum_{k=0}^{n} \alpha_{2k} \triangle \mathrm{lnGDP}^2{}_{t-k} \\ & + \sum_{k=0}^{n} \alpha_{3k} \triangle \mathrm{lnREW}_{t-k} \\ & + \sum_{k=0}^{n} \alpha_{4k} \triangle \mathrm{lnAGRI}_{t-k} + \lambda_1 \mathrm{lnCO}_{2t-1} + \lambda_2 \mathrm{lnGDP}_{t-1} \\ & + \lambda_3 \mathrm{lnGDP}^2{}_{t-1} + \lambda_4 \mathrm{lnREW}_{t-1} + \lambda_5 \mathrm{lnAGRI}_{t-1} + e_{1t}\end{aligned} \quad (2)$$

In Equation (2), $\triangle$ denotes the difference operator, α_0 the constant term, e_1t the error term, and $\alpha_{1\text{-}5}$ and $\lambda_{1\text{-}5}$ the coefficients. n shows the appropriate lag length. The cointegration relationship is tested using the F-test. The null hypothesis ($\lambda_1 = \lambda_2 = \lambda_3 = \lambda_4 = \lambda_5 = 0$) states that there is no cointegration. If the F-statistic is lower than the critical bounds, the null is accepted; if higher, it is rejected; if in between, no conclusion can be drawn [53]. If cointegration exists, lag lengths are determined using the AIC or SIC, and the model is estimated via OLS [52,54]. In the second stage, long-run coefficients are estimated [52]:

$$\begin{aligned}\mathrm{lnCO}_2 = \alpha_0 + \sum_{i=1}^{n1} & \alpha_{1i} \triangle \mathrm{lnCO}_{2t-i} \\ & + \sum_{k=0}^{n2} \alpha_{2k} \triangle \mathrm{lnGDP}_{t-k} \\ & + \sum_{k=0}^{n3} \alpha_{3k} \triangle \mathrm{lnGDP}^2{}_{t-k} + \sum_{k=0}^{n4} \alpha_{4k} \triangle \mathrm{lnREW}_{t-k} \\ & + \sum_{k=0}^{n5} \alpha_{5k} \triangle \mathrm{lnAGRI}_{t-k} + e_t\end{aligned} \quad (3)$$

The long-run coefficients help identify the direction and magnitude of the effect between variables [56,57]. The third stage involves estimating the short-run coefficients through the error correction model [52]:

$$\begin{aligned}\triangle lnCO_2 = \alpha_0 + & \sum_{i=1}^{n1} \alpha_{1i} \triangle lnCO_{2t-i} \\ & + \sum_{k=0}^{n2} \alpha_{2k} \triangle lnGDP_{t-k} \\ & + \sum_{k=0}^{n3} \alpha_{3k} \triangle GDP^2{}_{t-k} + \sum_{k=0}^{n4} \alpha_{4k} \triangle lnREW_{t-k} \\ & + \sum_{k=0}^{n5} \alpha_{5k} \triangle lnAGRI_{t-k} + \Psi COINTEQ_{t-k} + e_t\end{aligned} \quad (4)$$

In this equation, $\triangle$ is the first-difference operator, α_0 the constant, and e_t the error term. The coefficients $\alpha_{1\text{-}5}$ refer to the short-term dynamics. COINTEQ denotes the error correction term, and Ψ represents the speed of adjustment. Ψ is expected to be negative and significant. If between 0 and -1, shocks converge monotonically; between -1 and -2, convergence occurs with decreasing fluctuations; values beyond -2 imply divergence [58,59].

3.3. Robustness Check

In this study, the FMOLS (Fully Modified Ordinary Least Squares) and DOLS (Dynamic Ordinary Least Squares) methods were employed to assess the reliability of the results obtained from the ARDL model. Various studies in the literature have suggested that the FMOLS and DOLS approaches are appropriate for evaluating the robustness of long-run ARDL estimates [60]. Developed by [61], FMOLS is a semi-parametric regression method designed to produce consistent estimates in time series that are integrated of order one (I(1)) and cointegrated. This method corrects for common issues in time-series data, such as autocorrelation and endogeneity, thus providing more reliable results. Compared to the conventional OLS method, FMOLS yields more consistent estimates and facilitates inference due to its asymptotic distribution properties. Therefore, it is widely used in multivariate cointegration analyses [61]. The DOLS method, on the other hand, is a cointegration estimation technique that provides reliable inferences despite uncertainties regarding the integration order of the series and stands out for its asymptotic efficiency [62]. In other words, DOLS addresses the asymptotic inefficiency of OLS by incorporating short-run dynamics into the model when estimating long-run cointegration relationships. As a result, the distribution of parameter estimates becomes more concentrated around the true values, allowing for more reliable outcomes [63]. Additionally, DOLS effectively handles small-sample bias, autocorrelation, and deviations from normality [60].

4. Results

In the ARDL model, the descriptive statistics of the variables included in the analysis are presented in Table 2. The variable with the highest mean value is GDP per capita followed, respectively, by renewable energy consumption, agricultural value added, and carbon emissions. It was observed that the variable with the highest volatility is renewable energy consumption, while the variable with the lowest volatility is GDP per capita. According to the Jarque–Bera statistics, all variables were found to exhibit a normal distribution.

Table 2. Descriptive statistics.

	lnCO$_2$	**lnGDP**	**lnREW**	**lnAGRI**
Mean	0.106258	8.712519	7.089166	2.657173
Median	0.158690	8.655083	7.289702	2.724388
Maximum	0.702991	9.550741	8.359624	3.727795
Minimum	−0.892514	8.069710	5.388940	1.710866
Std. Dev.	0.445196	0.423007	0.783811	0.632076
Skewness	−0.392169	0.350538	−0.576747	0.173185
Kurtosis	2.112708	2.005981	2.631408	1.710175
Jarque–Bera	3.214005	3.390710	3.360517	4.087464
Jarque–Bera probability	0.200488	0.183534	0.186326	0.129544
Observations	55	55	55	55

In the ARDL model, the stationarity of the variables included in the analysis was first tested. For this purpose, the ADF and PP unit root tests, which are widely used in the literature, were applied. In both unit root tests, differencing was applied to the non-stationary series, and the differencing process continued until stationarity was achieved. According to the results obtained from the unit root tests, all variables were found to be non-stationary at level but stationary at the first difference (Table 3). These results indicate that the variables do not become stationary at the second difference, meaning that the ARDL model is applicable.

Table 3. ADF and PP unit root results.

Model	Variables	ADF	PP	ADF	PP	Decision
		I (0)		I (1)		
Fixed Model	$lnCO_2$	−2.13 [0] (0.23)	−0.23 [1] (0.19)	−8.14 [0] (0.00) *	−8.14 [2] (0.00) *	For I (1) H_0 refuse
	lnGDP ($lnGDP^2$)	0.89 [0] (0.99)	1.15 [4] (0.99)	−6.98 [0] (0.00) *	−6.98 [3] (0.00) *	
	lnREW	−1.13 [0] (0.69)	−0.94 [5] (0.76)	−8.27 [0] (0.00) *	−8.74 [5] (0.00) *	
	lnAGRI	−1.20 [0] (0.66)	−1.40 [7] (0.57)	−7.28 [0] (0,00) *	−7.46 [6] (0.00) *	
Fixed and Trendy Model	$lnCO_2$	−3.12 [0] (0.11)	−3.06 [2] (0.12)	−8.44 [0] (0.00) *	−8.44 [1] (0.00) *	
	lnGDP ($lnGDP^2$)	−1.66 [0] (0.75)	−1.74 [1] (0.71)	−7.09 [0] (0.00) *	−7.14 [4] (0.00) *	
	lnREW	−2.71 [0] (0.23)	−2.71 [0] (0.23)	−8.20 [0] (0.00) *	−8.67 [5] (0.00) *	
	lnAGRI	−2.37 [0] (0.38)	−2.37 [0] (0.38)	−7.31 [0] (0.00) *	−7.74 [7] (0.00) *	

Note: The values in the table represent the test statistics, [ ] the maximum lag length and bandwidth, and ( ) the probability value of the test statistic. *, indicates significance at the 1% level.

Following the stationarity tests of the variables, the model estimation was carried out. Before estimating the model, it is necessary to determine the appropriate lag lengths of the variables by taking into account the maximum lag length. In cases where the variables in the ARDL model consist of annual series, it is suggested that the maximum lag length be taken as two [52,55]. Accordingly, the maximum lag length was set to two, and the Akaike Information Criterion (AIC) was taken into consideration to determine the optimal lag length of the variables in the model. As a result of the evaluations, it was concluded that the most appropriate model is ARDL (1, 2, 1, 2, 2). In the ARDL model, the F bounds test was applied to examine the cointegration relationship between the variables. According to the findings obtained from the bounds test, the F-statistic value was found to be higher than the lower and upper bounds of the critical values in the table, indicating the existence of a cointegration relationship between the variables at the 1% significance level (Table 4).

Table 4. Bounds testing for long-run relationship.

Test Statistic	Value	Signif.	I (0)	I (1)
F-statistic	7.581204	10%	2.345	3.28
K	4	5%	2.763	3.813
		1%	3.738	4.947

Note: Case II restricted intercept and no trend model was used, and the critical values were taken from [20].

Following the determination of the cointegration relationship among the variables, the long-run coefficients of the model were estimated and are presented in Table 5. The estimation results of the model revealed that the GDP per capita, its squared term, and renewable energy consumption were statistically significant at the 1% significance level. According to the long-run estimation results, a 1% increase in the GDP per capita increases carbon emissions by approximately 19%, while renewable energy consumption reduces carbon emissions by 0.16%. The findings obtained are largely consistent with theoretical expectations. Indeed, renewable energy is considered one of the important and sustainable solutions to environmental problems caused by fossil fuels. The finding of this study that renewable energy consumption reduces CO_2 emissions in the long run is consistent with the results reported in [39,64]. In addition, the study in [37] also identifies renewable energy as a causal determinant of CO_2 emissions. On the other hand, the findings reported by [40] contradict the current results by stating that the long-run effect of renewable energy on the environment is statistically insignificant.

Table 5. Long-run coefficients based on ARDL model.

Variable	Coefficient	Std. Error	t-Statistic	Prob.
lnGDP	19.206	1.711	11.221	0.000 *
$lnGDP^2$	−1.011	0.091	−11.097	0.000 *
lnREW	−0.163	0.034	−4.807	0.000 *
lnAGRI	0.153	0.075	2.038	0.048 **
C	−89.533	8.011	−11.175	0.000 *
Turning point	$13,396			
Diagnostic Tests	**Test Statistic**		**p Values**	
BG-LM	1.108		0.340	
BPG	1.197		0.318	
Ramsey reset	0.011		0.990	
Jarque–Bera	0.264		0.876	
F-statistic	445.548		0.000 *	
Durbin–Watson	1.923			
R^2	0.992			
Adjusted R^2	0.990			

* and ** indicate that the coefficients are statistically significant at the 1% and 5% significance levels, respectively.

In addition, a 1% increase in agricultural value added was found to increase carbon emissions by 0.15%, and the coefficient of this variable was found to be statistically significant at the 5% level. This finding can be explained by the increasing dependence of the agricultural sector in Turkey on carbon-intensive production inputs such as mechanization, the use of chemical fertilizers and pesticides, and fossil fuel-based irrigation systems in the long term. In particular, the transition to modern agricultural techniques increases agricultural productivity while also leading to higher energy consumption and carbon emissions. This situation indicates that the growth of the agricultural sector may increase environmental costs in the long run. The findings obtained from this study are consistent with the results reported in [39,41] within the Turkish context but differ from those reported in [37,38].

In the long-run analysis of the ARDL model for Turkey, when the signs of the GDP per capita variables are examined, it is observed that the GDP per capita has a positive coefficient, while its squared term has a negative coefficient. This indicates that the EKC hypothesis is valid in the long run. Accordingly, it is predicted that an increase in the GDP per capita will reduce carbon emissions after reaching a certain threshold. This study's findings are generally consistent with those of [40]. Both studies confirm the validity of the EKC hypothesis for Turkey and show that agricultural value added increases CO_2 emissions. While this study finds a long-term emission-reducing effect of renewable energy consumption, [40] reports a significant reduction only in the short term and states that the effect is not statistically significant in the long term.

According to the model estimation results, the turning point predicted by the EKC hypothesis is estimated to occur when per capita income reaches approximately USD 13,396. Based on World Bank data, Turkey's real per capita income for the year 2022 was reported as USD 14,713 [45]. This suggests that the transition from increasing to decreasing environmental pressure may have already occurred in Turkey or is likely to occur in the very near future. The finding is largely consistent with various studies in the literature. For instance, in studies focusing on Turkey, [65] reported the EKC turning point at USD 9920, while [42] estimated it to be in the range of USD 13,523–14,077. Additionally, the study in [41], which used ecological footprint as the dependent variable, the turning point was calculated as USD 8184, and it was noted that environmental degradation began to decline during the analysis period. In this context, it is concluded that environmental policies in Turkey should be maintained decisively, and a balance between economic growth and environmental sustainability must be ensured.

In the estimated ARDL model, the BG-LM autocorrelation test and the Breusch–Pagan–Godfrey (BPG) heteroscedasticity test were applied. The test results indicated that there were no issues of autocorrelation or heteroscedasticity. Furthermore, it was determined that the Durbin–Watson statistic had a value close to two, confirming the absence of autocorrelation. The Jarque–Bera test revealed that the error terms were normally distributed. The Ramsey RESET test statistics also indicated that the model has the correct functional form. According to the R^2 values obtained from the model, it was determined that the model has high explanatory power. The probability value of the F-statistic showed that all independent variables in the model significantly affect the dependent variable.

Following the estimation of the long-run coefficients, an error correction model (ECM) based on the ARDL model was established within the scope of the third step in order to estimate the short-run coefficients. As seen in Table 6, all variables included in the model were found to be statistically significant at different significance levels. The results obtained from the error correction model show that the GDP per capita in the current period has a positive effect on carbon emissions, while the lagged GDP per capita has a reducing effect on carbon emissions. On the other hand, it was determined that the squared term of the GDP per capita in the current period has a negative effect on carbon emissions. This indicates that the EKC hypothesis is valid in the short term.

Table 6. Error correction for the selected ARDL model.

Variable	Coefficient	Std. Error	t-Statistic	*p* Values
D (lnGDP)	6.132	2.751	2.229	0.031 **
D (lnGDP (−1))	−0.395	0.138	−2.852	0.006 *
D ($lnGDP^2$)	−0.268	0.155	−1.726	0.091 ***
D (lnREW)	−0.110	0.029	−3.743	0.000 *
D (lnREW (−1))	0.083	0.029	2.887	0.006 *
D (lnAGRI)	0.179	0.071	2.491	0.017 **
D (lnAGRI (−1))	−0.152	0.069	−2.194	0.034 **
CointEq (−1) *	−0.752	0.105	−7.153	0.000 *
R-squared		0.781		
Adjusted R-squared		0.747		
Log likelihood		102.563		
Durbin–Watson stat		1.923		

*, **, and *** indicate that the coefficients of the variables are statistically significant at the significance levels of 1%, 5%, and 10%, respectively.

In addition, while carbon emissions increase with the lagged renewable energy consumption, they decrease with current renewable energy consumption. This finding suggests that renewable energy consumption may have temporary negative effects in the short term and may cause transition costs. It was found that increases in agricultural value added in the current period increase carbon emissions, whereas increases in the previous period have a reducing effect. This suggests that due to the nature of agricultural activities, they may trigger both short-term emission-generating and delayed emission-reducing processes. In particular, intensive agricultural activities increase emissions in the short term through energy use and input dependency; however, in the following periods, they may reduce emissions through mechanisms such as decreased activity volume after harvest, an increase in biomass, and carbon interaction with the soil.

The coefficient of the error correction term was found to be statistically significant at the 1% level. The fact that the coefficient lies between −1 and 0 indicates that short-term shocks can be eliminated in a monotonic manner. According to the value obtained, approximately 75% of the imbalances that occurred in the previous period can be corrected within one year; in other words, it is concluded that the short-term shocks can fully return to the long-term equilibrium level in about 1.3 years.

CUSUM and CUSUMQ tests, as proposed by [66], were applied to determine whether the long- and short-term coefficients obtained from the ARDL model were stable over time. In the CUSUM and CUSUMQ tests, the cumulative sums of error terms are iteratively examined on the graph, depending on time. In this way, it can be determined whether the structural breaks that may occur in the coefficients of the model and the predictive power of the model maintain their validity in different periods [66]. If the graphs obtained under the CUSUM and CUSUMQ tests lie within the critical value zone of 5% significance, it appears that the coefficients in the model are stationary and have no structural fracture [67]. The results of both tests in Figure 1 show that the null hypothesis is accepted, the graph is among the critical values, and the coefficients are stable.

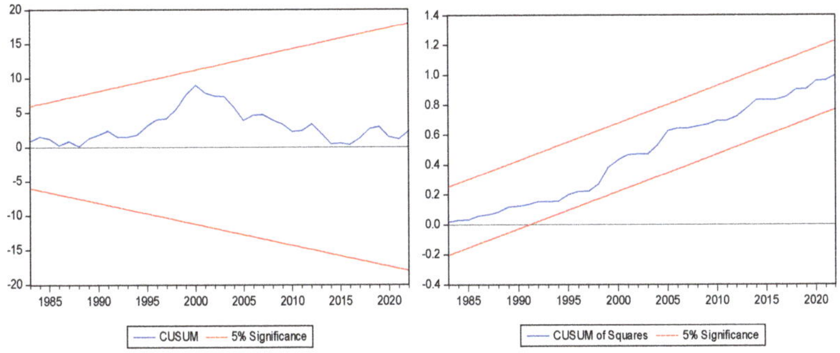

Figure 1. CUSUM and CUSUMQ tests results.

The robust model was estimated for the validity of the results obtained from the ARDL model. For this purpose, the FMOLS and DOLS methods were used, and the results obtained are presented in Table 7. The differences in the values of ARDL, FMOLS, and DOLS coefficients are due to the structural characteristics of the methods. The results from the FMOLS and DOLS regressions have been determined to be largely consistent with the results from the ARDL long-term model. As the regression results reveal, the EKC hypothesis is valid for Turkey. It has been found that renewable energy reduces carbon emissions, while agricultural added value has the effect of increasing carbon emissions. On the other hand, the turning point in per capita income predicted by the EKC hypothesis appears to be relatively consistent with the ARDL long-run results. The turning point is estimated at USD 13,583 in the FMOLS model and USD 13,687 in the DOLS model. Considering that Turkey's real per capita income is USD 14,713 [45], these values indicate that Turkey is either very close to or has just surpassed the turning point at which environmental pressures begin to decline. The results support the EKC hypothesis, which suggests that environmental degradation may decrease as income increases, highlighting the necessity of aligning economic growth with sustainability principles.

Table 7. FMOLS and DOLS results for robustness check.

FMOLS	Coefficient	Std. Error	t-Statistic	Prob.
lnGDP	19.183	1.235	15.521	0.000 *
$lnGDP^2$	−1.007	0.066	−15.235	0.000 *
lnREW	−0.137	0.024	−5.634	0.000 *
lnAGRI	0.173	0.050	3.453	0.001 *
C	−89.876	5.746	−15.639	0.000 *
Turning point	$13,853			
R-squared		0.988		
Adjusted R-squared		0.987		
DOLS	**Coefficient**	**Std. Error**	**t-Statistic**	**Prob.**
lnGDP	20.259	1.501	13.492	0.000 *
$lnGDP^2$	−1.064	0.079	−13.331	0.000 *
lnREW	−0.178	0.027	−6.486	0.000 *
lnAGRI	0.176	0.066	2.650	0.012 **
C	−94.594	7.045	−13.425	0.000 *
Turning point	$13,687			
R-squared		0.993		
Adjusted R-squared		0.990		

* and ** indicate that the coefficients for variables are significant at the significance levels of 1% and 5%, respectively.

5. Discussion

Differences in national production and consumption structures play a critical role in identifying sectoral sources of greenhouse gas emissions and in assessing the effectiveness of country-specific policy instruments. In emerging economies such as Turkey, sectoral analysis enables the formulation of targeted policy proposals for environmental sustainability. To empirically assess these sectoral dynamics in the context of Turkey, this study analyzes the validity of the EKC hypothesis over the period 1968–2022. In addition, it examines the impact of renewable energy consumption and agricultural value added on carbon emissions using the ARDL bounds testing approach. To ensure robustness, the results were also tested using the FMOLS and DOLS estimation methods. The empirical findings confirm a long-run cointegration relationship among carbon emissions, GDP per capita, renewable energy consumption, and agricultural value added.

In the long run, renewable energy consumption reduces emissions, while growth in agricultural value added contributes to increased emissions. The EKC hypothesis is found to be valid for Turkey in both the short and long runs. Short-run results indicate that the effects of renewable energy and agricultural value added on emissions vary over time, which may be explained by the costs of energy transition and agriculture's ongoing dependence on fossil fuels and chemical inputs. The error correction model shows that about 74% of short-term deviations are corrected annually, indicating a convergence to long-run equilibrium within approximately 1.3 years. Similarly, recent studies conducted in the European Union suggest that the impact of renewable energy on sustainability may be negative at lower levels and positive at higher levels of adoption, indicating that such effects can vary depending on scale and context [68].

The emission-increasing effect of agricultural value added is consistent with previous studies such as [29–31]. On the other hand, studies that report different results, such as [69], show that crop production and fisheries contribute to reducing CO_2 emissions, while the effect of livestock is statistically insignificant. Similarly, the long-term emission-reducing impact of renewable energy consumption reflects a widely observed trend in the literature [69–71]. However, divergent findings reported by studies such as [20,21] suggest that results in this field may vary depending on country-specific conditions and contextual factors. For instance, ref. [72] reported a negative effect of renewable energy consumption, suggesting that its current use in the agricultural sector may not be fully efficient or adequately integrated into productive activities. In line with the present study's findings, ref. [73] also reported that renewable energy consumption reduces the ecological footprint in Denmark, whereas value added from agriculture, forestry, and fisheries—as well as economic growth—increases environmental pressure.

According to the analysis findings, since Turkey has either reached or is very close to the EKC turning point, it is of great importance to maintain sustainable environmental policies with determination. The results from the study suggest the need to maintain a delicate balance between economic growth and environmental sustainability and emphasize the importance of supporting renewable energy policies with stronger incentives, particularly in Turkey, where reducing dependence on foreign fossil fuel sources could significantly benefit national economic growth.

In the case of Turkey, an increase in the real share of agricultural added value in the GDP, coupled with the low technology density of the production structure and traditional forms of production that create environmental pressure, leads to an increase in emissions. This shows that not only the volume of production, but also the mode of production and the technologies used, are decisive in terms of environmental impacts. In this context, it is necessary to accelerate structural transformation in the agricultural sector, adopt low-impact technologies, and activate financial and economic incentives to increase the use of clean energy. The adoption of technologies that enhance productivity—such as precision farming, low-emission equipment, and smart irrigation—can help alleviate the environmental burden of agricultural activities and should be accelerated through targeted subsidies and technical support.

Environmentally friendly and sustainable farming practices offer a promising path toward reducing emissions from the agricultural sector, particularly when supported by national policies aimed at increasing the sector's added value. In this respect, strategies such as improved livestock management, agroforestry, forest conservation, reforestation, sustainable fishing practices, and the adoption of energy-efficient technologies may be prioritized [21,74].

The use of renewable energy sources, which can reduce environmental impacts while lowering energy costs in agricultural production processes, should be expanded. Supporting energy efficiency and renewable energy investments will contribute to reducing environmental pressures while enabling Turkey to sustain its economic growth in a more environmentally sustainable manner. To this end, the adoption of renewable energy should be encouraged through various instruments such as green taxes, emissions trading schemes, R&D support, and efforts to increase public awareness. The steps taken towards a carbon credit trading platform in Turkey are promising in terms of developments in this field [75]. Establishing the necessary infrastructure for monitoring and reporting emissions in the agricultural sector and including this sector within the scope of the Emissions Trading System (ETS) may contribute to the reduction of carbon emissions. In addition to supporting the agricultural sector by providing additional income to farmers, such mechanisms can also contribute to the dissemination of sustainable practices such as reducing the use of chemical fertilizers. Existing agricultural subsidies should be restructured to discourage environmentally harmful practices and support sustainable production methods while complementing these reforms with green taxation instruments such as carbon taxes and border carbon adjustments to internalize environmental costs.

In order to reduce the restrictions encountered in the export of agricultural products to the EU, Turkey should enhance the compliance of its agricultural products with the EU's health and safety standards. As a matter of fact, in line with the European Green

Deal, important steps are being taken to integrate agricultural production in Turkey to fight climate change and meet emission reduction targets. In particular, limiting pesticide and fertilizer use, increasing organic farmland, and promoting agroecological methods are among the key strategies supporting both environmental sustainability and rural development [76].

6. Conclusions

In conclusion, the development of holistic and inclusive policies that aim to reduce environmental impacts while sustaining economic growth is of paramount importance. This study, based on national-level macroeconomic data, focuses on broad structural trends but does not account for micro-level diversity in agricultural practices or regional disparities in energy use. Additionally, certain structural factors—such as industrial policies, technological advancements, and urbanization—were excluded due to data limitations. Although this study offers sector-specific insights into the environmental impacts of agriculture and renewable energy, it does not account for institutional and policy-related uncertainties. As highlighted by [77], ESG-based sustainability uncertainties—such as those arising from fluctuating environmental, social, and governance conditions—can significantly influence both policy effectiveness and investment decisions.

Accordingly, it is recommended that future studies expand the analytical scope to include not only the renewable energy and agriculture sectors but also other emission-intensive sectors such as industry, transportation, and services. In addition, incorporating structural determinants—such as industrial policies, technological innovation, and energy consumption patterns—could enrich the explanatory capacity of future models. Evaluating the impact of market-based environmental instruments—such as carbon taxation, emissions trading schemes, and green finance—may also lead to more comprehensive findings. Micro-level analyses and institutional indicators could further support a multidimensional assessment of Turkey's emissions landscape.

In light of recent empirical studies, several promising directions emerge for future research, particularly regarding methodological diversity and variable selection in the analysis of the agriculture–energy–emissions nexus. Future studies aiming to deepen the Environmental Kuznets Curve (EKC) framework in the context of agriculture and renewable energy can be enriched through broader variable sets and methodological improvements. Firstly, the inclusion of green innovation indicators such as research and development (R&D) expenditures and the number of patents in the model may reveal the technological aspect of environmental improvements [78,79]. In addition, analyses can be conducted based on component indicators such as energy efficiency (e.g., energy intensity or energy productivity) and resource efficiency (e.g., productivity ratios based on DMC) [80]. On the regulatory dimension, it is possible to use indicators such as environmental governance quality and eco-efficiency scores [81]. Similarly, the effects of policy uncertainty on sustainability highlight the value of including indicators of institutional instability into the model [82]. From a financial perspective, green public expenditures and financial development indicators may help analyze how public investments and credit systems contribute to environmental goals [78,83]. Finally, the application of advanced econometric methods such as Fourier ARDL, NARDL, and wavelet coherence may allow for more effective identification of structural breaks, asymmetric effects, and time-varying relationships [73,79,84]. Such extensions will contribute to a more holistic and policy-oriented understanding of emission dynamics in the context of Turkey's evolving energy–agriculture structure.

Author Contributions: Conceptualization, N.K., Ö.E.K. and O.O.B.; Formal analysis, Ö.E.K., F.O.V., O.O.B., U.Ç., R.O.K., V.-A.V. and T.V.; Investigation, N.K., Ö.E.K., F.O.V., O.O.B., R.O.K. and T.V.;

Methodology, N.K., Ö.E.K., F.O.V., U.Ç. and T.V.; Resources, N.K., Ö.E.K., O.O.B., U.Ç. and V.-A.V.; Validation, N.K., O.O.B., U.Ç., R.O.K. and V.-A.V.; Writing—original draft, F.O.V. and U.Ç. All authors have read and agreed to the published version of the manuscript.

Funding: This research received no external funding.

Institutional Review Board Statement: Not applicable.

Informed Consent Statement: Not applicable.

Data Availability Statement: The data supporting the findings of this study are publicly available from the following sources: CO_2 emissions data were obtained from the Global Footprint Network (https://data.footprintnetwork.org/#/countryTrends?cn=223&type=BCpc,EFCpc, accessed on 10 January 2025), GDP and agricultural value-added data were retrieved from the World Bank World Development Indicators (https://databank.worldbank.org/source/world-development-indicators, accessed on 10 January 2025), and renewable energy consumption data were accessed from Our World in Data (https://ourworldindata.org/grapher/per-capita-renewables?tab=chart&country=~TUR, accessed on 10 January 2025). All datasets were accessed in January 2025. No new datasets were created nor were proprietary data used during this study.

Conflicts of Interest: The authors declare no conflicts of interest.

eferences

Grossman, G.M.; Krueger, A.B. Economic growth and the environment. *Q. J. Econ.* **1995**, *110*, 353–377. [CrossRef]

Virgens, N.d.; Silva, S.; Laranjeira, E. Applications of the circular economy to the second-hand textile and clothing market: The case of Humana in Portugal. *Int. J. Fash. Des. Technol. Educ.* **2023**, *16*, 214–223. [CrossRef]

Ahrens, D.; Benedikter, S.; Giessen, L. Rethinking synergies and trade-offs at the forest–sustainable development goals (SDGs) nexus—A systematic review. *Sustain. Dev.* **2025**. Early View. [CrossRef]

Emezirinwune, M.U.; Adejumobi, I.A.; Adebisi, O.I.; Akinboro, F.G. Synergizing hybrid renewable energy systems and sustainable agriculture for rural development in Nigeria. *Adv. Electr. Eng. Electron. Energy* **2024**, *6*, 100492. [CrossRef]

Paramati, S.R.; Sinha, A.; Dogan, E. The significance of renewable energy use for economic output and environmental protection: Evidence from the Next 11 developing economies. Environ. *Sci. Pollut. Res.* **2017**, *24*, 13546–13560. [CrossRef]

Panayotou, T. *Empirical Tests and Policy Analysis of Environmental Degradation at Different Stages of Economic Development*; International Labour Office Technology World Employment Programme Working Paper WP238; International Labour Office: Geneva, Switzerland, 1993.

Yıldırım, E.; Şükrüoğlu, D.; Aslan, A. Energy consumption and economic growth in the next 11 countries: The bootstrapped autoregressive metric causality approach. *Energy Econ.* **2014**, *44*, 14–21. [CrossRef]

Shahbaz, M.; Mahalik, M.K.; Shah, S.H.; Sato, J.R. Time-varying analysis of CO_2 emissions, energy consumption, and economic growth nexus: Statistical experience in next 11 countries. *Energy Policy* **2016**, *98*, 33–48. [CrossRef]

Inglesi-Lotz, R. The impact of renewable energy consumption to economic welfare: A panel data application. *Energy Econ.* **2014**, *53*, 58–63. [CrossRef]

10. Ministry of Commerce. Tarım Sektörü Raporu. 2023. Available online: https://www.ticaret.gov.tr/ihracat/sektorler/tarim-ve-gida (accessed on 19 February 2025).

11. Akca, H. Environmental Kuznets curve and financial development in Turkey: Evidence from augmented ARDL approach. Environ. *Sci. Pollut. Res.* **2021**, *28*, 69149–69159. [CrossRef]

12. Wang, Q.; Zhang, F.; Li, R. Revisiting the Environmental Kuznets Curve hypothesis in 208 countries: The roles of trade openness, human capital, renewable energy, and natural resource rent. *Environ. Res.* **2023**, *216*, 114637. [CrossRef]

13. Şerifoğlu, M.M.; Öge Güney, P. Is the environmental Kuznets curve (EKC) hypothesis still valid for OECD countries? A comprehensive analysis across multiple sources. *Qual. Quant.* **2025**, *59*, 547–573. [CrossRef]

14. Badeeb, R.A.; Lean, H.H.; Shahbaz, M. Are too many natural resources to blame for the shape of the environmental Kuznets curve in resource-based economies? *Resour. Policy* **2020**, *68*, 101694. [CrossRef]

15. Erdogan, S.; Okumus, I.; Guzel, A.E. Revisiting the environmental Kuznets curve hypothesis in OECD countries: The role of renewable, non-renewable energy, and oil prices. *Environ. Sci. Pollut. Res.* **2020**, *27*, 23655–23663. [CrossRef] [PubMed]

16. Jahanger, A.; Hossain, M.R.; Onwe, J.C.; Ogwu, S.O.; Awan, A.; Balsalobre-Lorente, D. Analyzing the N-shaped EKC among top nuclear energy generating nations: A novel dynamic common correlated effects approach. *Gondwana Res.* **2023**, *116*, 73–88. [CrossRef]

17. Biala, M.I.; Adeniyi, T.A. Economic growth and environmental sustainability: Assessing the validity of the environmental Kuznet curve in Nigeria. *J. Econ. Allied Res. (JEAR)* **2024**, *9*, 323–338. Available online: https://jearecons.com/index.php/jearecons article/view/480 (accessed on 10 January 2025).
18. Bölük, G.; Mert, M. Fossil & renewable energy consumption, GHGs (greenhouse gases) and economic growth: Evidence from a panel of EU countries. *Energy* **2014**, *74*, 439–446. [CrossRef]
19. Bekun, F.V.; Gyamfi, B.A.; Onifade, S.T.; Agboola, M.O. Beyond the environmental Kuznets curve in E7 economies: Accountin for the combined impacts of institutional quality and renewables. *J. Clean. Prod.* **2021**, *314*, 127924. [CrossRef]
20. Aguir Bargaoui, S. The impact of energy efficiency and renewable energies on environmental quality in OECD countries. *J. Know Econ.* **2022**, *13*, 3424–3444. [CrossRef]
21. Shahbaz, M.; Aydin, M.; Degirmenci, T.; Bozatli, O. The role of agriculture-forestry-fisheries sectors, biodiversity expenditures and renewable energy on environmental quality for The Netherlands: Evidence from novel Fourier ARDL. *Nat. Resour. Forum* **2024**, *2024*, 1–20. [CrossRef]
22. Voumik, L.C.; Hossain, M.I.; Rahman, M.H.; Sultana, R.; Dey, R.; Esquivias, M.A. Impact of Renewable and Non-Renewabl Energy on EKC in SAARC Countries: Augmented Mean Group Approach. *Energies* **2023**, *16*, 2789. [CrossRef]
23. Hayaloğlu, P. İklim değişikliğinin tarım sektörü ve ekonomik büyüme üzerindeki etkileri. *Gümüşhane Üniversitesi Sos. Bilim. Enstitüsü Elektron. Derg.* **2018**, *9*, 51–62. Available online: https://dergipark.org.tr/en/pub/gumus/issue/42294/45240 (accessed on 10 January 2025).
24. Afzal, M.; Ahmed, T.; Ahmed, G. Empirical Assessment of Climate Change on Major Agricultural Crops of Punjab, Pakista (MPRA Paper No. 70958). Munich Personal RePEc Archive. 2016. Available online: https://mpra.ub.uni-muenchen.de/70958 (accessed on 11 April 2025).
25. Başoğlu, A.; Telatar, O.M. İklim değişikliğinin etkileri: Tarım sektörü üzerine ekonometrik bir uygulama. *Karadeniz Teknik Üniversitesi Sos. Bilim. Enstitüsü Sos. Bilim. Derg.* **2013**, *3*, 7–25. Available online: https://web.archive.org/web/2018050811002 id_/http://www.ktu.edu.tr/dosyalar/sbedergisi_e88c5.pdf (accessed on 10 January 2025).
26. Belloumi, M. *Investigating the Impact of Climate Change on Agricultural Production in Eastern and Southern African Countries* AGRODEP Working Paper 0003; International Food Policy Research Institute (IFPRI): New Delhi, India, 2014; Available online https://cgspace.cgiar.org/items/1f266be5-21a1-4c5b-a272-9102548c43e0 (accessed on 24 February 2025).
27. Masud, M.M.; Rahman, M.S.; Al-Amin, A.Q.; Kari, F.; Filho, W.L. Impact of climate change: An empirical investigation o Malaysian rice production. *Mitig. Adapt. Strateg. Glob. Change* **2014**, *19*, 431–444. [CrossRef]
28. Raihan, A. An econometric evaluation of the effects of economic growth, energy use, and agricultural value added on carbo dioxide emissions in Vietnam. *Asia-Pac. J. Reg. Sci.* **2023**, *7*, 665–696. [CrossRef]
29. Pata, U.K. Linking renewable energy, globalization, agriculture, CO_2 emissions and ecological footprint in BRIC countries: A sustainability perspective. *Renew. Energy* **2021**, *173*, 197–208. [CrossRef]
30. Eştürk, Ö.; Aydın, F.F.; Levent, C. Tarım ve sanayi sektörlerinin çevre kirliliği üzerindeki etkisi: Seçilmiş OECD üyesi ülkelerde ekonometrik bir uygulama. *Ardahan Üniversitesi İktisadi Ve İdari Bilim. Fakültesi Derg.* **2023**, *5*, 109–116. [CrossRef]
31. Khurshid, N.; Khurshid, J.; Shakoor, U.; Ali, K. Asymmetric effect of agriculture value added on CO_2 emission: Does globalizatior and energy consumption matter for Pakistan. *Front. Energy Res.* **2022**, *10*, 1053234. [CrossRef]
32. Raihan, A.; Tuspekova, A. Dynamic impacts of economic growth, energy use, urbanization, tourism, agricultural value-added and forested area on carbon dioxide emissions in Brazil. *J. Environ. Stud. Sci.* **2022**, *112*, 794–814. [CrossRef]
33. Ahmed, S.; Abdi, A.H.; Sodal, M.; Yusuf, O.A.; Mohamud, M.H. Assessing the energy-economy-environment nexus in Somalia The impact of agricultural value added on CO_2 emissions. *Int. J. Energy Econ. Policy* **2025**, *15*, 221–232. [CrossRef]
34. Agboola, M.O.; Bekun, F.V. Does agricultural value-added induce environmental degradation? Empirical evidence from ar agrarian country. *Environ. Sci. Pollut. Res.* **2019**, *26*, 24733–24743. [CrossRef]
35. Ali, S.; Ying, L.; Shah, T.; Tariq, A.; Ali Chandio, A.; Ali, I. Analysis of the nexus of CO_2 emissions, economic growth, land under cereal crops and agriculture value-added in Pakistan using an ARDL approach. *Energies* **2019**, *12*, 4590. [CrossRef]
36. Wang, Y.; Ali, A.; Chen, Z. Dynamic relationships between environment-related technologies, agricultural value added, transpor infrastructure and environmental emissions in the five most populous countries. *Sci. Rep.* **2025**, *15*, 2308. [CrossRef] [PubMed]
37. Çetin, M.; Saygın, S.; Demir, H. Tarım sektörünün çevre kirliliği üzerindeki etkisi: Türkiye ekonomisi için bir eşbütünleşme ve nedensellik analizi. *Tekirdağ Ziraat Fak. Derg.* **2020**, *17*, 329–345. [CrossRef]
38. Baş, T.; Kara, F.; Alola, A.A. The environmental aspects of agriculture, merchandize, share, and export value-added calibrations in Turkey. *Environ. Sci. Pollut. Res.* **2021**, *28*, 62677–62689. [CrossRef]
39. Oğul, B. Tarım sektöründeki gelişmeler çevresel kirliliği etkiliyor mu? Türkiye üzerine ampirik bulgular. *MANAS Sos. Araştırmalar Derg.* **2023**, *12*, 1016–1026. [CrossRef]
40. Okumuş, İ. Türkiye'de yenilenebilir enerji tüketimi, tarım ve CO_2 emisyonu ilişkisi. *Eskişehir Osman. Üniversitesi İktisadi Ve İdari Bilim. Derg.* **2020**, *15*, 21–34. [CrossRef]

Bölük, G.; Karaman, S. The impact of agriculture, energy consumption and economic growth on ecological footprint: Testing the agriculture-induced EKC for Türkiye. *Environ. Dev. Sustain.* **2024**, *26*, 31817–31837. [CrossRef]

Pata, U.K. Renewable energy consumption, urbanization, financial development, income and CO_2 emissions in Turkey: Testing EKC hypothesis with structural breaks. *J. Clean. Prod.* **2018**, *187*, 770–779. [CrossRef]

Dinda, S. Environmental Kuznets curve hypothesis: A survey. *Ecol. Econ.* **2004**, *49*, 431–455. [CrossRef]

Global Footprint Network. National Footprint Accounts. Available online: https://data.footprintnetwork.org/#/countryTrends?cn=223&type=BCpc,EFCpc (accessed on 10 January 2025).

World Bank. *World Development Indicators Online Database*; World Bank Group: Washington, DC, USA, 2025; Available online: https://databank.worldbank.org/source/world-development-indicators (accessed on 10 January 2025).

Our World in Data. Per Capita Energy Consumption from Renewables, 1965 to 2023. 2025. Available online: https://ourworldindata.org/grapher/per-capita-renewables?tab=chart&country=~TUR (accessed on 10 January 2025).

Gujarati, D.N.; Porter, D.C. *Basic Econometrics*, 5th ed.; McGraw-Hill: New York, NY, USA, 2009.

Mushtaq, R.; SSRN. Augmented Dickey Fuller Test. 2011. Available online: https://ssrn.com/abstract=1911068 (accessed on 15 March 2025).

Phillips, P.C.; Perron, P. Testing for a unit root in time series regression. *Biometrika* **1988**, *75*, 335–346. [CrossRef]

Pata, U.K.; Yurtkuran, S.; Kalça, A. Türkiye'de enerji tüketimi ve ekonomik büyüme: ARDL sınır testi yaklaşımı. *Marmara Univ. J. Econ. Adm. Sci.* **2016**, *38*, 255–271. [CrossRef]

Pesaran, M.H.; Shin, Y. *Autoregressive Distributed Lag Modelling Approach to Cointegration Analysis*; Department of Economics Working Paper Series, No. 9514; University of Cambridge: Cambridge, UK, 1995.

Narayan, P.K. Fiji's tourism demand: The ARDL approach to cointegration. *Tour. Econ.* **2004**, *10*, 193–206. [CrossRef]

Gençoğlu, P.; Kuşkaya, S. Türkiye'de sağlığın eğitim üzerindeki etkileri: ARDL sınır testi yöntemi ile bir değerlendirilme. *İşletme ve İktisat Çalışmaları Dergisi* **2017**, *5*, 1–11. Available online: https://dergipark.org.tr/en/pub/iicder/issue/49867/639268 (accessed on 10 January 2025).

Narayan, P.K. The saving and investment nexus for China: Evidence from cointegration tests. *Appl. Econ.* **2005**, *37*, 1979–1990. [CrossRef]

Narayan, P.K.; Smyth, R. What determines migration flows from low income to high-income countries? An empirical investigation of Fiji-US migration 1972–2001. *Contemp. Econ. Policy* **2006**, *24*, 332–342. [CrossRef]

Sünbül, E. Dış ticaret hacmi ve döviz kuru ilişkisinin ARDL sınır testi ile analizi (Türkiye örneği). *J. Bank. Financ. Res.* **2021**, *8*, 1–16. Available online: https://dergipark.org.tr/tr/pub/jobaf/issue/59603/802918 (accessed on 10 January 2025).

Bıtrak, O.O. Türkiye kağıt sanayisinde selüloz ithalatı: Eşbütünleşme için artırılmış otoregresif dağıtılmış gecikmeli (ARDL) model ile sınır testi yaklaşımı. *Alanya Akad. Bakış* **2023**, *7*, 1043–1061. Available online: https://dergipark.org.tr/en/pub/alanyaakademik/issue/80135/1191253 (accessed on 10 January 2025).

Alam, I.; Quazi, R. Determinants of capital flight: An econometric case study of Bangladesh. *Int. Rev. Appl. Econ.* **2003**, *17*, 85–103. [CrossRef]

Gülmez, A. Türkiye'de dış finansman kaynakları ekonomik büyüme ilişkisi: ARDL sınır testi yaklaşımı. *Ekon. Ve Sos. Araştırmalar Derg.* **2015**, *11*, 139–152. Available online: https://dergipark.org.tr/en/pub/esad/issue/38966/456083 (accessed on 10 January 2025).

Raihan, A.; Voumik, L.C.; Mohajan, B.; Rahman, M.S.; Zaman, M.R. Economy-energy-environment nexus: The potential of agricultural value-added toward achieving China's dream of carbon neutrality. *Carbon Res.* **2023**, *2*, 43. [CrossRef]

Phillips, P.C.; Hansen, B.E. Statistical inference in instrumental variables regression with I (1) processes. *Rev. Econ. Stud.* **1990**, *57*, 99–125. [CrossRef]

Stock, J.H.; Watson, M.W. A simple estimator of cointegrating vectors in higher order integrated systems. *Econom. J. Econom. Soc.* **1993**, *61*, 783–820. [CrossRef]

Saikkonen, P. Asymptotically efficient estimation of cointegration regressions. *Econom. Theory* **1991**, *7*, 1–21. [CrossRef]

Dallı, T.; Kütükçü, E. Türkiye'de tarım, yenilenebilir ve yenilenemez enerji tüketimi, doğrudan yabancı sermaye girişi ve CO_2 emisyonu ilişkisi: ARDL analizi. *Osman. Korkut Ata Üniversitesi Fen Bilim. Enstitüsü Derg.* **2023**, *6*, 2154–2170. [CrossRef]

Bölük, G.; Mert, M. The renewable energy, growth and environmental Kuznets curve in Turkey: An ARDL approach. *Renew. Sustain. Energy Rev.* **2015**, *52*, 587–595. [CrossRef]

Brown, R.L.; Durbin, J.; Evans, J.M. Techniques for testing the constancy of regression relationships over time. *J. R. Stat. Soc. Ser. B Stat. Methodol.* **1975**, *37*, 149–192. [CrossRef]

Lawal, A.I.; Nwanji, T.I.; Asaleye, A.; Ahmed, V. Economic growth, financial development and trade openness in Nigeria: An application of the ARDL bound testing approach. *Cogent Econ. Financ.* **2016**, *4*, 1258810. [CrossRef]

Radulescu, M.; Simionescu, M.; Kartal, M.T.; Mohammed, K.S.; Balsalobre-Lorente, D. The impact of human capital, natural resources, and renewable energy on achieving sustainable cities and communities in European Union countries. *Sustainability* **2025**, *17*, 2237. [CrossRef]

69. Ridzuan, N.H.A.M.; Marwan, N.F.; Khalid, N.; Ali, M.H.; Tseng, M.-L. Effects of agriculture, renewable energy, and economi growth on carbon dioxide emissions: Evidence of the environmental Kuznets curve. *Resour. Conserv. Recycl.* **2020**, *160*, 10487 [CrossRef]
70. Li, R.; Wang, Q.; Li, L. Does renewable energy reduce per capita carbon emissions and per capita EF? New evidence from 13 countries. *Energy Strategy Rev.* **2023**, *49*, 101121. [CrossRef]
71. Akpanke, T.A.; Deka, A.; Ozdeser, H.; Seraj, M. EF in the OECD countries: Do energy efficiency and renewable energy matter *Environ. Sci. Pollut. Res.* **2024**, *31*, 15289–15301. [CrossRef] [PubMed]
72. Lamharher, R.; Ritahi, O.; Echaoui, A. The interconnected dynamics of agricultural growth, employment, renewable energy, an carbon emissions: Evidence from Morocco. *Afr. J. Commer. Stud.* **2025**, *6*, 174–186. [CrossRef]
73. Doğan, M.; Georgescu, I.; Çeştepe, H.; Sümerli Sarıgül, S.; Ergün Tatar, H. Renewable energy consumption and the ecologica footprint in Denmark: Assessing the influence of financial development and agricultural contribution. *Agriculture* **2025**, *15*, 83 [CrossRef]
74. Christensen, J.L.; Hain, D.S. Knowing where to go: The knowledge foundation for investments in renewable energy. *Energy Re Soc. Sci.* **2017**, *25*, 124–133. [CrossRef]
75. Tur, F.; Toprakçı, B. Karbondan Krediye: Onarıcı Tarım ve Karbon Kredileri. Türkiye Sınai Kalkınma Bankası. 2024. Avai able online: https://www.tskb.com.tr/uploads/file/karbondan-krediye-onarici-tarim-ve-karbon-kredileri.pdf (accessed on 1 April 2025).
76. Ağır, S.; Karakoç, U.; Topal, A. *Yeşil Dönüşüm ve Türkiye'de Gıda ve Tarım Sektörü*; Maliye Hesap Uzmanları Vakfı Yayınlar İstanbul, Turkey, 2023.
77. Ongan, S.; Gocer, I.; Işık, C. Introducing the new ESG-based Sustainability Uncertainty Index (ESGUI). *Sustain. Dev.* **2025**, *3* 4457–4467. [CrossRef]
78. Doğan, B.; Tiwari, A.; Bergougui, B.; Balsalobre-Lorente, D. Green innovation and fiscal spending: Decoding the path t sustainable development. *Sustain. Dev.* **2025**. Early View. [CrossRef]
79. Bergougui, B. Asymmetric impact of patents on green technologies on ecological footprint in Algeria: Evidence from NARD and wavelet coherence. *J. Environ. Manag.* **2024**, *355*, 120426. [CrossRef]
80. Bergougui, B. Algeria's pathway to COP28 and SDGs: Do environmental technology, energy productivity, and resource efficienc matter for sustainability? *Energy Strategy Rev.* **2024**, *55*, 101541. [CrossRef]
81. Bergougui, B.; Satrovic, E. Towards eco-efficiency of OECD countries: How does environmental governance restrain th destructive ecological effect of the excess use of natural resources? *Ecol. Inform.* **2025**, *87*, 103093. [CrossRef]
82. Bergougui, B.; Mehibel, S.; Boudjana, R.H. Asymmetric nexus between green technologies, policy uncertainty and environmenta sustainability in Algeria: Evidence from NARDL. *J. Environ. Manag.* **2024**, *360*, 121172. [CrossRef]
83. Bergougui, B. Investigating the relationships among green technologies, financial development, and ecological footprint i Algeria: Evidence from Fourier ARDL. *Sustain. Cities Soc.* **2024**, *112*, 105621. [CrossRef]
84. Bergougui, B. Moving toward environmental mitigation in Algeria: Do renewable energy and technological shocks matter fo CO_2 emissions? *Energy Strategy Rev.* **2023**, *51*, 101281. [CrossRef]

energies

MDPI

ticle

ustainable Mobility and Emissions: The Role of the Sale tructure in the Automotive Energy Transition

ga Orynycz [1,*], Ondrej Stopka [2], Anna Borucka [3,*], Ewa Kulesza [4], Jerzy Merkisz [5] and Petr Kolařík [2]

[1] Department of Production Management, Faculty of Engineering Management, Bialystok University of Technology, Wiejska Street 45A, 15-351 Bialystok, Poland
[2] Department of Transport and Logistics, Faculty of Technology, Institute of Technology and Business in České Budějovice, Okružní 517/10, 370 01 České Budějovice, Czech Republic; stopka@vste.cz (O.S.); kolarik@mail.vstecb.cz (P.K.)
[3] Faculty of Security, Logistics and Management, Military University of Technology, 00-908 Warsaw, Poland
[4] Department of Mechanics and Applied Computer Science, Faculty of Mechanical Engineering, Bialystok University of Technology, Wiejska Street 45A, 15-351 Bialystok, Poland; ewa.kulesza@pb.edu.pl
[5] Faculty of Civil Engineering and Transport, Poznań University of Technology, 60-965 Poznań, Poland; jerzy.merkisz@put.poznan.pl
* Correspondence: o.orynycz@pb.edu.pl (O.O.); anna.borucka@wat.edu.pl (A.B.)

Abstract

The aim of this article is to assess the sale structure impact of selected, popular brands of passenger vehicles on total CO_2 emissions in the context of the energy transition in the transport sector. A detailed analysis was conducted of the projected sales of gasoline-, diesel-, hybrid-, as well as electric-powered vehicles over the years 2021–2028. Based on the available empirical data, a mathematical model was developed to estimate emissions over the entire life cycle of vehicles, taking into account the unit carbon footprint of each type of drivetrain and the expected number of vehicles sold. The results indicate a gradual decline in total CO_2 emissions during the analyzed period, mainly due to the increasing share of alternative drivetrains. Despite the growth in electric vehicle sales, their impact on emission reductions remains limited due to the long lifespan of conventional vehicle fleets. The article concludes with a proposal to expand the LCA model to include regional, energy, and recycling components, which could help in formulating more effective climate policies.

Keywords: CO_2 emissions; sale forecast; life cycle analysis (LCA); energy transition; electric drivetrain; hybrid vehicle

1. Introduction

In the face of the growing global problem of greenhouse gas emissions and their impacts on climate change, the transport sector plays a crucial role [1–3]. It is estimated that transport accounts for a significant portion of total greenhouse gas (GHG) emissions, with carbon dioxide being one of the main contributors. Therefore, a precise assessment and comparison of CO_2 emissions related to different types of passenger vehicles is fundamental for both consumers and policymakers to make informed decisions. One of the comprehensive methodologies for evaluating the environmental impacts of products and services is life cycle assessment (LCA), which covers all stages of a product's life, from raw material extraction to its disposal.

The environmental advantage of electric vehicles (EVs) is intrinsically linked to the carbon intensity of the electricity grid they utilize. In regions with a high proportion of fossil fuels in their energy mix, the well-to-wheel emissions of EVs might be comparable to or even, in some specific analyses focusing on manufacturing, temporarily higher than those of efficient internal combustion engine vehicles (ICEVs), particularly when considering vehicle production. Conversely, as the energy sector decarbonizes through the increased adoption of renewable energy sources (solar, wind, and hydro), the life cycle emissions of EVs progressively decrease, solidifying their role as a key technology in mitigating transport-related CO_2 emissions [1,4].

Whereas dealing with CO_2 emissions generated by the transport sector is imperative, a comprehensive environmental assessment should also consider other pollutants and impacts. These include particulate matter (PM), nitrogen oxides (NO_x), and other volatile organic compounds emitted during vehicle operation and manufacturing, as well as the environmental burdens associated with battery production (resource depletion, water usage, and land use) and end-of-life management. A broader perspective might reveal trade-offs between different powertrain technologies across various environmental indicators, highlighting the need for multi-criteria decision-making [3,4].

The successful transition to a low-carbon transport sector is not solely dependent on technological advancements. Consumer acceptance, driven by factors such as purchase price, range anxiety, charging infrastructure availability, and individual environmental consciousness, plays a crucial role in shaping market trends. Policy interventions, including subsidies, tax incentives, and the development of a robust charging infrastructure, are essential to accelerate the adoption of cleaner vehicles and influence consumer preferences [1,2].

The energy transition in the transport sector is regarded a gradual process affected by technological innovation, infrastructure development, policy changes, and economic factors. Whilst the increasing market share of alternative drivetrains is a positive trend, the inertia of the existing vehicle fleet entails that the full benefits of this transition in terms of emission reductions will be realized over several years. Therefore, long-term policy frameworks and consistent signals are necessary to guide investment and innovation toward sustainable mobility solutions [1,3].

This article consists of two stages. The first is a comparison of the total CO_2 emissions in the life cycle of passenger vehicles powered by gasoline, diesel, and electricity, based on available studies and statistical data. The second stage focuses on the characterization and evaluation of the passenger vehicle market at the national level using a selected manufacturer as an example. Combining sales structure data with a discussion of the life cycle allows for a comprehensive analysis of the environmental impacts of dominant drivetrain (also known as powertrain or propulsion) technologies and an assessment of real market trends in terms of the energy transition in the transport sector.

This analysis will make it possible not only to determine which types of drivetrains have the lowest CO_2 emissions across their entire life cycle but also to indicate to what extent the current market offer and consumer preferences contribute to the reduction in greenhouse gas emissions. This analysis can also serve as a starting point for discussions on future directions for transport policy development and strategies for automotive manufacturers.

2. Literature Review

The issues of emissions and their environmental impacts in the context of market dynamics of sales across various energy technologies and the overarching global energy transformation are deemed critical areas of contemporary research and policy. Understanding the connections among these subjects is crucial for navigating the transition toward a sustainable energy future and mitigating the adverse effects of climate change and pollution [4]. Future research should yet focus on further elucidating the complex interconnections and addressing the identified gaps in the existing body of knowledge.

The topic of emissions and environmental impact of energy systems, particularly when quantifying the produced emissions associated with different energy sources and their resulting environmental consequences, has been addressed in a wide range of works, for instance [5–12]. Specifically, extensive research [5] details the greenhouse gas emissions (e.g., CO_2, CH_4, and N_2O) from the combustion of coal, oil, and natural gas, whilst [6] highlights air pollutants (e.g., SO_x, NO_x, and particulate matter) and their impacts on air quality, human health, and ecosystems. On the other hand, Sovacool focuses on low direct GHG emissions of nuclear power during operation but addresses concerns regarding radioactive waste disposal, potential accidents [7], and the nuclear fuel cycle, and in [8], research on solar, wind, hydro, geothermal, and biomass energy is discussed when emphasizing their potential for significantly lower GHG emissions compared to fossil fuels. However, several studies also investigate the life-cycle emissions associated with the manufacturing, deployment, and disposal of renewable energy technologies, e.g., Pehnt et al. focused on biomass [9] and Schleisner focused on wind [10]. In addition, environmental impacts beyond emissions, such as land use changes, habitat disruption, and water consumption, have been explored as well, e.g., by Godfray et al. on hydropower [11] and by Rykała et al. when using logistic regression [12].

As for the subject of the sales dynamics and market penetration of energy technologies with a special consideration of factors affecting the sales and market adoption of different energy technologies, including both fossil fuels and cleaner alternatives, multiple studies have been published as well [13–20]. While Acemoglu et al. deal with price competitiveness, investment costs, operational expenses, government subsidies, carbon pricing mechanisms, and tax incentives as key drivers of energy technology sales [13], Jaffe et al. analyze the price elasticity of demand for different energy sources and the impact of policy interventions on the market share [14]. On the other hand, innovation, performance improvements, technological maturity, infrastructure availability, and grid integration capabilities that significantly affect the marketability of energy technologies are dealt with by Foster [15], and Grübler et al. explore the diffusion of innovations and the role of learning curves in reducing the cost of new technologies [16]. Social and behavioral factors have been analyzed as well. For instance, Stern evaluates consumer preferences, awareness of environmental issues, social norms, and trust in new technologies as aspects that play crucial roles in adoption decisions [17], and Rogers provides insights into behavioral economics and sociology in terms of various energy technologies [18]. Last but not least, government policies, including renewable energy mandates, feed-in tariffs, emissions standards, and regulations on fossil fuels, having a profound impact on the sales and deployment of different energy technologies are discussed by Helm [19], and Vieira et al. compare the effectiveness of various policy instruments [20].

Apart from the above aspects and literature sources, the topic of the ongoing global energy transformation and its drivers (such as climate change, energy security, and technological advancements) have also been examined. For instance, Smil explores the pathways and challenges related to shifting away from fossil fuel-dominated energy systems. This includes analyses of stranded assets, the role of incumbent industries, and the need for new infrastructure [21]. Furthermore, a study [22] focused on the rapid growth of renewable energy technologies and their integration into existing energy systems, and Jacobson et al. address issues such as grid stability, energy storage, and the intermittency of renewable sources [23]. The increasing electrification of various sectors (e.g., transportation and heating) and the integration of different energy systems (e.g., power-to-gas) as key indicators of the energy transformation are discussed in [24]. This literature explores the potential for reducing emissions and increasing energy efficiency through these trends. And last but not least, Falkner investigates the role of international agreements (e.g., Paris Agreement), national energy strategies, and sub-national initiatives in driving the energy transition [25]. The effects of political economy, public opinion, and stakeholder engagement are analyzed as well.

While significant research exists in each area under investigation, a further in-depth and meticulous examination is needed to fully understand the comprehensive interconnections among emissions and environmental impacts, sales dynamics, and the broader energy transformation. For example, more research could explore

- The feedback loops between public awareness of environmental impacts and consumer demand for cleaner energy technologies;
- The influence of specific emission regulations on the market competitiveness of different energy sources;
- The roles of sales and market penetration in accelerating the learning curves and cost reductions of low-carbon technologies (so called "eco-friendly"), thus further driving down emissions;
- The social and economic consequences of different energy transition pathways, considering both environmental benefits and potential disruptions, and so forth.

The conducted research, in addition to scientific studies, was also based on available reports, analyses, and tests published by car manufacturers, specialist industry portals, and motoring enthusiasts. Such a wide range of sources allowed a more comprehensive and diverse picture of the subject matter to be obtained, taking into account both the theoretical foundations and practical aspects of use and the opinion of the automotive community.

3. Methodology

Life cycle assessment (LCA) is a well-established and standardized methodology for evaluating the environmental impacts of products, processes, and services, covering the full scope of their existence—from raw material extraction to final disposal [26]. The term "cradle to grave" means that LCA considers all stages of the life cycle: the extraction and processing of raw materials, component production and assembly, distribution of the finished product, its use by the consumer, as well as end-of-life waste management, including recycling and landfilling [27,28].

In the automotive sector, particularly for passenger vehicles, life cycle analysis enables a comprehensive assessment of the total environmental footprint of a given means of transport [29–31]. A particular emphasis is placed on estimating total greenhouse gas emissions, most commonly expressed in terms of the carbon dioxide equivalent (CO_2e), generated at each of the mentioned stages. This approach allows not only for a comparison of different drivetrain technologies—such as internal combustion engines (gasoline and diesel) and electric drivetrains—but also for the identification of life cycle stages with the highest contribution to emissions [32,33].

In practice, LCA takes into account numerous variables, such as the type and origin of raw materials, the source of energy used in the production process, the energy intensity of vehicle use (including the fuel type or energy mix), as well as end-of-life scenarios for vehicle disposal. As a result, the findings can serve as a basis for developing more sustainable strategies for technological development, climate policy, and consumer decision-making. Operational emissions include all energy-related emissions occurring during vehicle use, including fuel consumption during engine idling, warm-up, and stop-and-go conditions, as captured in the total lifetime fuel consumption estimates.

This study considers the following key stages of a vehicle's life cycle:

- Raw material extraction and processing—This stage includes emissions associated with obtaining the raw materials needed for vehicle production, such as steel, aluminum, plastics, and—in the case of electric vehicles—the materials used in battery production (lithium, cobalt, nickel, and manganese) [33]. These processes are often energy-intensive and generate significant amounts of CO_2. For example, meta-analyses of lithium-ion battery production show a global warming potential ranging from 12 to 313 kg CO_2e per kWh of battery capacity, with a global average around 187 kg CO_2e/kWh [34,35]. Another study reports 110 g CO_2e ($\approx$0.11 kg CO_2e) per Wh, i.e., 110 kg CO_2e/kWh [36]. Vehicle and component manufacturing—This stage relates to emissions from manufacturing the vehicle itself and its key components, including the engine, gearbox, and, in the case of electric vehicles, the battery. The production of lithium-ion batteries for electric vehicles is particularly energy-intensive and can significantly contribute to their initial carbon footprint, causing increases of 15% to 68% in vehicle manufacturing emissions and adding roughly 53–356 kg CO_2e per kWh of battery capacity (global average ~110–187 kg CO_2e/kWh) [33,37,38].

- Fuel/energy production and distribution—For gasoline and diesel vehicles, this stage includes crude oil extraction, transport, and the energy-intensive refining process to produce gasoline and diesel fuel. For electric vehicles, this stage involves the generation of electricity from various sources (fossil fuels, nuclear power, and renewable energy) and its transmission and distribution to charging stations. It is important to emphasize that the energy mix of a given region plays a crucial role in determining the CO_2 emissions associated with the use of electric vehicles [39].
- Vehicle use and maintenance—For gasoline and diesel vehicles, this stage focuses on direct CO_2 emissions from fuel combustion in the engine. Electric vehicles do not emit CO_2 directly during operation, but their use is indirectly associated with emissions from electricity generation. Indirect emissions related to the production and use of consumables such as tires and lubricants should also be considered.

End-of-life/recycling—This stage includes emissions related to the disposal of the vehicle and its components, as well as the potential environmental benefits from recycling materials (steel, aluminum, and battery components), and the energy savings and avoided emissions made possible by using recycled materials [33].

In the vehicle life cycle analysis, the well-To-wheels (WTW) approach is often used. It focuses on the life cycle of the energy carrier used to power the vehicle (liquid fuel or electricity), considering environmental protection aspects from the energy source to its use in the vehicle [33,40,41].

The methodology for calculating the linear regression applied to the sale forecasts is that the sale forecast was conducted independently for each drivetrain type using linear regression [42]:

$$S_i(t) = a_i \cdot t + b_i$$

where

$S_i(t)$ is the number of vehicles of type i (gasoline, diesel, hybrid, electric) sold in year t,
a_i, b_i is the regression coefficients for the drivetrain,
$t \in \{2021, \ldots, 2025\}$ are the observation years, and
predictions were made for the years $t = 2026, 2027, 2028$.

Based on the projected number of sold vehicles, CO_2 emissions over the full lifecycle of each vehicle were estimated using the following equation:

$$E_i(t) = S_i(t)\, \varepsilon_i$$

where

$E_i(t)$ is the CO_2 emissions for vehicle type i in year t, in tons, and
ε_i is the unit CO_2 emission per vehicle of type i.

4. Results Obtained

4.1. CO_2 Emissions of Petrol-Powered Passenger Cars

The life cycle of a petrol-powered car begins with the extraction of crude oil, its transportation, and the energy-intensive refining process required to produce petrol. Each of these stages contributes to CO_2 emissions. Subsequently, during the production of the vehicle itself, additional emissions are generated due to the extraction and processing of raw materials such as steel and aluminum [33], the production of various components, and the final assembly of the car.

The largest share of CO_2 emissions in the life cycle of a petrol car occurs during the usage phase [43,44]. Direct CO_2 emissions result from burning petrol in the engine. Average CO_2 emissions during the use phase of petrol cars are estimated at 120–150 g/km [45,46]. Studies show that the total life cycle CO_2 emissions of a petrol car in Europe in 2021 were just under 250 g CO_2/km. Forecasts for 2031, assuming the implementation of the Paris Agreement, predict only a slight decrease to around 240 g CO_2/km [6]. Another study by VDI estimated that a petrol car emits 37.1 tons ($\approx$186 gCO_2/km) of CO_2 over a lifetime distance of 200,000 km [47]. In comparison, the Renault Clio's average total life cycle CO_2 emissions in European use were estimated at approximately 47 tons ($\approx$235 gCO_2/km) [48].

The end-of-life phase of a petrol vehicle also generates CO_2 emissions, related to its disposal, including potential emissions from landfills and the energy needed to recycle materials such as steel and aluminum [33].

Analyzing the life cycle of a petrol vehicle shows that fuel combustion is the dominant phase in terms of emissions, although fuel production and vehicle manufacturing also make significant contributions. The limited forecasted reductions in emissions for petrol cars by 2031 suggest that achieving a substantial environmental impact reduction may be difficult through internal combustion engine improvements alone [49].

4.2. CO_2 Emissions of Diesel-Powered Passenger Cars

Similar to petrol vehicles, the life cycle of a diesel car begins with the extraction, transport, and refining of crude oil to produce diesel fuel. These processes generate CO_2 emissions, which may differ slightly from those involved in petrol production. Manufacturing a diesel vehicle also results in CO_2 emissions from material processing and assembly.

During operation, diesel vehicles emit CO_2 directly. Average tailpipe emissions are estimated at 130–170 g/km [50]. WHO data for Europe in 2021 indicate total emissions for diesel cars comparable to petrol cars, around 250 g CO_2/km [6]. A VDI study estimated emissions at 33.0 tons of CO_2 over 200,000 km, slightly lower than petrol vehicles in this specific analysis [47]. A noteworthy result from the Green NCAP study showed that the Škoda Octavia Combi 2.0 TDI consumes 164 MWh of energy over its lifetime, which is comparable to, or even better than, some electric and hybrid vehicles [48].

The end-of-life stage for diesel vehicles is similar to that of petrol vehicles.

Although diesel cars are commonly thought to emit more CO_2 during operation than petrol cars, some life cycle analyses suggest that due to potentially better fuel efficiency, their total emissions may be slightly lower [46,47,50]. However, these outcomes can vary depending on the adopted methodology and study assumptions. The Green NCAP study highlights the complexity of comparing different drivetrains, suggesting that in certain scenarios, modern diesel vehicles may have comparable or even lower life cycle energy consumption than some electric and hybrid vehicles [48]. This underscores the importance of the assumptions and methodologies used in LCA (life cycle assessment) studies [51].

4.3. CO_2 Emissions of Electric Passenger Cars

The life cycle of electric passenger cars differs significantly from combustion engine vehicles, particularly in the usage phase. However, they generate CO_2 emissions at other stages. The extraction and processing of raw materials necessary for battery production—such as lithium, cobalt, nickel, and manganese—are associated with significant CO_2 emissions [33]. These processes are energy-intensive and have notable environmental impacts. The production of lithium-ion batteries itself is highly energy-consuming and represents a major contribution to the initial carbon footprint of electric vehicles [1,33]. This stage is estimated to account for 50% to 90% of the total CO_2 emissions in the life cycle of an electric vehicle [38,52]. Nevertheless, due to technological advancements and the use of cleaner energy sources in manufacturing, CO_2 emissions per kWh of battery production show a decreasing trend [38,52].

The production of the electric vehicle itself, excluding the battery, involves emissions from material processing and assembly similar to those in combustion vehicles [33].

CO_2 emissions during the use phase of electric vehicles depend entirely on the source of electricity used for charging [53,54]. Assuming the average European energy mix in 2021, total CO_2 emissions for electric vehicles were about 80 g CO_2/km, with projections for 2030 indicating a decrease to around 65 g CO_2/km [55]. When powered entirely by renewable energy, emissions drop below 50 g CO_2/km [55]. In countries with a high share of fossil fuels in the energy mix, such as Poland, per-kilometer emissions during use can be higher compared to countries with cleaner energy sources [56].

The end-of-life stage of batteries and their recycling is critical for assessing the environmental impacts of electric vehicles. Battery recycling can help reduce CO_2 emissions and recover valuable raw materials [33]. Developing an efficient and sustainable battery recycling infrastructure is therefore crucial.

A VDI study found that the electric Volkswagen ID.3 emits 24.2 tons of CO_2 over 200,000 km [47]. For comparison, the Renault Zoe emits 32 tons of CO_2 over its entire life cycle in Europe [48]. The potential reduction in life cycle emissions for electric vehicles ranges from 17–30% when using non-renewable electricity to 60–80% when powered entirely by renewables compared to a petrol vehicle [57]. A WHO study suggests that electric vehicles in Europe generate 66–69% fewer greenhouse gas emissions than comparable petrol vehicles [6]. A study by Volvo indicates that manufacturing an electric vehicle may produce 70% more CO_2 than a combustion vehicle, but the total life cycle carbon footprint becomes lower over time and mileage [57].

The initial carbon footprint of electric vehicles is usually higher than that of petrol or diesel vehicles due to energy-intensive battery production [58]. However, operational emissions are significantly lower, especially when the electricity comes from renewable sources [59]. This leads to a lower overall life cycle emission level in many scenarios [58]. The point at which an electric vehicle's total emissions become lower than those of a comparable petrol or diesel car depends largely on the carbon intensity of the power grid. In regions with a high share of fossil fuels in the energy mix, this point may be reached after a longer period of use [56]. Studies such as those by WHO [6] suggest that even with the current European energy mix, electric vehicles already offer a significant reduction in life cycle greenhouse gas emissions compared to conventional vehicles. This advantage is expected to grow as the energy sector becomes more decarbonized [59].

4.4. Emission Comparisons and Statistics

Table 1 presents a comparison of the vehicle whole-life carbon emissions analysis of estimated lifecycle emissions (tons CO_2e) based on a 2015 vehicle in use for 150k KM using a 10% ethanol blend and 500 g/KWH grid electricity [60,61].

Table 1. Life cycle CO_2 emissions comparison: standard gasoline, hybrid, plug-in hybrid, and battery electric vehicles.

	Estimated Lifecycle Emissions (tons of CO_2e)	Proportion of Emissions in Production	Estimated Emissions in Production (tons of CO_2e)
Standard gasoline vehicle	24	23%	5.6
Hybrid vehicle	21	31%	6.5
Plug-in hybrid vehicle	19	35%	6.7
Battery electric vehicle	19	46%	8.8

Table 2 presents the life-cycle greenhouse gas (GHG) emissions (100-year global warming potential, in g CO_2e/MJ) for average fossil and biogenic blends of gasoline, diesel, and natural gas, as well as for grid electricity and hydrogen mixes in China, Europe (EU27+UK), India, and the United States. The data cover the 15–18 year lifetime of cars registered in 2021 and 2030, along with a sales-weighted global average across these regions. Emission values for gasoline, diesel, and natural gas are presented as (production emissions) + (consumption emissions) [60–62].

Table 2. Life-cycle greenhouse gas (GHG) emissions for average fossil and biogenic blends of gasoline, diesel, and natural gas, as well as for grid electricity and hydrogen mixes in China, Europe (EU27+UK), India, and the United States.

g CO_2e/MJ	Data	Gasoline (Including Biofuels)	Diesel (Including Biofuels)	Natural Gas (Including Biogas)	Average Grid Electricity	Hydrogen Mix
China	2021–2035	21.8 + 70.9	-	-	141.4	107.8
Europe	2021–2038	21.5 + 70.9	27.4 + 68.4	16.5 + 58.6	45.6	50.4
India	2021–2038	20.5 + 69.2	21.8 + 71.8	19.2 + 57.7	155.9	84.4
United States	2021–2035	22.2 + 69.7	-	-	66.5	64.3
Global average	2021 car	21.8 + 70.5	26.6 + 68.9	16.9 + 58.5	93.2	77.8

From the presented data, it is evident that CO_2 emissions over the vehicle life cycle vary depending on the type of vehicle and the study methodology. Gasoline (petrol) and diesel vehicles show similar levels of total CO_2 emissions per kilometer in many studies, especially when life-cycle emissions (including fuel production, vehicle manufacturing, and usage) are taken into account. For example, a comprehensive life cycle assessment by the International Council on Clean Transportation found that the total life cycle CO_2 emissions of diesel cars are only marginally lower (by about 3–5%) than those of gasoline cars, primarily due to slightly better fuel economy, which is largely offset by higher emissions during fuel production and vehicle manufacturing [61,63–65].

The energy mix has a significant impact on the life cycle emissions of electric vehicles. In countries where fossil fuels constitute a large share of the energy mix, the emission benefits of using electric vehicles may be reduced. However, as energy production shifts toward cleaner sources, the advantage of electric vehicles in terms of emissions will increase [66]. There is a trade-off between the higher initial emissions associated with the production of electric vehicles (especially batteries) and their lower operational emissions compared to gasoline and diesel vehicles.

This study is based on the assumption that renewable energy is low-emission in terms of CO_2, without considering the full life cycle of its production, which would require separate analyses. Nevertheless, many studies question the zero-emission status of renewable energy production and point out its potential environmental harm—which should also be highlighted in the context of the studies presented in this paper. It is commonly assumed that renewable energy production is climate-friendly and essentially free of CO_2 emissions, especially during the energy generation phase [67]. This argument is frequently raised, for example, regarding the direct operation of renewable energy installations, where solar panels convert sunlight and wind turbines convert wind energy into electricity without emitting CO_2 [67]. However, such a simplified perspective ignores the complexity of processes involved throughout the entire life cycle of renewable technologies. CO_2 emissions should be assessed at all stages of a renewable energy technology's life cycle, including raw material extraction, component production, transportation, installation, operation, and end-of-life disposal or recycling. Relying on an incomplete picture of the environmental impacts of renewable energy development may lead to misguided strategies and ultimately hinder the achievement of a truly sustainable energy system [68].

Nevertheless, this aspect will be analyzed in future studies by the authors as the next stage of the analyses presented in the paper. For now, it has only been signaled.

4.5. CO_2 Emissions Forecast Based on the Sales Structure of Selected Brands of Vehicles (2021–2028)

To expand the assessment of greenhouse gas emissions over the vehicle life cycle, an annual CO_2 emissions forecast was conducted based on the sales structure of vehicles from 2021 to 2025 and extrapolated to 2028. The analysis included passenger cars, classified by drivetrain type: gasoline, diesel, hybrid, and electric.

The collected sales data were used not only to determine direct emissions associated with vehicle operation but also as a foundation for further modeling of future development trends in electromobility and their potential environmental impacts. The assessment covered not only the magnitude of emissions but also the dynamics of change, including the gradual decline in the share of internal combustion engine (ICE) vehicles and the growing significance of alternative drivetrains.

Additionally, based on the sales data, the following charts were created [61]:

- A radar chart of selected cars' parameters (2020–2024), see Figure 1;
- A three-dimensional histogram of the number of vehicles sold by type from 2020 to 2025 (no. of obs), see Figure 2;
- An annual CO_2 emissions forecast by drivetrain type (2021–2028), see Figure 3.

For a comparative analysis, we selected five investigated vehicle models, each representing a different drivetrain type: x (electric drive), y (petrol engine), z (plug-in hybrid drive), k (petrol engine), and l (diesel engine).

Figure 1 above displays a radar chart that compares four key technical parameters: maximum power (kW), torque (Nm), 0–100 km/h acceleration (s), and range (km). To allow for a direct comparison despite varying units, these parameters were normalized within the (0, 1) range.

The values in the chart are sourced from Table 3 below and reflect actual technical data, consistent with official manufacturer specifications. The chart's calibration sets the circle thresholds at normalized levels of 0.2, 0.4, 0.6, 0.8, and 1.0, thereby maintaining proper proportions between the models. Table 3 presents the actual values for each parameter.

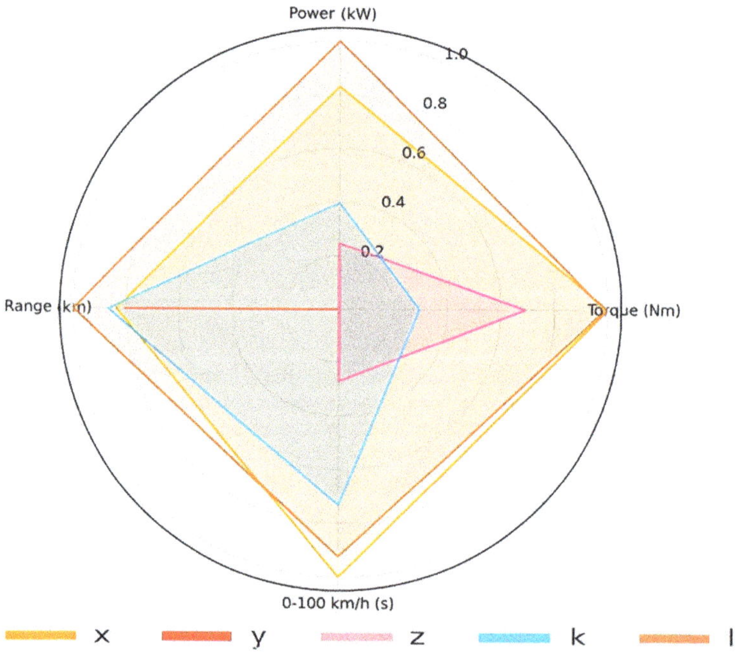

Figure 1. Comparison of the technical parameters of vehicles (2020–2024).

Table 3. Actual, catalog values of selected car parameters.

Symbol	Power (kW)	Torque (Nm)	Range (km)	Acceleration 0–100 km/h (s)	Powertrain Type
x	385	855	679	4.3	Electric
y	430	850	500	4.5	Petrol
z	230	550	117	6.7	Hybrid
k	270	500	590	5	Petrol
l	110	320	600	8.6	Diesel

An analysis of the chart reveals that model x an electric vehicle, excels in range and torque. In contrast, model *y* demonstrates the highest power output. Model *z*, a plug-in hybrid, shows significantly lower power and range values, but it offers a good compromise between performance and efficiency.

Model *k* represents a premium car with high power and dynamic performance, making it suitable for fast highway driving, though it has moderate fuel efficiency.

Finally, model *l* is a more economical and practical choice that is well-suited for daily use, especially in urban and suburban conditions, with a focus on fuel efficiency and low emissions.

Figure 1 presents four key technical parameters of the selected models from the years 2020–2024: maximum power (in kW), maximum torque (Nm), acceleration from 0 to 100 km/h (s), and total driving range (km). The data for each parameter were normalized to standardize the scale and allow for direct comparison of values originally expressed in different units.

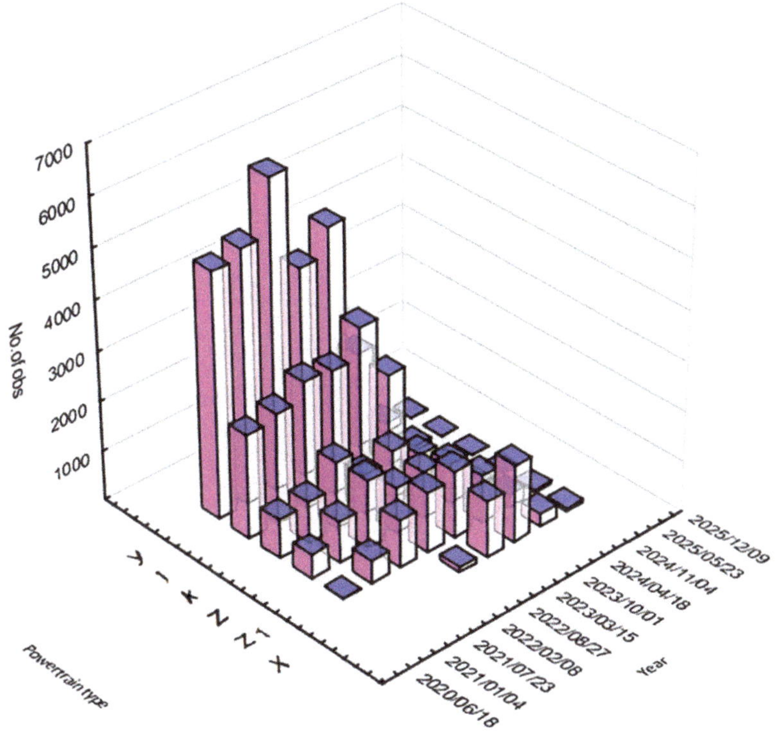

Figure 2. Histogram of the number of vehicles sold by type (2020–2025).

Figure 2 presents the number of vehicles sold, with the axes representing the following [62]:

The *X*-axis represents the sales date, which covers the period from June 2020 to December 2025. Dates are marked at aggregate points (e.g., every few months), allowing an analysis of sales seasonality and dynamics over time.

The *Y*-axis represents the vehicle drivetrain type, where x denotes electric vehicles, y and k correspond to petrol-powered vehicles (with k referring specifically to an exclusive convertible model), z and z_1 denote hybrid vehicles, and l represents diesel-powered vehicles.

The *Z*-axis represents the number of vehicles sold in a given period and for a given type/model (No. of obs = number of observations).

The results reveal that throughout the analyzed time frame, petrol-powered vehicles of type y consistently exhibited the highest sales volumes, with a pronounced peak observed between 2022 and early 2024. While a gradual decline is visible in the most recent data (2025), petrol engines still maintained a dominant market share.

The k variant, characterized as a premium convertible, showed significantly lower sales figures across the entire period, which aligns with its exclusive market positioning. Despite this, its relatively stable sales trajectory may suggest the presence of a steady, niche consumer base.

Electric vehicles, x, demonstrated a clear upward trend over time, with a noticeable acceleration in sales from 2023 onward. This pattern reflects the increasing adoption of electromobility, likely driven by broader regulatory, economic, and technological developments across the industry.

The hybrid category, z, experienced substantial sales during the mid-2021 to mid-2023 window, after which a modest decline occurred. In contrast, the alternative hybrid model z_1

maintained a steady yet comparatively lower sales volume, suggesting consistent interest within a targeted segment of the market.

Finally, diesel-powered vehicles, 1, consistently recorded the lowest sales levels across all time points. Their marginal and declining trend further supports the broader observation of their decreasing relevance in light of global decarbonization policies and shifting consumer preferences.

In summary, the data presented in Figure 2 reveal a progressive transformation in the automotive market, characterized by a gradual shift away from traditional internal combustion engines—particularly diesel—toward electrified powertrains. While petrol engines remain prevalent, the growing uptake of electric and hybrid vehicles indicates a transition toward more sustainable mobility solutions.

Calculation Methodology (see Section 3)

Assumed Unit Emission Values

The values ε_i were adopted from the sources [61,69,70] (see Table 4).

Table 4. The values ε_i adopted for individual vehicle types.

Vehicle Type	Unit Emissions ε_i [t CO_2/vehicle]
Gasoline	37.0
Diesel	33.0
Hybrid	28.0
Electric	22.0

To better illustrate the changes, the relative change in emissions compared to the base year (2021) was introduced:

$$\Delta E_i(t) = ((E_i(t) - E_i(2021))/(2021)) \cdot 100\%$$

Assuming an average vehicle lifetime mileage of 200,000 km and average emissions per km, emissions can also be calculated with

$$\varepsilon_i = e_i \cdot d$$

where

e_i is the average emissions per 1 km,

d = 200,000 km and is the assumed mileage.

Results and Interpretation

For example, in 2026, the forecasted sales are

- 4527 gasoline vehicles;
- 3704 diesel vehicles;
- 100 hybrid vehicles;
- 893 electric vehicles.

This translates into

$$Ebenz(2026) = 4527 \cdot 37.0 = 167,499\, t\, CO_2$$

$$Ediesel(2026) = 3704 \cdot 33.0 = 122,232\, t\, CO_2$$

$$Ehyb(2026) = 100 \cdot 28.0 = 2800\, t\, CO_2$$

$$Eelek(2026) = 893 \cdot 22.0 = 19,646\, t\, CO_2$$

$$Etotal(2026) = 167,499 + 122,232 + 2800 + 19,646 = 312,177\,t\,CO_2$$

These calculations were repeated for each year, creating an annual emissions summary (see Table 5), which can serve as a basis for assessing the impacts of sales trends on emissions in the automotive sector.

Table 5. Results for 2021–2028.

Year	Gasoline	Diesel	Hybrid	Electric	Total Emissions (t)
2021	6300	4500	150	300	312,300
2022	5200	4200	130	450	282,170
2023	5000	4000	120	550	270,400
2024	4700	3800	110	675	256,350
2025	4600	3750	105	800	251,950
2026	4527	3704	100	893	248,177
2027	4450	3650	95	975	244,180
2028	4375	3600	90	1050	240,185

Figure 3 depicts a forecast of the vehicles under investigation according to the drivetrain type and year (2021–2028).

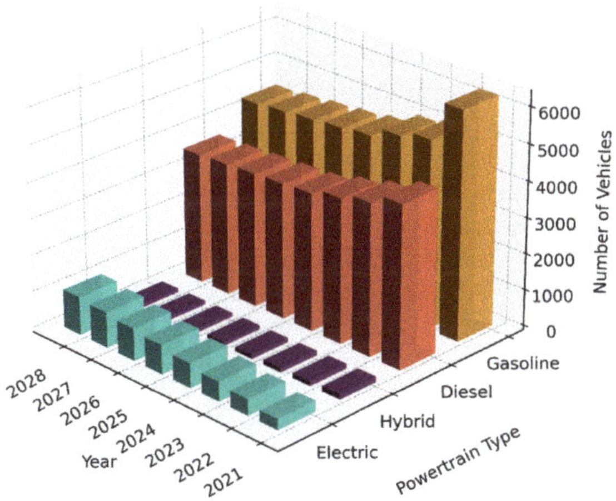

Figure 3. Forecast of vehicles by type and year (2021–2028).

Figure 3 above shows the number of vehicles sold between 2021 and 2028, broken down by drivetrain type: gasoline, diesel, hybrid, and electric. A clear reduction in the sales of internal combustion engine vehicles and an increase in the sales of electric vehicles are evident during the analyzed period.

5. Discussion

Based on the results obtained (see Section 4 above), the estimated CO_2 emissions (in tons) for four types of drivetrains: gasoline (orange line), diesel (red line), hybrid (dark pink line), and electric (light pink line) are depicted in Figure 4.

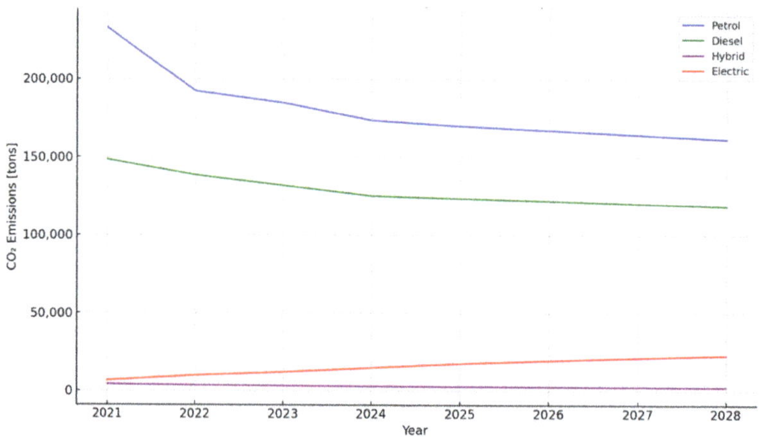

Figure 4. Annual CO_2 emissions forecast by drivetrain type (2021–2028).

A downward trend in CO_2 emissions from gasoline vehicles is evident, from over 210,000 tons in 2021 to approximately 165,000 tons in 2028. Similarly, emissions from diesel vehicles declined from around 145,000 tons to just above 120,000 tons over the same period.

Hybrids generate significantly lower emissions than internal combustion engine (ICE) vehicles, on the order of several tens of thousands of tons annually, and show a stable, slow upward trend, which corresponds to an increasing number of hybrid vehicle sales.

Additionally, although CO_2 emissions associated with electric vehicles remain low (mainly due to electricity production and charging), the chart indicates a growing share of this group: emissions increased from approximately 6600 tons in 2021 to nearly 23,000 tons in 2028. It is worth noting that even though the emissions generated by electric vehicles are significantly lower, their growing number may, in the coming years, contribute to a shift in the emissions profile from the usage phase to the electricity production phase.

Figure 4 illustrates the evolution of emission structures in the premium private transport sector in the context of the energy transition. It also confirms that the continuous increase in the share of electric and hybrid drivetrains contributes to reducing total CO_2 emissions in the long term, although achieving this requires maintaining a high pace of technological transformation.

The results obtained confirm that, despite the growing share of electric vehicles, total CO_2 emissions remain significant due to the dominance of internal combustion engine vehicles in the sales structure. Only a substantial change in the sales structure (a mass transition to zero-emission drivetrains) will result in a real reduction in emissions in the post-2030 period [71]. The presented methodology can be applied to a scenario-based analysis of climate policies and the modeling of potential benefits from accelerating the energy transformation in the transport sector.

Although the sales of electric vehicles are growing dynamically, their impact on the overall reduction of CO_2 emissions by 2028 is still limited. A key factor that restricts the environmental effect is the high share of internal combustion vehicles from previous years, which remain in operation for many years [72,73].

The presented model also allows for the simulation of the impacts of decarbonization policies, changes in the energy mix, and tax regulations. In the next stages of the study, it is planned to introduce emission weights depending on the country of vehicle use and account for the effect of material recycling in the product's end-of-life phase [74].

In the context of further development of the proposed model, it will also be crucial to consider the differentiated emission intensity depending on the vehicle segment and its lifecycle mileage. Commercial vehicles, such as delivery and heavy trucks, have significantly higher unit emissions than passenger cars, meaning that their energy transformation could result in relatively larger environmental benefits. Expanding the model to include these aspects will allow for a more accurate assessment of the potential impacts of different policy scenarios. This will be part of the authors' further research [75,76].

It is also worth noting that the transport transformation is not limited to changes in the vehicle drivetrain itself. An important element will also be the development of the charging infrastructure, intelligent traffic management systems, and the promotion of alternative forms of mobility, such as public transport and micromobility. Only the synergistic actions of all these elements will provide a comprehensive and lasting reduction in emissions in the transport sector.

From the perspective of policymakers, it is therefore essential to develop integrated strategies that combine financial incentives, environmental regulations, and infrastructure investments. The presented model can support the decision-making process by identifying the most effective directions for action, considering both implementation costs and the expected effects in terms of emission reduction and economic impact.

In future research, it is also planned to include the social and economic aspects of the transformation, such as the availability of electric vehicles for different social groups, the impact on employment in the automotive sector, and changes in energy consumption patterns. Such an extension would enable a more comprehensive understanding of the consequences of transport decarbonization and assist in designing a just energy transition [77].

6. Conclusions

The conducted study aimed to evaluate the impact of the sales structure of the selected brands of passenger cars on total CO_2 emissions with regard to the ongoing global energy transformation in the transport sector. By analyzing sales forecasts for gasoline, diesel, hybrid, and electric vehicles over the years 2021–2028, and employing the LCA-based mathematical model when incorporating the carbon footprint of each powertrain type (called also as engine propulsion or drivetrain type) and projected sales figures, this research provides valuable insights into the emission trends of a major automotive manufacturer.

Following the outcomes of the analysis obtained, the findings indicate a gradual decline in total CO_2 emissions over the analyzed period. This reduction is primarily attributed to the increasing market share of alternative drivetrains, in particular hybrid and electric vehicles. However, this research highlights that despite the growth in electric vehicle sales, their immediate impact on overall emission reductions is somewhat constrained by the long lifespan of the existing fleet of conventionally powered vehicles. This suggests that a sustained and accelerated transition toward alternative drivetrains is inevitable to achieve significant emission reductions in the short to medium term.

Nevertheless, this study acknowledges several ensuing limitations that should be taken into consideration when interpreting the findings.

- First, this analysis is based on sales forecasts, which inherently carry a degree of uncertainty. Thus, deviations in actual sales figures could affect the projected emission reduction level.
- Second, the LCA model applied relies on mean emission factors for each drivetrain type. Variations in manufacturing processes, energy sources for electricity generation, and end-of-life treatment could affect the actual life cycle emissions.
- Third, this research focuses solely on the sales structure of passenger cars of one brand within a specific timeframe and does not account for the broader vehicle fleet or the

impact of external factors (such as evolving energy policies, consumer behavior beyond purchasing decisions, and so on). To enhance the understanding of the automotive sector's role in the energy transformation and to inform more effective climate policies, several avenues for further research are recommended, as follows:

- Expanding the LCA model to incorporate regional variations in energy grids, the environmental impacts of battery production and recycling, and the emissions associated with vehicle manufacturing would provide a more comprehensive evaluation.
- In addition, future research could explore the effects of government incentives, charging infrastructure development, and consumer adoption rates on the pace of electric vehicle penetration and its subsequent impact on emission reduction as well.
- Furthermore, analyzing the emission trends of other major automotive manufacturers and across different vehicle segments would offer a broader perspective on the sector's overall contribution to the energy transition.
- Last but not least, investigating the potential of other alternative fuels and drivetrain technologies beyond electrification, such as hydrogen fuel cells and sustainable biofuels, within the same analytical framework could provide a more holistic understanding of future mobility scenarios and their environmental implications

Author Contributions: Conceptualization, O.O., O.S. and A.B.; methodology, O.O., O.S. and A.B.; software, A.B., E.K. and J.M.; validation, E.K., J.M. and P.K.; formal analysis, A.B. and P.K.; investigation, O.S. and A.B.; resources, O.O. and A.B.; data curation, A.B., J.M. and P.K.; writing—original draft preparation, O.O., O.S. and A.B.; writing—review and editing, O.S., A.B. and P.K.; visualization, O.O. and E.K.; supervision, O.S. and A.B.; project administration, O.S., A.B. and E.K. All authors have read and agreed to the published version of the manuscript.

Funding: This study was financed by the Faculty of Logistics and Security Management Military University of Technology, Research grant no. 065/2025.

Data Availability Statement: An anonymized, aggregated version of the sales data (source [62]) can be provided by the author upon reasonable request, subject to the data owner's (Manufacturer X) approval. The data are not publicly available due to the confidentiality of sales data and business issues.

Conflicts of Interest: The authors declare no conflicts of interest.

References

1. Niewczas, A.; Mórawski, Ł.; Rymarz, J.; Dębicka, E.; Hołyszko, P. Operational Risk Assessment Model for City Buses. *Eksp. Niezawodn. Maint. Reliab.* **2023**, *25*, 14. [CrossRef]
2. Yun, M.; Ren, J.; Zhang, L.; Chen, X. Fracture Process and Instability Precursor Determination of Freeze-thaw Red Sandston Based on Acoustic Emission Monitoring. *Ekspl. Niezawodn. Maint. Reliab.* **2024**, *26*, 191193. [CrossRef]
3. Folęga, P.; Burchart, D.; Kubik, A.; Turoń, K. Application of the life cycle assessment method in public bus transport. *Eksp. Niezawodn. Maint. Reliab.* **2025**. *online first*. [CrossRef]
4. Jaroń, A.; Borucka, A.; Deliś, P.; Sekrecka, A. An Assessment of the Possibility of Using Unmanned Aerial Vehicles to Identify an Map Air Pollution from Infrastructure Emissions. *Energies* **2024**, *17*, 577. [CrossRef]
5. IPCC. *Climate Change 2014: Synthesis Report*; IPCC: Geneva, Switzerland, 2014; ISBN 978-92-9169-143-2. Available online https://www.ipcc.ch/site/assets/uploads/2018/05/SYR_AR5_FINAL_full_wcover.pdf (accessed on 7 April 2025).
6. WHO. *WHO Global Air Quality Guidelines: Particulate Matter (PM2.5 and PM10), Ozone, Nitrogen Dioxide, Sulfur Dioxide and Carbo Monoxide*; World Health Organization: Geneva, Switzerland, 2021; ISBN 978-92-4-003422-8.
7. Sovacool, B.K. Contesting the Science of Nuclear Power: A Critical Review of the Field. *Energy Policy* **2011**, *39*, 7263–727 [CrossRef]
8. IEA. *World Energy Outlook 2023*; International Energy Agency: Paris, France, 2023; ISBN 978-92-64-99157-8.
9. Pehnt, M.; Oeser, T.; Scholz, R.W. Life Cycle Assessment of Biomass Electricity. *Energy* **2006**, *31*, 553–563. [CrossRef]
10. Schleisner, L. Life Cycle Assessment of a Wind Farm. *Renew. Energy* **2000**, *20*, 129–142. [CrossRef]
11. Godfray, C.; Beddington, J.; Crute, I.; Haddad, L.; Lawrence, D.; Muir, J.; Pretty, J.; Robinson, S.; Thomas, S.; Toulmin, C. Fooc Security: The Challenge of Feeding 9 Billion People. *Science* **2010**, *327*, 812–818. [CrossRef]

Rykała, M.; Grzelak, M.; Borucka, A. Analysis of the Impact of Selected Dynamic Parameters of a Motor Vehicle on CO_2 Emissions Using Logistic Regression. *Appl. Sci.* **2024**, *14*, 10349. [CrossRef]

Acemoglu, D.; Aghion, P.; Bursztyn, L.; Hemous, D. The Environment and Directed Technical Change. *Am. Econ. Rev.* **2012**, *102*, 131–166. [CrossRef]

Jaffe, A.B.; Newell, R.G.; Stavins, R.N. A Tale of Two Market Failures: Technology and Environmental Policy. *Ecol. Econ.* **2005**, *54*, 164–174. [CrossRef]

Foster, R.N. *Creative Destructional Innovation*; Berrett-Koehler Publishers: San Francisco, CA, USA, 2010; ISBN 978-1-60509-677-8.

Grübler, A.; Nakićenović, N.; Victor, D.G. Dynamics of Energy Technologies and Global Change. *Energy Policy* **1999**, *27*, 247–280. [CrossRef]

Stern, P.C. Toward a Coherent Theory of Environmentally Significant Behavior. *J. Soc. Issues* **2000**, *56*, 407–424. [CrossRef]

Rogers, E.M. *Diffusion of Innovations*, 5th ed.; Free Press: New York, NY, USA, 2003; ISBN 978-0-7432-5809-9.

Helm, D. Climate Change Policy: Why Has So Little Been Achieved? *Oxf. Rev. Econ. Policy* **2010**, *26*, 78–91. [CrossRef]

Vieira, A.S.; Stewart, R.A.; Lamberts, R.; Beal, C.D. Renewable Energy and Energy Conservation Area Policy (REECAP) Framework: A Novel Methodology for Bottom-Up and Top-Down Principles Integration. *Energy Strategy Rev.* **2020**, *32*, 100544. [CrossRef]

Smil, V. *Energy Transitions: Global and National Perspectives*, 2nd ed.; Praeger: Santa Barbara, CA, USA, 2017; ISBN 978-1-4408-5405-4.

IRENA. *Renewable Energy Statistics 2023*; International Renewable Energy Agency: Abu Dhabi, United Arab Emirates, 2023; ISBN 978-92-9260-537-7. Available online: https://www.irena.org/Publications/2023/Jul/Renewable-energy-statistics-2023 (accessed on 7 April 2025).

Jacobson, M.Z.; Delucchi, M.A.; Bazouin, G.; Bauer, Z.A.F.; Heavey, C.C.; Fisher, E.; Morris, S.B.; Piekutowski, D.J.; Vencill, T.A.; Yeskoo, T.W. 100% Clean and Renewable Wind, Water, and Sunlight (WWS) All-Sector Energy Roadmaps for the 50 United States. *Energy Environ. Sci.* **2015**, *8*, 2093–2117. [CrossRef]

IEA. *The Future of Electrification*; International Energy Agency: Paris, France, 2024; ISBN 978-92-64-86291-5.

Falkner, R. The Paris Agreement and the New Logic of International Climate Politics. *Int. Aff.* **2016**, *92*, 1107–1125. [CrossRef]

EUR-Lex. Reduction of CO_2 Emissions from New Passenger Cars and New Light Commercial Vehicles. Available online: https://eur-lex.europa.eu/PL/legal-content/summary/reduction-in-co2-emissions-of-new-passenger-cars-and-of-new-light-commercial-vehicles.html?fromSummary=32 (accessed on 7 April 2025).

Bandekar, P.A.; Putman, B.; Thoma, G.; Matlock, M. Cradle-to-Grave Life Cycle Assessment of Production and Consumption of Pulses in the United States. *J. Environ. Manag.* **2022**, *302*, 114062. [CrossRef]

Gursel, I.V.; Moretti, C.; Hamelin, L.; Jakobsen, L.G.; Steingrimsdottir, M.M.; Junginger, M.; Høibye, L.; Shen, L. Comparative Cradle-to-Grave Life Cycle Assessment of Bio-Based and Petrochemical PET Bottles. *Sci. Total Environ.* **2021**, *793*, 148642. [CrossRef]

Buberger, J.; Kersten, A.; Kuder, M.; Eckerle, R.; Weyh, T.; Thiringer, T. Total CO_2-Equivalent Life-Cycle Emissions from Commercially Available Passenger Cars. *Renew. Sustain. Energy Rev.* **2022**, *159*, 112158. [CrossRef]

Wang, N.; Tang, G. A Review on Environmental Efficiency Evaluation of New Energy Vehicles Using Life Cycle Analysis. *Sustainability* **2022**, *14*, 3371. [CrossRef]

Oda, H.; Noguchi, H.; Fuse, M. Review of Life Cycle Assessment for Automobiles: A Meta-Analysis-Based Approach. *Renew. Sustain. Energy Rev.* **2022**, *159*, 112214. [CrossRef]

Koroma, M.S.; Costa, D.; Philippot, M.; Cardellini, G.; Hosen, M.S.; Coosemans, T.; Messagie, M. Life Cycle Assessment of Battery Electric Vehicles: Implications of Future Electricity Mix and Different Battery End-of-Life Management. *Sci. Total Environ.* **2022**, *831*, 154859. [CrossRef]

Verma, S.; Dwivedi, G.; Verma, P. Life Cycle Assessment of Electric Vehicles in Comparison to Combustion Engine Vehicles: A Review. *Mater. Today Proc.* **2022**, *49*, 217–222. [CrossRef]

Li, J.; Li, L.; Yang, R.; Jiao, J. Assessment of the lifecycle carbon emission and energy consumption of lithium-ion power batteries recycling: A systematic review and meta-analysis. *J. Energy Storage* **2023**, *65*, 107306. [CrossRef]

Zhao, E.; Walker, P.D.; Surawski, N.C.; Bennett, N.S. Assessing the life cycle cumulative energy demand and greenhouse gas emissions of lithium-ion batteries. *J. Energy Storage* **2021**, *43*, 103193. [CrossRef]

Peters, J.F.; Baumann, M.; Zimmermann, B.; Braun, J.; Weil, M. The environmental impact of Li-Ion batteries and the role of key parameters–A review. *Renew. Sustain. Energy Rev.* **2017**, *67*, 491–506. [CrossRef]

Bouter, A.; Guichet, X. The greenhouse gas emissions of automotive lithium-ion batteries: A statistical review of life cycle assessment studies. *J. Clean. Prod.* **2022**, *344*, 130994. [CrossRef]

Li, P.; Xia, X.; Guo, J. A review of the life cycle carbon footprint of electric vehicle batteries. *Sep. Purif. Technol.* **2022**, *296*, 121389. [CrossRef]

39. Aguilar, P.; Groß, B. Battery electric vehicles and fuel cell electric vehicles, an analysis of alternative powertrains as a mean t decarbonise the transport sector. *Sustain. Energy Technol. Assess.* **2022**, *53*, 102624.
40. Edwards, R.; Larivé, J.-F.; Rickeard, D. Well-to-Wheels Analysis of Future Automotive Fuels and Powertrains in the Europea Context. *SAE Trans.* **2004**, *113*, 1072–1084.
41. Orsi, F.; Muratori, M.; Rocco, M.; Colombo, E.; Rizzoni, G. A Multi-Dimensional Well-to-Wheels Analysis of Passenger Vehicle in Different Regions: Primary Energy Consumption, CO_2 Emissions, and Economic Cost. *Appl. Energy* **2016**, *169*, 197–20 [CrossRef]
42. Zhang, Y.; Zhao, L. Linear regression applications for environmental emissions trend forecasting. *Environ. Model. Softw.* **202** *150*, 105318. [CrossRef]
43. Andrych-Zalewska, M.; Chłopek, Z.; Pielecha, J.; Merkisz, J. Investigation of Exhaust Emissions from the Gasoline Engine of Light Duty Vehicle in the Real Driving Emissions Test. *Ekspl. Niezawodn. Maint. Reliab.* **2023**, *25*, 165880. [CrossRef]
44. Furch, J.; Jelínek, J. Design of a Tribotechnical Diagnostics Model for Determining the Technical Condition of an Interna Combustion Engine during Its Life Cycle. *Ekspl. Niezawodn.–Maint. Reliab.* **2022**, *24*, 5. [CrossRef]
45. Tucki, K. A computer tool for modelling CO_2 emissions in driving tests for vehicles with diesel engines. *Energies* **2021**, *14*, 26 [CrossRef]
46. Benga, G.C.; Savu, S.V.; Stefan, I. Analysis of CO_2 Emissions for a Supercharged Engine and a Series-Parallel Hybrid System. I Proceedings of the International Conference on Mechanical Engineering (ICOME 2022); Springer Nature: Berlin/Heidelber Germany, 2023; Volume 15, p. 331.
47. Scherelis, G.; Ludewig, C. *Volkswagen Group, Media Information NO. 128/2019 Electric Vehicles with Lowest CO_2 Emissions*; Volkswage Group: Wolfsburt, Germany, 2019.
48. Green NCAP. Life Cycle Assessment Results & Fact Sheets. 2024. Available online: https://www.greenncap.com/european-lca results/ (accessed on 4 June 2025).
49. Joshi, A. Review of vehicle engine efficiency and emissions. *SAE Int. J. Adv. Curr. Pract. Mobil.* **2022**, *4*, 1704–1733. [CrossRef]
50. Chandrashekar, C.; Rawat, R.S.; Chatterjee, P.; Pawar, D.S. Evaluating the real-world emissions of diesel passenger Car in India heterogeneous traffic. *Environ. Monit. Assess.* **2023**, *195*, 1248. [CrossRef]
51. Gentil, E.C.; Damgaard, A.; Hauschild, M.; Finnveden, G.; Eriksson, O.; Thorneloe, S.; Kaplan, P.O.; Barlaz, M.; Muller, O.; Matsu Y.; et al. Models for waste life cycle assessment: Review of technical assumptions. *Waste Manag.* **2010**, *30*, 2636–2648. [CrossRef
52. Baars, J.; Domenech, T.; Bleischwitz, R.; Melin, H.E.; Heidrich, O. Circular economy strategies for electric vehicle batteries reduc reliance on raw materials. *Nat. Sustain.* **2021**, *4*, 71–79. [CrossRef]
53. Kostopoulos, E.D.; Spyropoulos, G.C.; Kaldellis, J.K. Real-world study for the optimal charging of electric vehicles. *Energy Re* **2020**, *6*, 418–426. [CrossRef]
54. Cronshaw, M. *Energy in Perspective*; Springer: Berlin/Heidelberg, Germany, 2021. [CrossRef]
55. Papadimitriou, D.; Duysinx, P. *Future Move: A Review of the Main Trends in the Automotive Sector at Horizon 2030 in the Great Regio* University of Liège: Liège, Belgium, 2022.
56. Kawecka-Wyrzykowska, E. Financing Energy Transition in Poland: Possible Contribution of EU Funds. *Eur. Integr. Stud.* **2022**, *1* 65–77. [CrossRef]
57. Malik, F.H.; Ayadi, W.; Hussain, G.A.; Haider, Z.M.; Alkhatib, F.; Lehtonen, M. Evaluating Carbon Emissions: A Lifecycl Comparison Between Electric and Conventional Vehicles. *World Electr. Veh. J.* **2025**, *16*, 287. [CrossRef]
58. Prystrom, J. Eco-innovation as a factor of competitiveness on the example of Volvo Car Corporation. *Mark. Rynek* **2023**, *4*, 35–4 [CrossRef]
59. Razmjoo, A.; Kaigutha, L.G.; Rad, M.V.; Marzband, M.; Davarpanah, A.; Denai, M.J.R.E. A Technical analysis investigating energ sustainability utilizing reliable renewable energy sources to reduce CO_2 emissions in a high potential area. *Renew. Energy* **2021** *164*, 46–57. [CrossRef]
60. Savage, J.; Esposito, G. Vehicle Life Cycle GHG Emissions Study to Show the Role of Renewable Fuels in Meeting Net Zer Zemo Partnership. 2024. Available online: https://www.zemo.org.uk/assets/reports/Vehicle%20life%20cycle%20GHG%2 emissions%20study%202024.pdf (accessed on 7 April 2025).
61. Manufacturer, X. Sales Summary of Brand X Vehicles, 2020–2024 nationwide data, Poland, 2020–2024, available in 2025 (interna data, unpublished).
62. Bieker, G. *A Global Comparison of the Life-Cycle Greenhouse Gas Emissions of Combustion Engine and Electric Passenger Cars*; ICCT Berlin, Germany, 2021. Available online: https://theicct.org/wp-content/uploads/2021/12/zevtc-decarbonizing-by-2050-Jul20 1%E2%80%AF.pdf (accessed on 7 April 2025).
63. Searle, S.; Bieker, G.; Baldino, C. *Decarbonizing Road Transport by 2050 Zero-Emission Pathways for Passenger Vehicles*; Internationa Council on Clean Transportation: Washington, DC, USA, 2021. Available online: https://theicct.org/publication/decarbonizing road-transport-by-2050-zero-emission-pathways-for-passenger-vehicles/ (accessed on 5 May 2025).

International Council on Clean Transportation. *A Global Comparison of the Life-Cycle Greenhouse Gas Emissions of Combustion Engine and Electric Passenger Cars*; International Council on Clean Transportation: Washington, DC, USA, 2022. Available online: https://theicct.org/publication/a-global-comparison-of-the-life-cycle-greenhouse-gas-emissions-of-combustion-engine-and-electric-passenger-cars/ (accessed on 7 April 2025).

Adamiak, B.; Andrych-Zalewska, M.; Merkisz, J.; Chłopek, Z. The uniqueness of pollutant emission and fuel consumption test results for road vehicles tested on a chassis dynamometer. *Eksploat. Niezawodn. Maint. Reliab.* **2025**, *27*, 195747. [CrossRef]

Stan, C. Flora and fauna have inversed carbon dioxide cycles. In *Energy Versus Carbon Dioxide*; Springer: Berlin/Heidelberg, Germany, 2022. [CrossRef]

Sinn, H.W. *The Green Paradox: A Supply-Side Approach to Global Warming*; MIT Press: Cambridge, MA, USA, 2012.

Maser, C.; Pollio, C.A. *Resolving Environmental Conflict Towards Sustainable Community Development*; CRC Press: Boca Raton, FL, USA, 1995.

EUR-Lex. *Regulation (EU) 2019/631: CO2 Emission Performance Standards for New Passenger Cars and for New Light Commercial Vehicles*; European Commission: Brussels, Belgium, 2023. Available online: https://eur-lex.europa.eu/legal-content/EN/TXT/?uri=CELEX:32019R0631 (accessed on 7 April 2025).

Umweltbundesamt. *CO2 Emissions from Transport—Methodological Framework and Results from a Life Cycle Perspective*; German Federal Environment Agency: Dessau-Roßlau, Germany, 2023. Available online: https://www.umweltbundesamt.de/en/publikationen/co2-emissions-from-transport (accessed on 7 April 2025).

Turek, M.; Kalupová, B.; Cempírek, V.; Tůmová, K.; Abramović, B. Electromobility and the Environment in the Czech Republic. *LOGI Sci. J. Transp. Logist.* **2024**, *15*, 201–212. [CrossRef]

Christoforidis, C.; Christakis, I.; Kotzasavva, F.; Fousekis, G.; Rimpas, D. Electric Vehicles Charging Stations and Their Influence on the Electricity Utility Grid. *Arch. Motoryz.* **2025**, *107*, 102–126. [CrossRef]

Bhat, M.Y.; Sofi, A.A.; Ganie, J.A. Green Wheels in Motion: Electric Vehicle Sales in the Path to Decarbonization. *Transp. Res. D Transp. Environ.* **2025**, *142*, 104704. [CrossRef]

Paisarnvirosrak, N.; Rungrueang, P. Firefly Algorithm with Tabu Search to Solve the Vehicle Routing Problem with Minimized Fuel Emissions: Case Study of Canned Fruits Transport. *LOGI Sci. J. Transp. Logist.* **2023**, *14*, 263–274. [CrossRef]

Tilly, N.; Yigitcanlar, T.; Degirmenci, K.; Paz, A. How Sustainable Is Electric Vehicle Adoption? Insights from a PRISMA Review. *Sustain. Cities Soc.* **2024**, *117*, 105950. [CrossRef]

Pečman, J.; Šarkan, B.; Ližbetinová, L.; Ľupták, V.; Loman, M.; Bartuška, L. Impact of Acceleration Style on Vehicle Emissions and Perspectives for Improvement through Transportation Engineering Solutions. *Arch. Motoryz.* **2024**, *104*, 48–62. [CrossRef]

Hiller, D. Savings of Material Resources and Carbon Emissions When Converting Fossil Fuel Car to Electric: A Case Study for Sweden. Master's Thesis, University of Stuttgart, Institute of Automotive Engineering, Stuttgart, Germany, 2022. Available online: https://urn.kb.se/resolve?urn=urn:nbn:se:kth:diva-311253 (accessed on 7 April 2025).

energies

MDPI

Article

Occupant-Centric Load Optimization in Smart Green Townhouses Using Machine Learning

Seyed Morteza Moghimi [1,*], Thomas Aaron Gulliver [1,*], Ilamparithi Thirumarai Chelvan [1,†] and Hossen Teimoorinia [2,3,†]

[1] Department of Electrical and Computer Engineering, University of Victoria, STN CSC, P.O. Box 1700, Victoria, BC V8W 2Y2, Canada
[2] Department of Physics and Astronomy, University of Victoria, Victoria, BC V8P 5C2, Canada
[3] NRC Herzberg Astronomy and Astrophysics, 5071 West Saanich Road, Victoria, BC V9E 2E7, Canada
* Correspondence: seyedmortezamoghimi@uvic.ca (S.M.M.); agullive@uvic.ca (T.A.G.)
† These authors contributed equally to this work.

Abstract

This paper presents an occupant-centric load optimization framework for Smart Green Townhouses (SGTs). A hybrid Long Short-Term Memory and Convolutional Neural Network (LSTM-CNN) model is combined with real-time Internet of Things (IoT) data to predict and optimize energy usage based on occupant behavior and environmental conditions. Multi-Objective Particle Swarm Optimization (MOPSO) is applied to balance energy efficiency, cost reduction, and occupant comfort. This approach enables intelligent control of HVAC systems, lighting, and appliances. The proposed framework is shown to significantly reduce load demand, peak consumption, costs, and carbon emissions while improving thermal comfort and lighting adequacy. These results highlight the potential to provide adaptive solutions for sustainable residential energy management.

Keywords: occupant satisfaction; green townhouse; smart townhouse; machine learning; load demand; optimization

1. Introduction

The increasing demand for energy efficiency and sustainability in the building sector has led to substantial advances in smart building technologies. Smart Green Townhouses (SGTs) integrate Renewable Energy Sources (RESs), Internet of Things (IoT) devices, and advanced control systems to optimize energy consumption, reduce costs, and lower environmental impact. However, effective load management in residential buildings remains challenging due to the influence of occupant behavior on energy use patterns [1,2]. Occupant energy behavior is shaped by psychological and economic factors such as daily routines, comfort preferences, and decision-making habits [3]. These factors are captured in the proposed framework by modeling occupancy patterns based on presence and usage tendencies. This enables real-time occupant-centric optimization for improved energy efficiency in dynamic residential environments.

Machine Learning (ML) has emerged as a powerful tool for managing the complexity of load forecasting and control in smart buildings [4]. ML models allow for accurate prediction and effective control in real-time systems [5,6]. For example, hybrid deep learning architectures, such as the combination of Convolutional Neural Networks (CNNs) and Long Short-Term Memory (LSTM) networks, have been used for robust and accurate forecasting [7]. CNN-LSTM models have been combined with metaheuristic algorithms

https://doi.org/10.3390/en1813332

like the Coati Optimization Algorithm for renewable energy forecasting [8]. While these methods have proven effective, existing research does not sufficiently address the role of real-time occupant behavior in residential load optimization [5,6,9].

Deep learning has been considered for equipment scheduling and energy control in dynamic systems [10,11]. However, few approaches provide real-time adaptability while balancing energy savings, cost efficiency, and occupant comfort. While occupancy-based HVAC prediction and control have been employed [12,13], integration with occupant-aware load forecasting has not been considered. Thus, a real-time, occupant-centric load optimization framework is proposed. This framework integrates a hybrid LSTM-CNN model with Multi-Objective Particle Swarm Optimization (MOPSO) to dynamically balance load demand, cost, emissions, and comfort. Public datasets and real-time IoT sensor data [14–16] are used to enable intelligent control of systems such as HVAC and lighting based on actual occupancy and preferences. While LSTM-CNN models have been employed in energy forecasting, the integration with real-time occupant-centric data has not been considered. The results presented for Connected Smart Green Townhouses (CGSTs) demonstrate the applicability, scalability, and performance of the proposed framework in realistic residential scenarios.

The contributions of this work are as follows.

- A hybrid LSTM-CNN model is proposed to predict load demand, cost, and emissions.
- Occupant behavior is integrated into a dynamic multi-objective optimization framework.
- The proposed framework is evaluated for four connected townhouses in Burnaby, British Columbia (BC), Canada. The results obtained indicate significant load reduction and cost savings while ensuring occupant satisfaction.

The remainder of this paper is structured as follows. Section 2 presents the methodology and optimization algorithms. The performance is evaluated in Section 3 and the results are discussed. Section 4 provides some concluding remarks including the implications for residential load optimization.

2. Methodology

A hybrid LSTM-CNN model is employed for dynamic load, cost, and carbon emission prediction [5,6]. It was implemented using Python with Pandas for data manipulation, NumPy for calculations, and Matplotlib for visualization [5]. Although the framework has been designed for real-time applications, model execution time and update frequency depend on available computational resources and sensor sampling rates. In the experiments, the hybrid LSTM-CNN model was used for prediction and optimization for a 24 h horizon within 12–18 s on a standard desktop computer which is sufficient for real-time hourly operation. The framework can be configured to run at hourly intervals for day-ahead optimization and at shorter intervals (e.g., 15 min) for finer control. An LSTM is used due to its ability to capture long-term dependencies and temporal patterns in sequential data, which is critical for accurate forecasting of loads influenced by occupant behavior and weather. A CNN is employed to extract local features and spatial correlations from multi-dimensional inputs such as occupancy and environmental data. In this paper, occupant behavior refers to measurable actions and patterns that affect residential energy consumption. This includes real-time presence detection (e.g., motion and door sensors), appliance usage habits (e.g., when and how long devices are used), HVAC and lighting preferences (e.g., thermostat setpoints, lighting usage), and feedback interactions with control systems. This behavior is inferred from IoT sensor data and smart devices to support dynamic load optimization.

The MOPSO algorithm is used to optimize the tradeoff between load efficiency, cost savings, and carbon emission reduction while ensuring occupant comfort. It was chosen considering the advantages and applicability to SGT systems [5]. It is a well-established algorithm that has excellent convergence [17], diversity preservation, and suitability for nonlinear, multi-objective problems. Here, MOPSO is used to effectively balance load, cost, and emissions while ensuring occupant comfort which is particularly challenging in real-time residential applications. This illustrates the practicality of MOPSO for the optimization of complex, dynamic building energy systems.

Real data from several sources is employed including occupancy and load demand data from [14,15], IoT sensor data from [16], weather data from [18,19], and energy prices from [18]. Data from IoT sensors (e.g., motion detectors, door sensors) and smart devices (e.g., thermostats) are integrated with utility data to infer occupancy patterns [20]. The AMPds dataset [14,15] provides high-resolution time-series data on electricity, water, and natural gas consumption. This facilitates load demand modeling under seasonal and weather variations. IoT sensors support real-time adjustments due to changes in load [16]. Energy-efficient technology such as heat pumps [21] ensure efficient and sustainable SGTs [22].

Although the focus here is on short-term, real-time occupant behavior, the proposed framework can easily be adapted to long-term patterns. Seasonal and holiday-related behavior can be captured using historical data to improve load prediction while maintaining adaptability to dynamic occupant profiles. However, the goal here is real-time prediction in a dynamic environment that is more challenging due to the very short time scale.

2.1. The Proposed Framework

The proposed framework uses publicly available datasets [14–16] and real-time IoT data collected from SGTs [5,6]. To ensure consistency across diverse features, the data are normalized to the range [0, 1]. The normalization for load type j (e.g., electricity, gas, water) and townhouse type i is

$$L_{\text{normalized},i,j}(t) = \frac{L_{i,j}(t) - L_{\text{min},i,j}}{L_{\text{max},i,j} - L_{\text{min},i,j}}, \tag{1}$$

where $L_{i,j}(t)$ is the load for townhouse type i and load type j at time t, $L_{\text{max},i,j}$ is the maximum load for townhouse type i and load type j across the dataset, and $L_{\text{min},i,j}$ is the corresponding minimum load. This ensures that normalization is performed independently for each load and townhouse type to prevent any single parameter from dominating the total load profile due to differences in magnitude or units such as electricity (kWh), gas (m^3), and water (liters). The normalization for the ith bedroom townhouse is

$$L_{\text{normalized},i}(t) = \frac{1}{3}\sum_{j=1}^{3} L_{\text{normalized},i,j}(t), \tag{2}$$

Normalization is critical for the performance of ML models so that features with a large numerical range do not disproportionately influence model learning. It also improves numerical stability and accelerates convergence, particularly for gradient-based optimizers such as ADAM used in the proposed LSTM-CNN model.

The electricity load in kilowatt-hours (kWh) is

$$\text{Electricity Load (kWh)} = L_{\text{appliance}} + L_{\text{lighting}} + L_{\text{HVAC}}, \tag{3}$$

where $L_{\text{appliance}}$, L_{lighting}, and L_{HVAC} represent the demand from electrical appliances, lighting, and HVAC systems, respectively. The gas load is

$$\text{Gas Load (kWh)} = \text{Gas Volume} \cdot C_{\text{gas}}, \tag{4}$$

where Gas Volume (m^3) is the volume of natural gas consumed and C_{gas} is the calorific value of gas (kWh/m^3). The water load is

$$\text{Water Load (kWh)} = \text{Water Volume} \cdot C_{water}, \tag{5}$$

where Water Volume (m^3) is the volume of water used and C_{water} is the energy required to pump, heat, and treat water (kWh/m^3).

The base (unoptimized) load demand at time t is given by

$$L_{base}(t) = L_{demand}(t) - L_{renewable}(t), \tag{6}$$

where $L_{demand}(t)$ is the total unoptimized load demand (kW) and $L_{renewable}(t)$ is the unoptimized renewable energy (e.g., PV power). The optimized load without occupant data at time t is

$$L_{opt,noOcc}(t) = L_{base}(t) - (L_{renewable,opt}(t) + L_{battery}(t)), \tag{7}$$

where $L_{renewable,opt}(t)$ is the optimized renewable energy (kW) and $L_{battery}(t)$ is the battery energy used to offset grid demand (kW). The maximum limits for battery and grid power ($L_{battery,max}$ and $L_{grid,max}$) are based on practical system design assumptions supported by real-world parameters. $L_{battery,max}$ is set in the range 5–10 kW. This aligns with commercially available residential-scale lithium-ion battery systems, and is consistent with industry-standard systems used in residential microgrids in Canada [5,6]. $L_{grid,max}$ is set in the range 12–15 kW. This reflects typical urban service capacity in a townhouse unit in BC and aligns with empirical values used in predicting load management for smart grid applications [9,13]. These values are not dictated by specific regulations but are based on realistic operating conditions and previous research.

The optimized load with occupant data at time t considers occupant-driven factors and is given by

$$L_{opt,occ}(t) = L_{opt,noOcc}(t) - L_{occ,adj}(t), \tag{8}$$

where $L_{occ,adjustment}(t)$ accounts for real-time occupancy-based load optimization including HVAC demand adjustments considering presence, load shifting to avoid peak times, and adaptive appliance control (e.g., delayed washing machine cycles). Substituting (7) in (8) gives

$$L_{opt,occ}(t) = L_{base}(t) - (L_{renewable,opt}(t) + L_{battery}(t) + L_{occ,adj}(t)). \tag{9}$$

2.2. LSTM-CNN Model

The LSTM is used to capture temporal patterns in the input data to learn dependencies over time. The hidden state at time t is

$$h_t = \sigma(W_x x_t + W_h h_{t-1} + b_l), \tag{10}$$

where x_t is the input at time t which includes features such as load demand and occupancy data, W_x and W_h are weight matrices for the input data and previous hidden state, respectively, b_l is the bias term, and σ is the sigmoid activation function

$$\sigma(x) = \frac{1}{1 + e^{-x}}. \tag{11}$$

This function limits values to the range $[0, 1]$ and helps the network learn complex, non-linear relationships by introducing smooth gradients and preventing large outputs.

The CNN extracts spatial features from multi-dimensional data. The CNN output is

$$y = \sigma(\text{Conv2D}(x, W) + b_c), \tag{12}$$

where x is the input data structured as a two-dimensional (2D) matrix, Conv2D() is 2D convolution given by

$$\text{Conv2D}(x, W) = \sum_i \sum_j W[i, j] \cdot x[i, j],$$

where W is the convolutional filter (kernel) used to extract local patterns, and b_c is the bias term. The kernel is a small matrix of learnable weights used to extract spatial features from the input data. During training, it slides across the input matrix to learn local patterns such as occupancy and appliance usage. This helps improve the prediction capability of the model. The bias terms are randomly initialized and updated each iteration via backpropagation using the ADAM optimizer.

The combined LSTM and CNN output is

$$\text{Output} = g(h_t, y), \tag{13}$$

where $g()$ is the fusion function which here is concatenation. The average training time was 4.6 h. A dropout rate of 0.2 is applied after each LSTM and CNN layer to prevent overfitting. The model was trained using the Adam optimizer with a learning rate of 0.001 and batch size of 64.

The MOPSO algorithm objective function is

$$F = w_1 f_1 + w_2 f_2 + w_3 f_3, \tag{14}$$

where

$$f_1 = \sum_{t=1}^{N} L_{\text{demand}}(t), \tag{15}$$

is the total load demand in kWh that includes all electrical appliances, HVAC systems, and other energy-consuming devices within the building

$$f_2 = \sum_{t=1}^{N} L_{\text{demand}}(t) \cdot C_{\text{electricity}}, \tag{16}$$

is the operational costs

$$f_3 = \sum_{t=1}^{N} L_{\text{demand}}(t) \cdot E_{\text{factor}}, \tag{17}$$

is the carbon emissions, N is the number of time steps, $C_{\text{electricity}}$ is the electricity cost per kWh, E_{factor} is the carbon emission factor in kg CO_2/kWh, and w_1, w_2, and w_3 are weights representing the importance of each objective. The weights used here are $w_1 = 0.3$, $w_2 = 0.4$, and $w_3 = 0.3$. They were selected empirically to balance the three goals: load efficiency, operational cost reduction, and carbon emission reduction and are the result of extensive evaluation across many scenarios. Parameter tuning was conducted to ensure that no single objective disproportionately dominates the optimization. The corresponding constraints are

$$L_{\text{grid}}(t) + L_{\text{renewable}}(t) + L_{\text{battery}}(t) \geq L_{\text{demand}}(t), \quad \forall t \tag{18}$$

$$0 \leq L_{\text{grid}}(t) \leq L_{\text{grid,max}}, \tag{19}$$

$$0 \leq L_{\text{battery}}(t) \leq L_{\text{battery,max}}, \tag{20}$$

where $L_{grid}(t)$ is the load supplied by the grid (kWh), $L_{renewable}(t)$ is the load met by renewable sources (kWh), and $L_{battery}(t)$ is the energy supplied from battery storage (kWh).

To maintain occupant comfort, the deviation between the actual and desired indoor temperature should be within an acceptable range

$$\sum_{t=1}^{N} L_{demand}(t) \cdot |T_{actual}(t) - T_{desired}(t)| \leq \text{Threshold}, \tag{21}$$

where $T_{actual}(t)$ is the actual indoor temperature at time t, $T_{desired}(t)$ is the corresponding desired temperature, and Threshold is the maximum acceptable cumulative deviation over the N time steps (optimization period). This ensures energy savings do not compromise thermal comfort so load scheduling decisions account for occupant preferences.

2.3. Performance Metrics

Occupant comfort satisfaction is based on thermal comfort, lighting adequacy, and feedback adherence. The thermal comfort measures how close the indoor temperature is to the desired temperature and is given by

$$\text{Thermal Comfort}(t) = 1 - \min\left[1, \frac{|T_{actual}(t) - T_{desired}(t)|}{5}\right], \tag{22}$$

where 5 °C is considered the maximum difference. The lighting adequacy assesses how well lighting levels meet occupant needs. It is based on the percentage of time that the lighting intensity stays within the desired range and is expressed as

$$\text{Lighting Adequacy}(t) = \frac{\sum_{i=1}^{r} \beta(L_{min} \leq L_i(t) \leq L_{max})}{r}, \tag{23}$$

where $L_i(t)$ is the actual lighting intensity in room i at time t, L_{min} and L_{max} are the minimum and maximum acceptable light levels, $\beta()$ is an indicator function that returns 1 if the lighting is within the desired range, and 0 otherwise, and r is the number of rooms. Feedback adherence measures how well the building systems (e.g., HVAC, lighting) respond to occupant feedback and is given by

$$\text{Feedback Adherence}(t) = \frac{F_{implemented(t)}}{F_{submitted(t)}}, \tag{24}$$

where $F_{implemented}(t)$ is the number of feedback requests that were implemented by the system at time t and $F_{submitted}(t)$ is the corresponding number of feedback requests submitted by occupants.

The occupant comfort satisfaction at time t is a weighted sum of thermal comfort, lighting adequacy, and feedback adherence and is expressed as

$$\begin{aligned}\text{Occupant Comfort Satisfaction}(t) = w_T &\cdot \text{Thermal Comfort}(t) \\ + w_L &\cdot \text{Lighting Adequacy}(t) \\ + w_F &\cdot \text{Feedback Adherence}(t),\end{aligned} \tag{25}$$

where $w_T = 0.4$, $w_L = 0.3$, and $w_F = 0.3$. These weights are based on the relative importance of thermal comfort, lighting adequacy, and feedback adherence in the overall comfort of the occupants.

The performance evaluation metrics are

$$\text{Mean Absolute Error (MAE)} = \frac{1}{n}\sum_{i=1}^{n} |\text{Predicted}_i - \text{Actual}_i|, \tag{26}$$

$$\text{Root Mean Squared Error (RMSE)} = \sqrt{\frac{1}{n}\sum_{i=1}^{n} (\text{Predicted}_i - \text{Actual}_i)^2}, \tag{27}$$

$$\text{Coefficient of Determination } (R^2) = 1 - \frac{\sum_{i=1}^{n} (\text{Predicted}_i - \text{Actual}_i)^2}{\sum_{i=1}^{n} (\text{Actual}_i - \overline{\text{Actual}})^2}, \tag{28}$$

where n is the number of data values, Predicted_i is the ith predicted value from the ML model, Actual_i is the corresponding actual value, and $\overline{\text{Actual}}$ is the mean of the actual values given by

$$\overline{\text{Actual}} = \frac{1}{n}\sum_{i=1}^{n} \text{Actual}_i. \tag{29}$$

The MAE is the average magnitude of the prediction errors so smaller values indicate better accuracy. The RMSE measures the standard deviation of prediction errors, penalizing larger errors more heavily than the MAE. Lower values indicate higher accuracy. R^2 indicates how well the model explains the variations in the actual data. Values closer to 1 signify better model performance.

3. Performance Results

The results presented in this section were generated using OpenStudio (v3.4.0) for load simulation with and without occupant input. Data analysis and optimization were performed using Python v3.11.5 with Pandas v2.1.1 and Matplotlib v3.8.0 [5,6]. Real-time occupancy data was collected with ThingSpeak v2.0 [23].

We consider a CSGT complex with 1-bedroom, 2-bedroom, 3-bedroom, and 4-bedroom units as shown in Figure 1. Connected townhouses include two key components: connected water systems [14,24] and party walls [25,26]. Party walls are shared walls between adjacent properties and are jointly owned and maintained by property owners. Party wall agreements outline the responsibilities for maintenance, repairs, alterations, and dispute resolution.

Figure 1. Four connected SGTs as a townhouse complex.

Table 1 gives the SGT parameters including the number of residents, occupancy, hot water usage, lighting, EV charging, HVAC consumption, shared wall insulation, and fire safety. Table 2 presents the base and optimization results with and without occupant data for the CSGT complex. The percentage reduction compared to the base results is also given. This indicates a significant improvement in load efficiency, cost savings, and carbon emission reduction. The load is decreased by 10.6% without occupant data and 14.3% with occupant data, confirming the effectiveness of real-time adaptive load management. Operational costs are reduced by up to 13.0% and carbon emissions are 11.0% and 15.5% lower, respectively. Peak load is reduced by 10.5% and 14.0% which helps grid stability and lowers peak-hour costs. Overall, occupant-centric optimization increases load efficiency, cost savings, and environmental performance, making CSGTs a practical solution for the future.

Table 1. CSGT Parameters.

eature	1-Bedroom CSGT	2-Bedroom CSGT	3-Bedroom CSGT	4-Bedroom CSGT	CSGT
esidents	2 (Couple)	3 (Couple + Child)	4 (Couple + 2 Children)	5 (Couple + 3 Children)	Multiple Units
eak Load (kW)	7	9	12	14	36
ccupancy (Hours/Day)	8–12	10–16	14–18	16–20	Distributed
Iot Water Usage Liters/Day)	180	270	400	460	1020
ighting (Hours/Day)	3–5	5–7	6–9	6–10	N/A
V Charging	N/A	N/A	1–2 EVs (2–3 h/day)	2 EVs (2–4 h/day)	N/A
IVAC Consumption kWh/m^2/Year)	26	28	30	32	N/A
hared Wall Insulation	R-22	R-22	R-22	R-22	Energy-Efficient Walls
ire Safety	2 h Fire Rated Walls	2 h Fire Rated Walls	2 h Fire Rated Walls	2 h Fire Rated Walls	Fire-resistant Construction

Table 2. CGST Complex Optimization Results.

Parameter	Base	Optimized Without Occupant Data	Optimized With Occupant Data
Load (kWh)	20,000	17,880 (10.6%)	16,950 (14.3%)
Operational Costs (CAD)	2500	2250 (10.0%)	2175 (13.0%)
Carbon Emissions (kg CO_2)	5000	4450 (11.0%)	4225 (15.5%)
Peak Load (kW)	15.0	13.4 (10.5%)	12.9 (14.0%)

While the proposed framework benefits significantly from real-time occupant data, there are practical challenges in data acquisition. These include sensor inaccuracies, intermittent connectivity, and privacy concerns related to monitoring occupant presence and preferences. In this study, these issues were mitigated by using anonymized data collection protocols, leveraging non-intrusive IoT sensors (e.g., motion detectors, smart thermostats), and implementing local edge-processing to minimize data transmission risks.

Figures 2–5 present the base load, optimized load without occupant data, and optimized load with occupant data for the individual CSGTs over a 24 h period. Figure 6 gives the corresponding results for the CSGT complex. Occupant data includes real-time and historical information on occupant presence, behavior, and preferences collected via sensors, IoT devices, smart meters, and user inputs [5,6,14]. These results indicate that occupant-aware optimization improves load efficiency and reduces peak demand. For example, the 1-bedroom CSGT has a base load peak of 14.3 kWh, and this decreases to 13.9 kWh without occupant data and 13.3 kWh with occupant data. The 2-bedroom unit has a peak load reduction from 15.7 kWh to 15.2 kWh and 14.7 kWh. The 3-bedroom and 4-bedroom CSGTs have base peak loads of 16.8 kWh and 18.1 kWh, respectively, and they decrease to 15.3 kWh and 16.4 kWh without occupant data and 13.8 kWh and 14.5 kWh when occupant data is incorporated. Figure 6 indicates that the cumulative effect is a significant decrease in the complex peak load from nearly 67 kWh to 61 kWh without occupancy data and 54 kWh with this data. These results demonstrate the value of real-time occupancy data in dynamic energy management, enabling control strategies that adapt load profiles to actual usage patterns and occupancy conditions. They also confirm that the proposed framework effectively improves load efficiency and lowers operational costs and carbon emissions while ensuring occupant comfort.

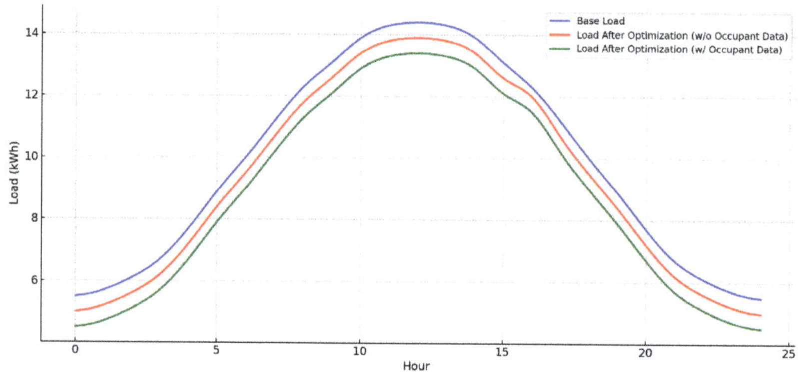

Figure 2. Base load, optimized load without occupant data, and optimized load with occupant data for the 1-bedroom CSGT over a 24 h period.

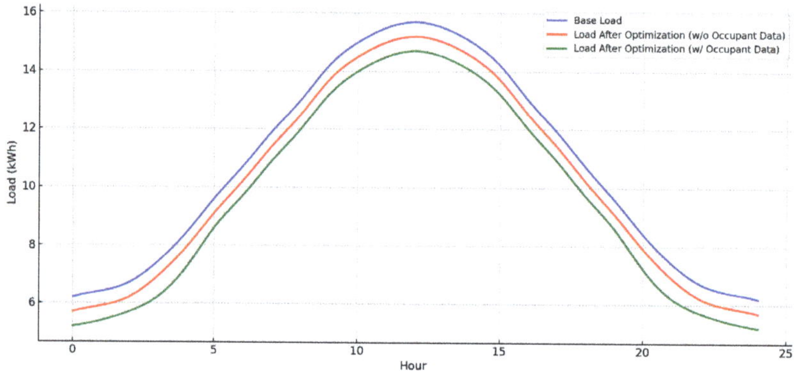

Figure 3. Base load, optimized load without occupant data, and optimized load with occupant data for the 2-bedroom CSGT over a 24 h period.

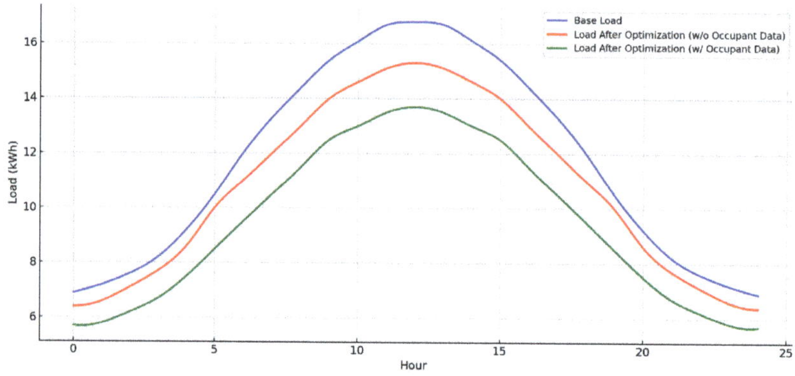

Figure 4. Base load, optimized load without occupant data, and optimized load with occupant data for the 3-bedroom CSGT over a 24 h period.

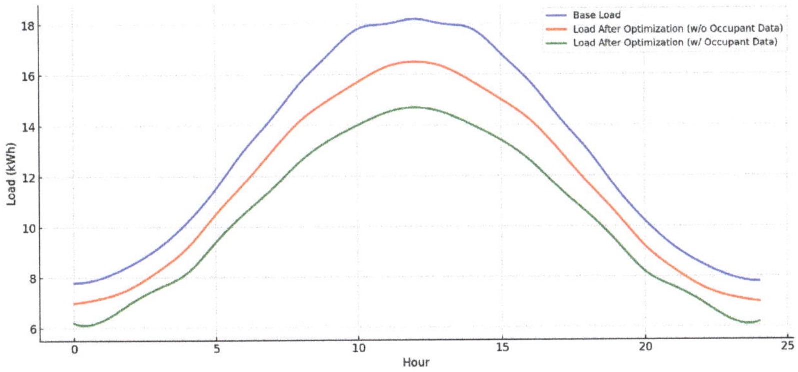

Figure 5. Base load, optimized load without occupant data, and optimized load with occupant data for the 4-bedroom CSGT over a 24 h period.

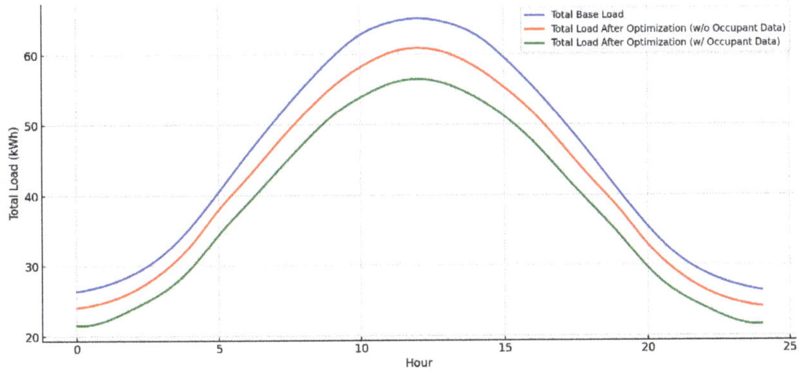

Figure 6. Base load, optimized load without occupant data, and optimized load with occupant data for the CSGT complex over a 24 h period.

Figure 7 presents the townhouse complex base, optimized without occupant data, and optimized with occupant data HVAC and peak load reductions, cost savings, carbon emissions reduction, and occupant comfort satisfaction compared to the historical data in [14–16]. The base results are the worst as optimization improves all five parameters. For example, optimization without occupant data provides an improvement in HVAC and peak loads of about 18%, and with occupant data there is an additional 2–6% improvement for all parameters. In particular, occupant comfort satisfaction is improved to over 85%. These results show the effectiveness of incorporating real-time occupant data into load management to improve efficiency, reduce costs, and lower environmental impact while maintaining a high level of occupant satisfaction.

Figure 8 gives the cost savings versus emissions reduction for the townhouse complex optimized with occupant data. This shows the tradeoff between economic and environmental benefits with cost savings between 10% and 17%, and emission reductions between 10% and 17%. This reflects optimization considering load demand and occupant behavior and indicates an inverse relationship between the two parameters. Thus, reducing the environmental impact increases costs.

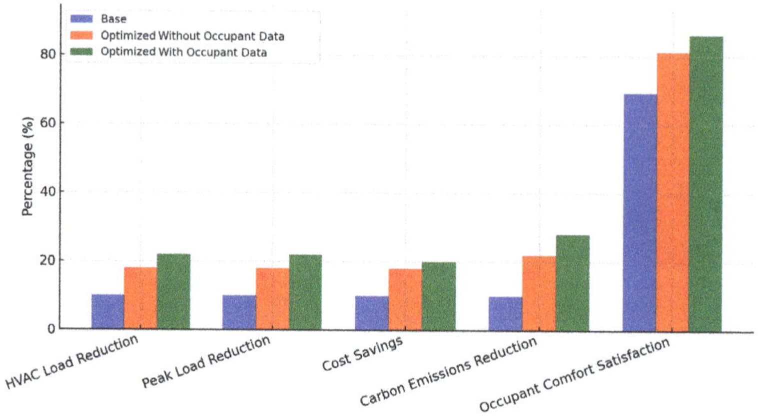

Figure 7. Performance improvement relative to historical data for the base load, optimized load without occupant data, and optimized load with occupant data.

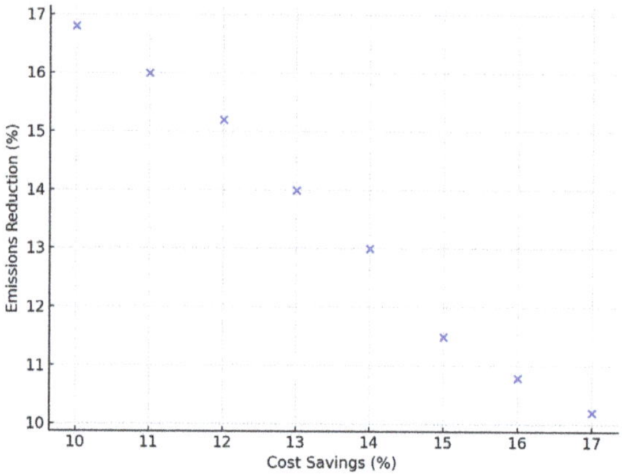

Figure 8. Cost savings versus emissions reduction for the CSGT complex optimized with occupant data.

Figure 9 presents the townhouse complex occupant satisfaction in terms of thermal comfort, lighting adequacy, and feedback adherence for the base, optimized without occupant data, and optimized with occupant data cases. These results indicate optimization increases all three parameters. For example, thermal comfort increased from a base of approximately 67% to 80% optimized without occupant data and 86% optimized with occupant data. The corresponding lighting adequacy improved from 75% to 83% and 89%, and the feedback adherence from 60% to 72% and 82%. This confirms the effectiveness of occupant-aware energy optimization in improving occupant satisfaction.

Figure 10 illustrates the impact of occupancy on the CSGT complex optimized load with occupant data over a 24 h period. Full occupancy indicates all residents are present which results in significant HVAC, lighting, and appliance use. With partial occupancy, there are fewer residents so energy consumption is lower. No occupancy means the building is unoccupied so only essential systems are running such as standby appliances, HVAC with setback control, and water heating. The results in Figure 10 show that occupancy has a significant effect on load demand. With full occupancy, the peak load is about 66 kW at midday due to increased appliance and HVAC usage. Partial occupancy has a lower peak load of about 61 kW, reflecting moderate demand due to fewer residents. No occupancy has the lowest peak load which is below 57 kW. Thus, occupancy-driven load optimization is important to reduce peak demand and overall CSGT load. The performance improvements observed, i.e., reductions in load, operational cost, and carbon emissions, are a result of the integration of the prediction capability of the hybrid LSTM-CNN model and the dynamic capability of the MOPSO algorithm. The LSTM effectively captures time-dependent occupancy and load trends, while the CNN identifies spatial usage patterns from sensor inputs across multiple zones and townhouses. This reduces grid dependence during peak hours and aligns energy use with occupant presence, which lowers energy demand, utility costs, and greenhouse gas emissions.

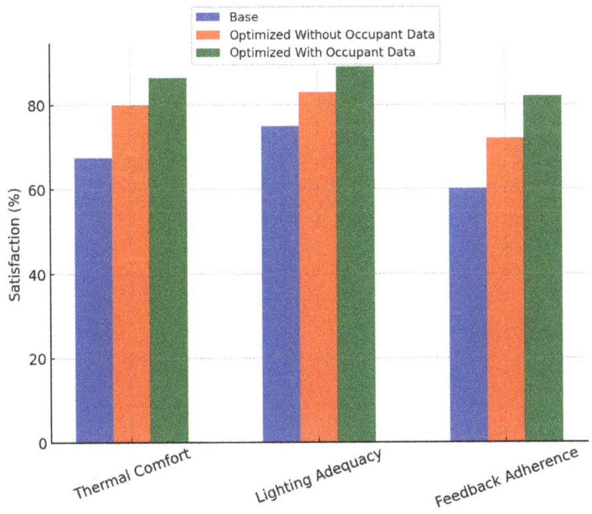

Figure 9. Base, without occupant data, and optimized with occupant data CGST complex occupant satisfaction over a 24 h period.

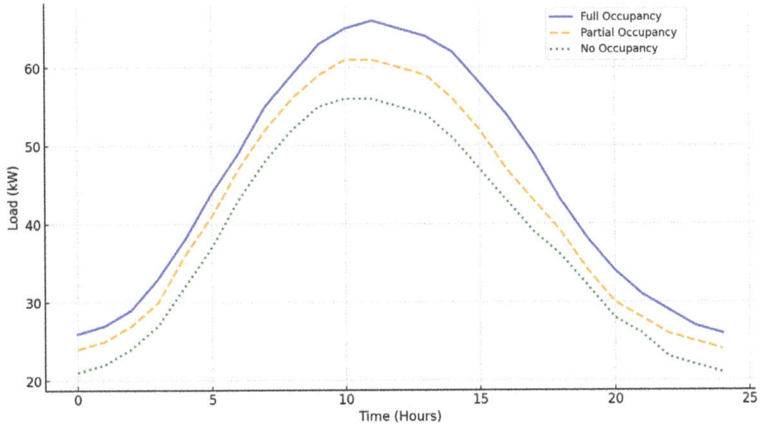

Figure 10. CSGT complex optimized load with full, partial, and no occupancy.

Table 3 presents the MAE, RMSE, and R^2 for the CSGT complex load optimization with the Linear Regression (LR), LSTM, CNN, and proposed hybrid LSTM-CNN models.

The LR model has the worst performance with an MAE of 0.80 kWh and RMSE of 1.20 kWh, indicating poor prediction accuracy. The LSTM model has an MAE of 0.60 kWh and an RMSE of 0.72 kWh, which is lower. The CNN model improves on these results with an MAE of 0.56 kWh and an RMSE of 0.75 kWh. Further, R^2 is 0.93 which is better than with the LR and LSTM models. The proposed model provides the best overall performance with the lowest MAE (0.47 kWh) and RMSE (0.68 kWh), and the highest R^2 (0.95), indicating superior prediction accuracy and reliability. These results validate the effectiveness of the model in energy load forecasting. While the proposed model outperforms the others, its decisions are less interpretable due to the deep learning architecture. Incorporating explainable AI techniques can improve understanding of the input-output relationships, particularly for stakeholders seeking clarity in operational decisions.

Table 3. CSGT performance with four models.

Model	MAE (kWh)	RMSE (kWh)	R^2
Linear Regression (LR)	0.80	1.20	0.85
LSTM	0.60	0.72	0.89
CNN	0.56	0.75	0.93
Proposed	0.47	0.68	0.95

Sensitivity of Key Parameters

The impact of key parameters is now considered.

1. **Occupant Behavior Weights (w_T, w_L, w_F):** Increasing the weight assigned to thermal comfort (e.g., w_T from 0.4 to 0.6) will improve temperature satisfaction but also increase HVAC usage, potentially raising energy consumption by up to 8% [12]. Similarly, a higher feedback adherence weight (w_F) will improve personalization but may introduce variations that will reduce energy efficiency.
2. **Optimization Objective Function Weights (w_1, w_2, w_3):** Adjusting the MOPSO weights shifts the balance between load, cost, and emissions. Prioritizing emissions ($w_3 > 0.4$) will reduce carbon emissions but increase reliance on storage and renewable energy, resulting in higher costs [18,19]. On the other hand, emphasizing cost ($w_2 > 0.4$) will improve affordability but may lower occupant satisfaction due to less flexible HVAC control.
3. **MOPSO Parameters:** As observed in [17], larger swarm sizes or increased cognitive/social weights improve convergence but require longer computation times which may not be feasible for real-time applications.
4. **Model Hyperparameters (e.g., depth, learning rate):** Deep architectures such as LSTM and CNN can provide better accuracy but risk overfitting, especially with small datasets [7,8]. A properly tuned architecture balances prediction accuracy and generalizability. The hyperparameters used here are based on the results in [5,6].

4. Conclusions

This paper introduced a scalable occupant-centric load optimization framework for Connected Smart Green Townhouses (CSGTs), integrating a hybrid Long Short-Term Memory-Convolutional Neural Network (LSTM-CNN) with Multi-Objective Particle Swarm Optimization (MOPSO). Real-time occupant data was used to dynamically optimize energy loads, resulting in substantial performance improvements. The results obtained indicate reductions in load by 12.7%, operational costs by 13.0%, and carbon emissions by 15.5%. Furthermore, peak load demand was reduced by up to 12.7% which helps improve grid stability. In addition, occupant satisfaction was improved with thermal comfort increasing by 19%, lighting adequacy by 14%, and feedback adherence by 22%. The tradeoff between cost savings and emissions reduction indicates the proposed framework can be used in real-world applications. Future research will consider Renewable Energy Systems (RESs) and system scalability to ensure sustainable and adaptive energy management for CSGTs considering occupant satisfaction. Further, Transformer-based architectures and statistical time-series models such as Prophet can be employed to improve long-term forecasting and model interpretability. Parameters such as occupant comfort weights and the cost-emission tradeoff can be investigated to assess their impact on model performance in residential scenarios considering occupant behavior.

Author Contributions: Conception and design, S.M.M., T.A.G., I.T.C. and H.T.; preparation and analysis, S.M.M., I.T.C. and T.A.G.; writing—original draft, T.A.G., S.M.M. and I.T.C.; writing—review and editing, T.A.G., S.M.M., I.T.C. and H.T.; supervision, T.A.G. All authors have read and agreed to the published version of the manuscript.

Funding: This research received no external funding.

Data Availability Statement: The original contributions presented in the study are included in the article, further inquiries can be directed to the corresponding author.

Conflicts of Interest: The authors declare no conflicts of interest.

Abbreviations

The following abbreviations are used in this manuscript:

Symbol	Description	Symbol	Description
BC	British Columbia	CAD	Canadian Dollar
COA	Coati Optimization Algorithm	CNN	Convolutional Neural Network
CSGT	Connected Smart Green Townhouse	HVAC	Heating, Ventilation, and Air Conditioning
IoT	Internet of Things	LR	Linear Regression
LSTM	Long Short-Term Memory	MAE	Mean Absolute Error
ML	Machine Learning	MSE	Mean Squared Error
PV	Photovoltaic	R^2	Coefficient of Determination
RES	Renewable Energy Source	RMSE	Root Mean Squared Error
		SGT	Smart Green Townhouse

eferences

Bäcklund, K.; Molinari, M.; Lundqvist, P.; Palm, B. Building occupants, their behavior and the resulting impact on energy use in campus buildings: A literature review with focus on smart building systems. *Energies* **2023**, *16*, 6104. [CrossRef]

Mylonas, A.; Tsangrassoulis, A.; Pascual, J. Modelling occupant behaviour in residential buildings: A systematic literature review. *Build. Environ.* **2024**, *256*, 111959. [CrossRef]

D'Oca, S.; Chen, C.F.; Hong, T.; Belafi, Z. Synthesizing building physics with social psychology: An interdisciplinary framework for context and occupant behavior in office buildings. *Energy Res. Soc. Sci.* **2017**, *34*, 240–251. [CrossRef]

Moghimi, S.M.; Gulliver, T.A.; Chelvan, I.T. Energy management in modern buildings based on demand prediction and machine learning—A review. *Energies* **2024**, *17*, 555. [CrossRef]

Moghimi, S.M.; Gulliver, T.A.; Chelvan, I.T.; Teimoorinia, H. Resource optimization for grid-connected smart green townhouses using deep hybrid machine learning. *Energies* **2024**, *17*, 6201. [CrossRef]

Moghimi, S.M.; Gulliver, T.A.; Chelvan, I.T.; Teimoorinia, H. Load optimization for connected modern buildings using deep hybrid machine learning in island mode. *Energies* **2024**, *17*, 6475. [CrossRef]

Agga, A.; Abbou, A.; Labbadi, M.; El Houm, Y.; Ali, I.H.O. CNN-LSTM: An efficient hybrid deep learning architecture for predicting short-term photovoltaic power production. *Electr. Power Syst. Res.* **2024**, *208*, 107908. [CrossRef]

Abou Houran, M.; Bukhari, S.M.S.; Zafar, M.H.; Mansoor, M.; Chen, W. COA-CNN-LSTM: Coati optimization algorithm-based hybrid deep learning model for PV/wind power forecasting in smart grid applications. *Appl. Energy* **2023**, *349*, 121638. [CrossRef]

Moghimi, S.M.; Gulliver, T.A.; Chelvan, I.T.; Teimoorinia, H. Adaptive machine learning for automatic load optimization in connected smart green townhouses. *Algorithms* **2025**, *18*, 132. [CrossRef]

10. Zhang, J.; Qian, K.; Luo, H.; Liu, Y.; Qiao, X.; Xu, X.; Tian, J. Process monitoring for tower pumping units under variable operationa conditions: From an integrated multitasking perspective. *Control Eng. Pract.* **2025**, *156*, 126229. [CrossRef]
11. Zhang, J.; Tian, J.; Alcaide, A.M.; Leon, J.I.; Vazquez, S.; Franquelo, L.G.; Luo, H.; Yin, S. Lifetime extension approach base on the Levenberg–Marquardt neural network and power routing of DC–DC converters. *IEEE Trans. Power Electron.* **2023**, *3* 10280–10291. [CrossRef]
12. Meftah, U. Smart Strategies for Building Energy Efficiency: Integrating Occupancy-Based HVAC Control and Machine Learnin Predictions. Master's Thesis, University of Missouri-Columbia, Columbia, MO, USA, 2024. Available online: https://www.proques com/openview/80467fcdfaef57fe457c5f5ee3b925c3/1?pq-origsite=gscholar&cbl=18750&diss=y (accessed on 31 July 2024).
13. Wang, Z.; Calautit, J.; Wei, S.; Tien, P.W.; Xia, L. Real-time building heat gains prediction and optimization of HVAC setpoint: A integrated framework. *J. Build. Eng.* **2022**, *49*, 104103. [CrossRef]
14. Makonin, S. Electricity, water, and natural gas consumption of a residential house in Canada from 2012 to 2014. *Sci. Data* **2016**, 160037. [CrossRef] [PubMed]
15. Makonin, S.; Popowich, F.; Bartram, L.; Gill, B.; Bajić, I.V. AMPds: A public dataset for load disaggregation and eco-feedbac research. In Proceedings of the IEEE Electrical Power & Energy Conference, Halifax, NS, Canada, 21–23 August 2013. [CrossRef
16. Gaur, M.; Makonin, S.; Bajić, I.V.; Majumdar, A. Performance evaluation of techniques for identifying abnormal energy consumptio in buildings. *IEEE Access* **2019**, *7*, 62721–62733. [CrossRef]
17. Chen, X.; Xiao, S. Multi-objective and parallel particle swarm optimization algorithm for container-based microservice schedulin *Sensors* **2021**, *21*, 6212. [CrossRef]
18. Yuan, J.; Zeng, X.; Zhou, J.; Li, J.; Lv, J.; Chen, R.; Chen, K.; Yang, W.; Zhang, Y. Data-driven real-time home energy managemen system based on adaptive dynamic programming. *Electr. Power Syst. Res.* **2025**, *238*, 111055. [CrossRef]
19. Da, T.N.; Cho, M.Y.; Thanh, P.N. Hourly load prediction based on feature selection and a hybrid CNN-LSTM method for building' smart solar microgrid. *Expert Syst.* **2024**, *41*, e13539. [CrossRef]
20. Ren, J.; Zhou, X.; Jin, X.; Ye, Y.; Causone, F.; Ferrando, M.; Li, P.; Shi, X. A systematic review of occupancy pattern in urban buildin energy modeling: From urban to building-scale. *J. Build. Eng.* **2024**, *95*, 110307. [CrossRef]
21. Decuypere, R.; Robaeyst, B.; Hudders, L.; Baccarne, B.; Van De Sompel, D. Transitioning to energy-efficient housing: Drivers an barriers of intermediaries in heat pump technology. *Energy Policy* **2021**, *156*, 112709. [CrossRef]
22. Benavente-Peces, C. On the energy efficiency in the next generation of smart buildings—Supporting technologies and technique *Energies* **2019**, *12*, 4399. [CrossRef]
23. Madhan Raj, B.; Ashwin, S.; Harish, R.; Manishwar, M.; Sophia Jasmine, G. Occupancy-based cost-efficient campus energ management system. In Proceedings of the International Conference on Inventive Computation Technologies, Lalitpur, Nepa 24–26 April 2024; pp. 2047–2051. [CrossRef]
24. Jangsten, M.; Lindholm, T.; Dalenbäck, J.O. Analysis of operational data from a district cooling system and its connected building *Energy* **2020**, *203*, 117844. [CrossRef]
25. Agugiaro, G.; Zwamborn, A.; Tigchelaar, C.; Matthijssen, E.; León-Sánchez, C.; van der Molen, F.; Stoter, J. On the influence of part walls for urban energy modelling. *Int. Arch. Photogramm. Remote Sens. Spat. Inf. Sci.* **2022**, *XLVIII-4/W5-2022*, 9–16. [CrossRef]
26. Palmer, J.; Terry, N. Looking critically at heat loss through party walls. *Sustainability* **2022**, *14*, 3072. [CrossRef]

energies

MDPI

·ticle

;ravity Energy Storage and Its Feasibility in the Context of ustainable Energy Management with an Example of the 'ossibilities of Mine Shafts in Poland

atarzyna Tobór-Osadnik [1], Jacek Korski [2], Bożena Gajdzik [3,*], Radosław Wolniak [4,*] and Wieslaw Grebski [5]

[1] Department of Safety Engineering, Silesian University of Technology, 44-100 Gliwice, Poland; katarzyna.tobor-osadnik@polsl.pl
[2] KOMAG Institute of Mining Technology, 44-100 Gliwice, Poland; jkorski@komag.eu
[3] Department of Industrial Informatics, Silesian University of Technology, 44-100 Gliwice, Poland
[4] Faculty of Organization and Management, Silesian University of Technology, 44-100 Gliwice, Poland
[5] Penn State Hazleton, Pennsylvania State University, 76 University Drive, Hazleton, PA 16802, USA; wxg3@psu.edu
* Correspondence: bozena.gajdzik@polsl.pl (B.G.); radoslaw.wolniak@polsl.pl (R.W.)

Abstract

This paper discusses the viability and efficiency of gravity energy storage (GES) systems utilizing abandoned coal mine shafts in Poland as a new frontier of energy management within the broader framework of sustainable energy transition. After a thorough analysis of shaft infrastructure, economic factors, and regulatory environment, the research demonstrates how GES is in line with circular economy and sustainability principles yet there are certain technical and financial limitations—smaller lifting capacities and expensive adjustments, for instance—that are currently stalling its large-scale adoption. The results highlight the importance of harmonizing such repurposing efforts with the available renewable energy infrastructure and call for aggressive policy, technological, and funding efforts to sustain the conceptual promise with actual fulfilment.

Keywords: gravity energy storage; mines; coal mining; mine shafts; sustainable energy; energy transition

1. Introduction

The concept of storing electrical energy has long been understood, initially applied in systems aimed at stabilizing electrical performance, such as those in spark-ignition engine vehicles—later adapted to hybrid drive technologies. Building on this foundation, the notion of storing already produced energy in dedicated energy storage systems began to take shape. For years, pumped-storage hydroelectric power plants have been implemented in large-scale energy systems [1], serving the role of stabilizing the grid by capturing surplus energy during periods of low demand and releasing it when the system experiences an energy deficit [2].

Since 2015, the idea of utilizing gravity for electrical energy storage has been increasingly discussed in the literature [3,4]. Gravity-based energy storage systems function by capturing the potential energy that arises when an object is elevated, with gravity acting as the driving force behind this energy accumulation. These systems use surplus electricity from the grid to raise a heavy mass, thereby storing energy in the form of gravitational potential, which can later be released and transformed back into electrical energy through a generator [5]. This system can be used in mines when their basic function is stopped, including closed coal mines [6]. The technology of gravity-based energy storage in mining environments is still in the development phase, although initial implementations have already begun to emerge [7]. One such solution is the gravity energy storage system (GESS), which operates on a similar principle to that of pumped hydro storage. However, rather than relying on water reservoirs, the GESS utilizes columns of loose bulk materials—such as sand, gravel, or other granular substances—positioned at varying heights to store and release energy [8].

The main goal of this article is to investigate the feasibility and functionality of energy storage in mine shafts using gravity energy storage technology in the context of sustainable energy management and the ongoing energy transition.

In this paper, the following research questions (RQs) were formulated:

- RQ1: What are the technical and economic limitations and the potential opportunities for implementing gravity energy storage in mine shafts in Poland?
- RQ2: To what extent can gravity energy storage in mine shafts represent a competitive solution compared to other energy storage methods, such as pumped-storage power plants or battery systems?
- RQ3: Is the concept of using mine shafts as gravity energy storage facilities consistent with the assumptions of sustainable energy transition and the plans for restructuring the mining industry in Poland?

Subsequent parts of this paper present theory background, contextual applicability, the research method, and findings from the analysis of gravity energy storage (GES) in Polish mine shafts. In the Background section, how energy storage is conducted in the context of overall energy transition is explained wherein EU policy action and technology innovation are the context where GES is an applied solution. The Study Method section distinguishes between theory and practical application according to rudimentary physics formulae and environmental limitations—legal, economic, and infrastructural—to analyse mine shaft conversion practicability. In the Results of Analysis section, the authors show quantitative results for which they prove that although theoretically energy can be conserved by utilizing gravity in mine shafts, their efficacy is rather low due to lifting capacity, shaft geometry, and maintenance. The Discussion section places these findings within the context of policy alignment, circular economy potential, and relative performance versus competing technologies and concludes that GES is a slightly better technology at scale. The Conclusion then combines these findings to ensure that although GES aligns with sustainability goals and policy instruments, its application in Poland requires focused, hybrid solutions based on focused technological and financial support.

2. Background

Energy storage systems have been known and used for several centuries. In recent years, the growing momentum of global climate policies combined with the expansion of renewable energy sources (RES) has significantly accelerated interest in energy storage technologies. These storage systems are essential to the ongoing energy transition, as they facilitate the effective integration of RES into power systems while contributing to grid stability and reliability [9–12].

The importance of ensuring access to energy is underscored by the United Nations 2030 Agenda for Sustainable Development, which identifies it as the seventh of the 17 Sustainable Development Goals [13]. Energy storage is seen as the future of zero-emission energy systems and forms the foundation for implementing the green transition in economies and industries. The effective regulation of energy storage in the European Union requires balancing technological innovations with appropriate legal frameworks [14]. International cooperation and strong partnerships are essential for developing sustainable technologies [15]. Many regions of the world, particularly the EU, are striving to create an internal electricity market based on non-fossil fuel energy sources. Directive (EU) 2019/944 [16] sets out rules for the generation, transmission, distribution, supply, and storage of electricity, as well as consumer protection measures, with the aim of creating integrated, competitive, consumer-centred, flexible, fair, and transparent electricity markets within the European Union. It was amended in 2024 by Directive (EU) 2024/1711.

It is estimated that over 75% of greenhouse gas emissions in the EU currently come from energy production and use across various sectors of the economy, which underscores the conviction that achieving the ambitious climate targets for 2030 and 2050 is a critical EU strategy [17]. Achieving net zero emissions by 2050 is seen as a key objective for limiting global warming and meeting the goals of the Paris Agreement [18]. The new energy policy is only possible thanks to technological breakthroughs, including new storage possibilities such as pumped-storage hydropower, battery and hydrogen storage, and, notably, gravity storage in mine shafts (following the decommissioning of coal mines) [3,19–23].

Gravity storage systems are part of the broader category of mechanical energy storage systems. According to the principles of their operation, storage systems can be classified as chemical, thermal, or mechanical. Among mechanical storage systems, those that utilize potential energy, kinetic energy, or compression energy are in use (see Figure 1). Within the category of potential energy storage systems are the gravity-based storage systems in mine shafts. The storage capacity of these systems is determined by two parameters: the mass of the load (or loads) and the available height difference between the resting levels of the load [3,24–28].

Energy storage in former mining sites—particularly through the innovative approach known as underground gravity energy storage (UGES)—offers a promising solution for long-term energy retention. This technique repurposes decommissioned mines into storage facilities by using gravitational potential energy, thereby contributing to the advancement of a more sustainable and resilient energy system [UGES]. It is considered in the restructuring scenarios for the mining industry alongside the management of methane [29–31]. In EU countries, including Poland, the decommissioning of hard coal mines has been an ongoing process for years, aimed at gradually closing the mines in line with energy transition objectives and market changes. In Poland, there is a plan to close 12 mines by the end of 2049. On 28 May 2021, in Katowice, the government, trade unions, mining municipalities, and mining companies signed a social agreement [32]. According to the Energy Policy of Poland (PEP2040), the coal phase-out rate from 2030 to 2040 is projected to be 65% (2025), 56% (2030), 38% (2035), and 28% (2040) [33]. According to INSTRAT data from December 2024, there were 19 hard coal mines and four lignite mines operating in Poland [34]. The transition of mining in Poland is the main climate policy in this country [35]. Of the innovative energy storage solutions in the European (EU) new energy policy, pumped-storage hydropower plants could be used with closed mines. In Poland, existing facilities of this type currently offer an energy storage capacity of around 8 GWh, with projections indicating that this capacity could potentially be expanded, although only by a few times at most. Following the planned cessation of lignite extraction at the Turów open-pit mine, there has been a proposal to construct a pumped-storage plant with an installed capacity of 2300 MW and a total storage potential of up to 160 GWh over the next three decades—an amount still considered modest in scale. The above limitations mean that other methods and facilities for storing the electricity are being sought, and mainly gravitational, pressure, and electrochemical ones are considered [36].

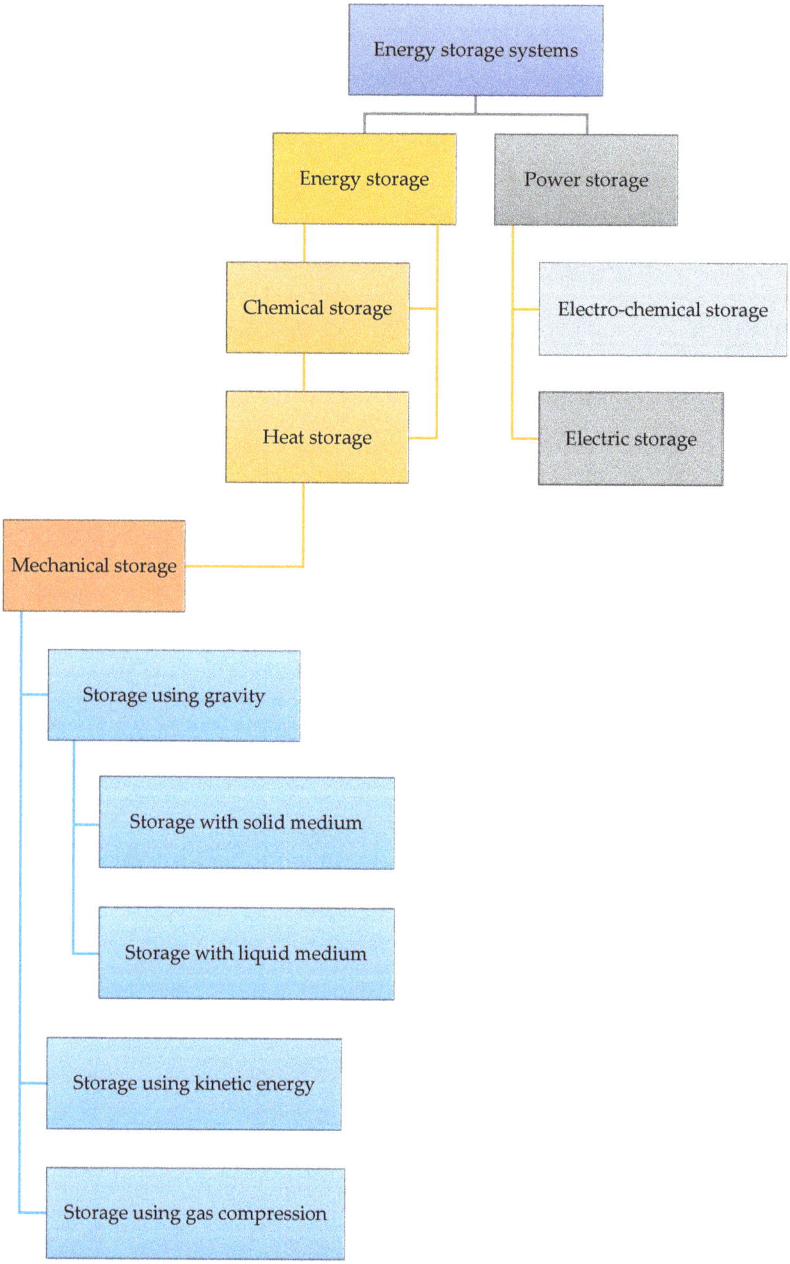

Figure 1. Gravity storage systems as a part of mechanical energy storage typology [3].

Energy storage facilities could present an opportunity for decommissioned mines, fulfilling the primary function of energy storage while also offering additional, secondary functions, such as the utilization of existing electrical infrastructure within the closed mine, the reuse of shaft and hoisting equipment, the retention of some mining-specialized jobs at the site of the closed mine, and the deferral of shaft decommissioning costs until after the storage facility has completed its operation. Of course, establishing such storage facilities requires significant direct investment costs, including purchasing and installing technology, draining water from decommissioned mine shafts, maintaining shaft ventilation for monitoring and operational needs, and covering the expenses of maintenance and repair work within the shaft. In the case of storage systems with multiple (divided) loads, it is also necessary to build loading and unloading systems for the weights and masses both at the surface and underground [25,37–42].

The gravity-based solution (GES) is comparable to a pumped-storage power plant, except that water is replaced with composite blocks or "mobile masses" made from inexpensive, locally sourced materials (including waste materials such as discarded wind turbine blades). The composite blocks and the lifting mechanism for storing and releasing energy are central to gravity-based storage systems. The operational process involves transferring energy from renewable sources to motors that raise the blocks (the Gravitricity technology). When electricity demand arises, the elevated blocks are lowered, and the kinetic energy generated during their descent is converted into electrical power. Gravity-based energy systems are particularly suitable for locations where traditional water-pumping methods are impractical. They are also applicable in post-mining areas that offer appropriate vertical elevation and sufficient quantities of granular bulk materials [43]. One such example is the Gravitricity technology, which operates by raising and lowering heavy weights—up to a combined mass of 12,000 tons—within deep mine shafts to store and release energy. Recent research by Imperial College London has shown that the electricity discharged by a commercially available 10 MW lithium-ion battery system would cost USD 367 (283 GBP)/MWh over its lifetime, compared to USD 171/MWh for the Gravitricity system. According to Recharge, the gravity-based technology is currently being piloted in the United Kingdom, Finland, Poland, the Czech Republic, and South Africa, where mine shafts can reach depths exceeding 2000 m [43–45].

Among the concepts for building gravity energy storage systems in the workings of inactive mines are the following solutions: underground pumped-storage hydropower plants using mine workings as, at minimum, the lower water reservoir; gravity storage systems with a single fixed weight; gravity storage systems with multiple fixed weights; gravity storage systems with containers filled with a working liquid as the load; gravity storage systems with containers for loose ballast material; gravity storage technologies utilizing the gravity of water flow; gravity storage technologies based on pressure gravity; and others that are currently in the technological development phase [3,43].

Here is an overview of international examples and projects related to various types of gravity energy storage, particularly those making use of the infrastructure of inactive mines:

- (i) Underground pumped-storage power plants in mine workings: In Finland, in the inactive Pyhäsalmi mine, the deepest mine shaft in Europe (with a depth of almost 1.5 km), a project is underway to build an underground pumped-storage power plant. The planned installation will have a capacity of 75 MW and an energy storage capacity of 530 MWh, utilizing the existing shafts and workings as the lower water reservoir. The project has received financial support from the European Union amounting to EUR 26.3 million [44].
- (ii) Gravity energy storage systems with a single or multiple weights: Energy storage technology based on lifting and lowering heavy weights in mine shafts (research on adapting this technology is currently underway in inactive mines such as Velenje in Slovenia, Pyhäsalmi in Finland, Darkov in the Czech Republic, and Grube Teutschenthal in Germany) [45].
- (iii) The world's first gravity energy storage systems utilizing gravity blocks—constructed from waste materials such as construction debris, coal ash, or fiberglass—and designed for an operational lifespan of up to 50 years, have been established

in China. A notable example is the China Tianying Rudong Gravity Energy Storage Project, located in Yangkou town, Rudong county, Nantong. This pioneering facility, with an initial capacity of 26 MW, is expected to reach a total energy storage capability of 100 MWh once it becomes fully operational by the end of 2025 [46].

- (iv) In Australia, the difference in height between two water reservoirs is used to store and generate electricity. (Snowy 2.0 project: The Snowy 2.0 project involves tunnelling through 27 km (17 miles) of rock to create a huge pipeline connecting two bodies of water. The elevation difference is 700 m (2296 feet). The energy gravity system would provide about 10% of Australia's electricity production at full output [47].)
- (v) A recent study by the International Institute for Applied Systems Analysis (IIASA) introduced a novel concept for energy storage, involving the transportation of sand into decommissioned underground mines. This approach, known as underground gravity energy storage (UGES), generates electricity by lowering sand into the mine and harnessing its gravitational potential through regenerative braking systems [26]. According to the findings, a single mine could offer an energy storage capacity ranging from 10 to 100 GWh, depending on the use of 5 to 50 million tonnes of sand. The deeper and wider the shaft, the greater the potential for power generation, while the overall size of the mine determines the maximum energy that can be stored [26,48].

The growth in the use of renewable energy sources has demonstrated the need to build and develop efficient and large-scale energy storage solutions. The extensive infrastructure of coal mines can provide the basis for repurposing mine facilities as energy storage. Several equivalent solutions are under development.

2.1. Examples of Energy Storage in Coal Mines

2.1.1. Pumped Hydro Energy Storage (PHES)

One of the most established methods of large-scale energy storage is pumped storage (PHES). In PHES systems, energy storage is achieved by pumping water from the lower reservoir to the upper reservoir during off-peak hours and releasing it during peak demand hours. A PHES power plant has been proposed at the Bełchatów opencast lignite mine in Poland. Pits serving as the lower reservoir are used and the upper reservoir is to be on heaps. Key parameters affecting the efficiency of this investment include reservoir volume, useful head, and energy losses. Evaporation and groundwater recharge rates also affect the efficiency of this system. GIS tools [49,50] and the Colebrook–White [51] formula were used to measure and determine the key parameters. The estimated energy storage potential is between 16.4 and 36.2 GWh per cycle, with an efficiency of around 75%. This could generate between 9.7 and 13.7 TWh of electricity per year, equivalent to 45% of the output of a nearby power station [52].

Underground pumped hydro storage power plants (UPHS) can be used in underground coal mines. These use the voids in closed coal mines as lower reservoirs. This method is particularly effective when combined with renewable energy sources such as wind or solar power. A case study in Asturias, Spain, demonstrated the feasibility of integrating UPHS with a 60 MW wind farm. The system, which had an investment cost of EUR 193 M, could generate EUR 54 M in benefits over 40 years, while reducing CO_2 emissions by 29,000 tonnes per year [53]. In China, the suitability of a combination of wind farms and coal mines in pumped hydro storage (PHS) was analysed. Storage capacities and efficiencies were analysed. The results presented indicate that such systems are technically feasible in regions rich in wind and solar energy [54].

Underground gravity energy storage (UGES) is an innovative method using sand as a storage medium. During off-peak hours, sand is lifted and stored above ground, and during peak demand hours it is lowered into underground tanks to generate electricity. A major advantage of UGES is the relatively long energy storage time. At the same time, the use of sand eliminates the risk of water pollution and provides a cost-effective solution with relatively low investment costs. The global potential of UGES is estimated at 7 to 70 TWh [26].

2.1.2. Compressed Air Energy Storage (CAES)

Compressed air energy storage (CAES) involves compressing air during off-peak hours and storing it in underground caverns. During peak demand, the compressed air is expanded to generate electricity. Coal mines, with their existing underground voids, are ideal locations for CAES systems. The Adiabatic CAES (A-CAES) method stores both the compressed air and the heat generated during compression, enabling more efficient energy recovery. A study in a Spanish coal mine showed that A-CAES systems can work. However, there may be minor geo-mechanical problems. These systems are estimated to have a fairly high capacity of more than 10,000 cycles [55,56]. Combining CAES with thermal energy storage (TES) systems can increase efficiency. The hybrid system in the post-mining shaft achieved an energy output of 140 MWh at 5 MPa, with a round-trip efficiency of 70.44% [57].

2.1.3. Battery Storage Systems

Lithium–iron–phosphate batteries are increasingly used in coal mines for energy storage [58]. These batteries offer a high performance and long life, making them suitable for both energy storage and generation. Research has focused on optimising battery configurations, charging mechanisms, and battery management systems. Experimental data have shown that these systems can effectively improve charging performance and extend battery life [59].

2.1.4. Hybrid Systems and Integration

Integrating multiple energy storage technologies can increase the overall efficiency and reliability of coal mine-based systems. For example, combining CAES with TES or UPHS with wind power can create hybrid systems that optimise energy storage and generation. A study on industrial energy management systems highlighted the benefits of integrating large-scale energy storage with smart grid concepts [60,61]. This approach improves power supply reliability, operational efficiency, and environmental friendliness [62]. In summary, each of the methods described offers unique advantages, and their integration with renewable energy sources and smart grid systems can further enhance their efficiency and environmental benefits. As the energy transition [63] continues, repurposing coal mines for energy storage is a sustainable and economically viable solution.

The use of underground energy storage is also associated with a number of environmental risks. In the event of leaks of the chemicals used, groundwater can be contaminated [64]. Greenhouse gas emissions are no less of a threat. The use of CAES systems using compressed air heating promotes CO_2 emissions [65]. This places an additional burden on the efficiency of these installations. Using the underground spaces of mines for energy storage can result in damage to geological structures, micro-impacts, and the consequent disturbance of the land surface. These effects are often irreversible and involve not only environmental but also financial and social costs [37]. This can lead to damage to surface infrastructure. When adapting existing mine workings for energy storage, the environmental effects associated with the business must also be taken into account, such as noise or the disturbance of existing ecosystems. Therefore, when developing technologies using the existing underground infrastructure of mines, these constraints must be kept in mind.

As important as the environment is, the safety of these installations must also be considered. Underground storage facilities are less vulnerable to fire, terrorist attack, or warfare [66].

3. Study Method

Feasibility was analysed in two areas:

- Theoretical feasibility refers to the compliance of a proposed project with fundamental natural laws—such as those of physics, mechanics, or thermodynamics—indicating that the concept is scientifically sound and does not violate established principles [67].
- Practical or situational feasibility, on the other hand, indicates that the project or process can be realistically implemented within a given context, taking into account specific local conditions such as time, location, and available resources.

A practically feasible project must also be theoretically feasible (Figure 2).

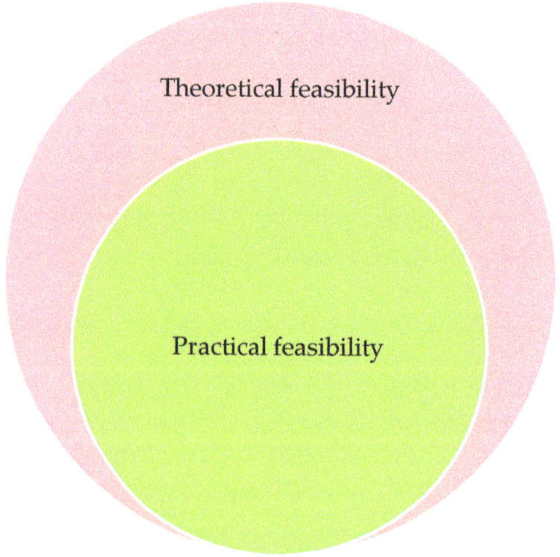

Figure 2. Theoretical and practical feasibility. Source: Authors' own presentation based on [27].

The practical feasibility of theoretically feasible projects is influenced by restrictions occurring at the time and place of the intended implementation of this project in the form of the following:

- Financial or, more broadly, economic constraints;
- Restrictions resulting from applicable standards and legal regulations at the time and place of project implementation;
- Limitations resulting from the availability and skills of human resources along with the existing patterns of social culture in the place and time of the project;
- Local constraints such as terrain, climate, geological structure or locally available infrastructure.

A practically feasible solution must take into account and not contradict the existing constraints. This means that the range of acceptable (practically feasible) solutions must not conflict with any of the groups of restrictions (Figure 3).

Calculations were based on basic formulas for calculating potential energy and kinetic energy as well as the dependence of the path on time and the speed of movement, assuming that the mass of the weight, the permissible speed of the weight, and the depth of the shaft are known, i.e., using the following [68]:

$$E = m \times g \times h \tag{1}$$

$$E = \frac{m \times v^2}{2} \tag{2}$$

$$h = v \times t \text{ where} : (t = \frac{h}{v}) \tag{3}$$

where m—mass, v—speed, h—shaft working height, g—gravity, and E—Energy.

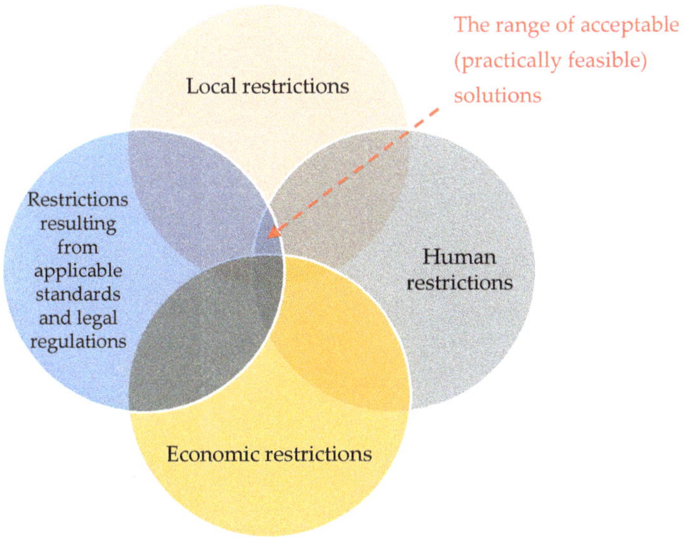

Figure 3. The range of acceptable solutions in light of different restrictions. Source: Authors' own presentation based on [27].

Due to the lack of extensive experience with the use of gravitational energy storage, especially in mine shafts, it is important to determine the expected technical parameters of the project in order to assess the purposefulness of such a project and its functionality. The technical parameters have a direct bearing on the industrial application of the technology and its cost-effectiveness. Compared to the locally obtainable parameters in the project, it is possible to obtain information about one of the important areas of functionality and the purposefulness of undertaking the project.

Local constraints may make it impossible to achieve an economically effective scale for the project.

It should be emphasized that the lack of practical experience with gravitational energy storage makes it difficult to use common tools in performing functionality studies.

4. Results of Analysis

The vast majority of ideas assume the use of the shaft as a hollow tube and the existing shaft hoist drive as part of the energy storage installation. Mine shafts used to transport people, raw material, and other materials with a high efficiency and speed of movement require the forced movement of conveyances (cages or skips), which results in equipping them with steel structures or ropes to guide these conveyances (Figure 4).

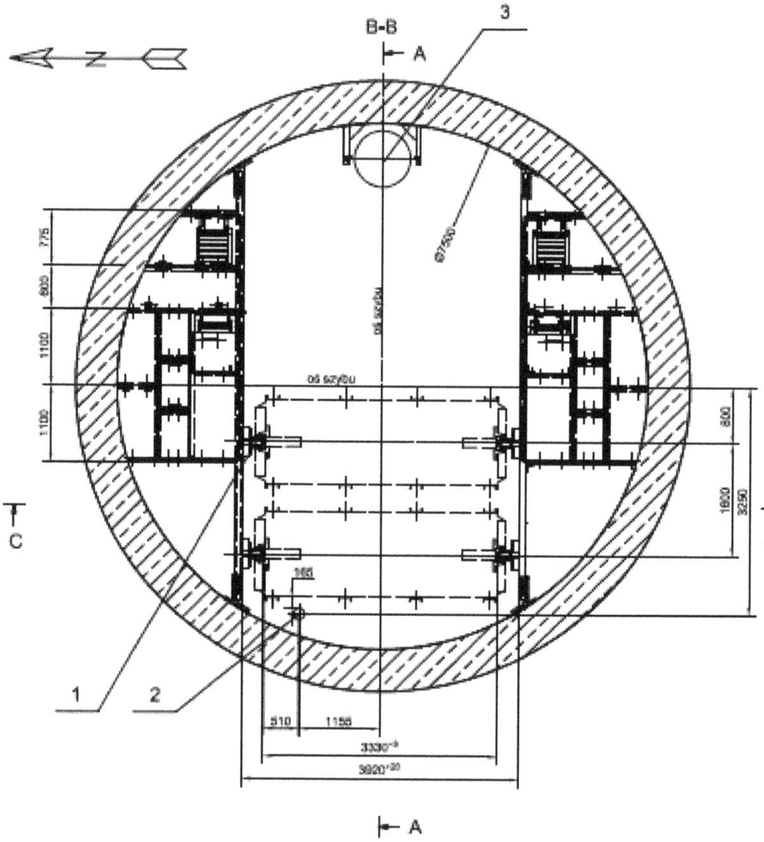

Figure 4. Exemplary shaft tube cross-section with equipment (temporary one-hoist system with two cages: 1—cage leading system, 2—cable, and 3—pipeline). Source: Author's own study (co-author: J.K).

The shaft reinforcement limits the dimensions of the weight, which translates into the capacity of energy storing. Moreover, shaft hoists used in the Polish mines use Koeppe pulleys, the design of which are optimized for pulling large loads upwards or balanced loads in both directions. An additional complication in such a system is the use of stabilizing ropes, which makes it impossible to lower heavy loads to the shaft bottom [69].

The main argument for building gravity energy storage facilities in shafts of liquidated mines is the low investment costs for setting up such a storage facility because the shaft exists and there is an electric reversible drive that can be used. The low expected costs of using such an installation are notable in light of the long expected life cycle—about 50 years (Figure 5).

Shaft hoists have structural limitations on the transported weight (in the conditions of Polish mines these are a maximum of 80 tons, and most often only 30–35 t) and, in the case of simple shaft adaptation shaft energy storage, this is a critical value.

This significantly limits the amount of energy stored in such conditions, the increase of which will require significant investment costs. Moreover, the existing shafts require constant service and maintenance, and in the long expected period of their further use, i.e., 50 years, they will also require additional costs related to their natural wear and tear. For the operation of the gravity energy storage system in the mine shaft, it will be necessary to secure the possibility of inspection and repair, and thus ventilation of all necessary workings. Water flows into each mine, even after its closure, until the water table equalizes the natural water level in the ground. This means that it will also be necessary to drain all workings and installations necessary for the operation of the gravity storage facility.

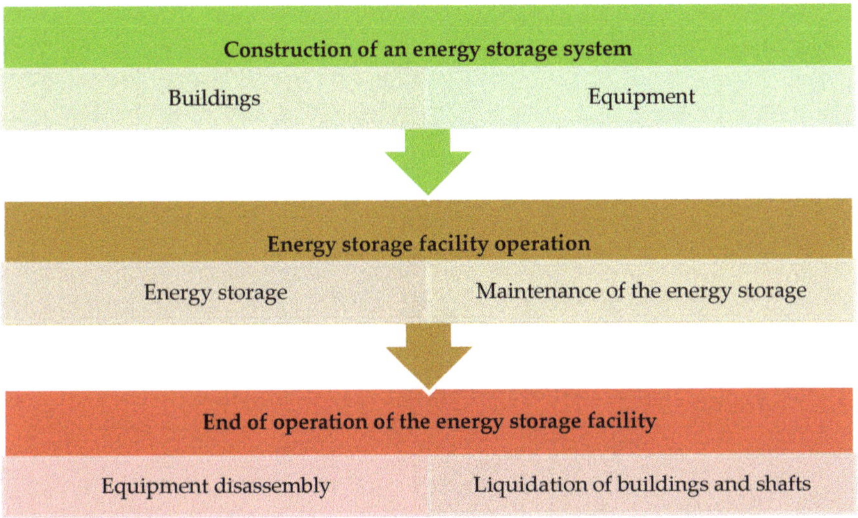

Figure 5. Gravity energy storage system in mine shaft lifetime-cycle. Source: Author's own study (JK) based on [26].

The calculations showed that the power obtained for a 70 Mg load when lowering at a speed of 5.0 m/s and with 100% efficiency would be approximately 2750 MW for approximately 100 s, and the amount of theoretically produced energy would be less than 80 kWh (0.08 MWh). To obtain storage capacity as stated in the description of the surface energy storage facility, it would be necessary to lift and lower thousands of tons of masses in a single shaft—this requires completely different technical solutions. Simply, mine shaft adaption as a gravity energy storage system is feasible but ineffective in capacity and energy volume (Figure 6).

Taking into account the use of hard coal mine shafts in the Upper Silesian Coal Basin as gravitational energy storage, their basic utility parameter in the form of useful depth was analysed. It should be noted that this depth of lowering the weight and its mass determine the amount of energy stored in the shaft. In Figure 7, the quantitative division of shafts in active hard coal mines according to the active depth in the ranges of 120 m between the shallowest and the deepest shaft is shown.

It should be noted that the depths of mine shafts in Polish mines are insufficient for long-term operation with energy return. As an example, it can be pointed out that the deepest shaft at the weight-lowering speed v = 1 m/s will generate power for only 1290 s (21.5 min) It should be emphasized that less than 10% of these shafts are not equipped with hoisting devices and therefore do not have any steel structures in their cross-section. This is beneficial for building a gravity storage facility from scratch. It also does not have permanent hoisting devices that could be adapted to the energy storage system. In shafts that are equipped with hoisting devices (sometimes several in one shaft), most shaft hoists are driven by a drive wheel (Koeppe wheel). This makes it difficult to adapt the hoisting machine to work as a gravitational energy storage system due to the risk of rope slippage on the drive wheel. In the drive wheel hoisting machines (Koeppe), working weight in Polish coal mines does not exceed 50 tons (Figure 8).

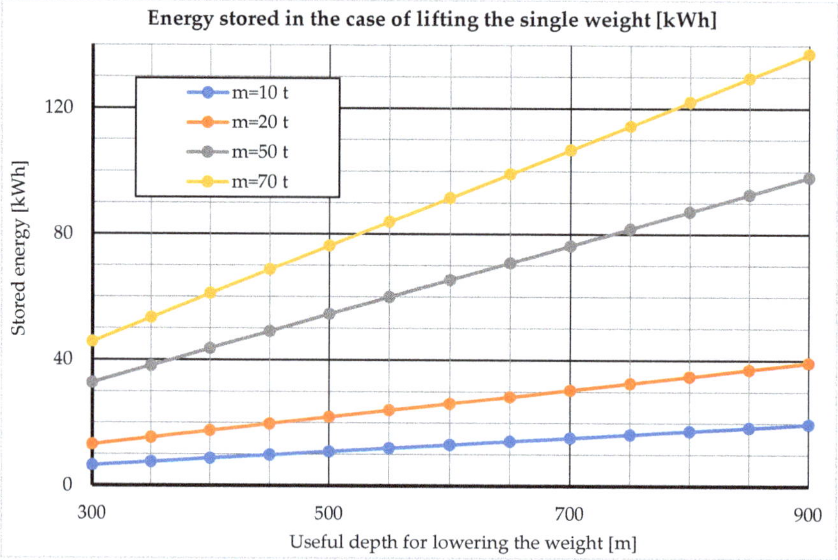

Figure 6. Possibility of gravitational energy storage in a mine shaft for weights with typical masses for active shafts. Source: Author's own analysis (co-author J.K.).

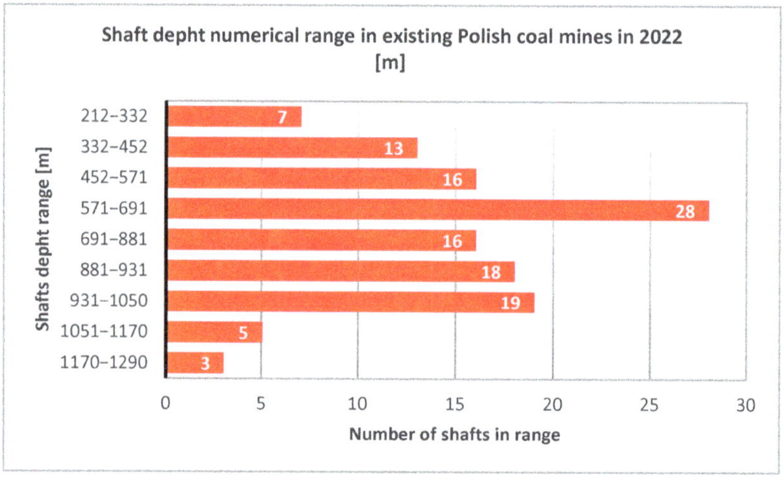

Figure 7. Number of shafts in Polish hard coal mines in different active depth ranges. Source: [Data ARP SA].

With the high technical complexity of the gravitational energy storage in the mine shaft, not only with the use of existing shaft hoists, the amount of energy that can be stored is very small. Therefore, it should be pointed out that this idea, although logically correct, is not scalable.

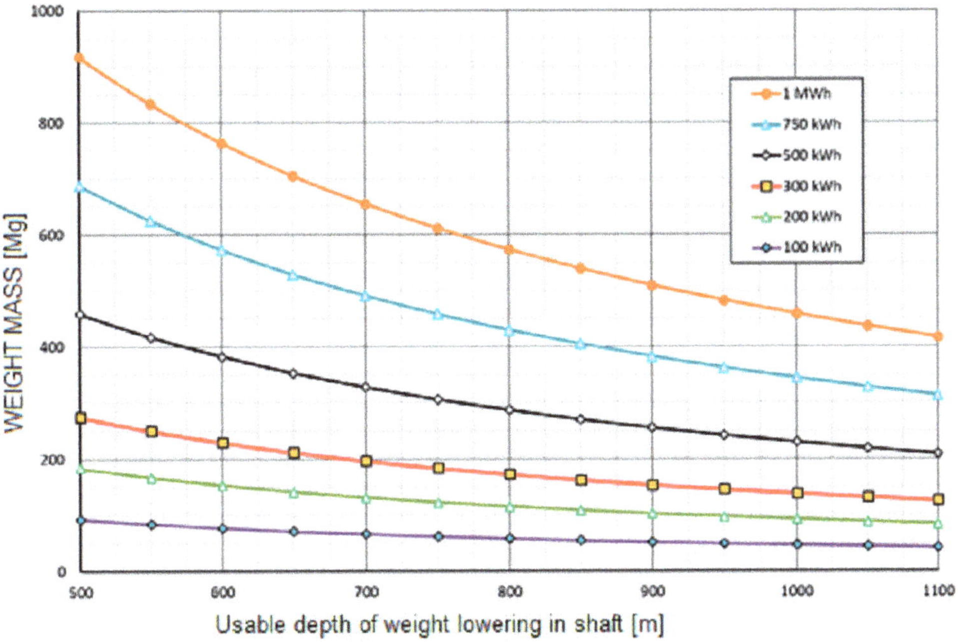

Figure 8. Relation between weight mass and active shaft depth for several expected values of the amount of energy stored in the shaft. Source: Author's own analysis (co-author J.K).

5. Discussion

The findings of this study provide an overall picture of the viability of gravity energy storage (GES) in mine shafts, showing both the potential and the constraints of this new energy storage technology under Poland's energy transformation context. The results of this analysis support those of Aneke and Wang [1] and Hossain et al. [9], who refer to the fact that energy storage systems play a significant role in stabilizing the power grid, particularly with the expanding uses of variable renewable energy sources. Nevertheless, aside from overall compatibility with the objectives of sustainable development and the theoretical possibility of Polish energy storage, the technical and cost feasibility of practical application within existing mine shafts is low.

Technically, studies identify substantial constraints in terms of the ability of existing shaft infrastructure to provide the massive loads necessary for optimal gravity storage. In particular, the existing shaft hoisting systems, as configured for use in mining operations, are not capable of lifting and lowering the enormous composite blocks or "mobile masses" contemplated for gravity storage use [3,24]. This agrees with the work of Siostrzonek [24] and Kula et al. [25], where they show that shaft equipment limitations—e.g., restricted lifting capacity and structure design—overwhelmingly limit the volume of energy which can be stored by methods based on gravity. Moreover, the analysis conducted in this research supports Hunt et al.'s [26] discovery, that considering the low theoretical cost savings from the reuse of mine infrastructure and the extended operating life, practical figures are negligible compared to other electricity storage technologies.

Policy and sustainability-wise, the plan to reuse abandoned mine shafts as a facility for gravity energy storage is extremely aligned with the general goals of the EU climate and energy policies, as outlined in the European Commission's Clean Planet for All strategy [17] and the directives for the internal electricity market [16]. In addition, it aligns with the ideals of the circular economy since it seeks to reuse existing industrial structures to facilitate the uptake of renewable energy systems and the shift towards a low-carbon economy [3,28–31]. This strategic alignment is most apposite to Poland, where the phased closure of the mines, as in Poland's Energy Policy 2040 (PEP2040) [33], provides room for a reengineering of utilizing current mining infrastructure.

Its broader significance is that although gravity energy storage in mine shafts is theoretically possible as a novel, site-specific store, it would not be competitive with tried-and-tested systems like PHES or contemporary BESSs (battery energy storage systems). PHES is the yardstick against which mechanical energy storage is measured [1,3,26] with its established capability for large-scale delivery and established dependability. In like manner, lithium-ion and other batteries, more expensive in the initial instance, are more flexible and capable of accommodating variable renewable sources of power [59]. Therefore, gravity storage in mine shafts is better described as a complementary technology that might find application in special cases where attributes of the mine shaft and local grid load make it economically feasible.

Gajdzik et al. [70,71] offer context to this evaluation that is useful, as they mention Poland's comparatively high electricity and heat demand in its steel sector, a pre-eminent industrial energy consumer. Their research illustrates that gravity storage, in theory, would be capable of functioning as a balancing agent for such energy-intensive processes, especially within a strategy of energy management. The current energy sources that can be realized in existing mine shafts would fall short of the requirements of large industrial consumers, hence casting aspersions on the efficacy of GES as a standalone source to feed heavy industry. Stecuła et al. [72] highlight the disruptive potential of digital management and urban AI in the implementation of city energy solutions. This is consistent with our findings that, although GES in mine shafts has modest technical potential, its application would heavily be facilitated by smart digital control systems to maximize energy flows, regulate loads, and incorporate storage activities into overall grid requirements. These technological solutions would assist in alleviating, in part, the operation inefficiencies indicated in our study, most importantly in relation to the finite time and low capacity of energy yield feasible in the mine shaft-based GES systems considered.

The low scalability and low yield nature of GES results in our study are also reflected in the treatment of Gajdzik et al. [73], who studied the CO_2 emission intensity of the steel process. Their research indicates that the decarbonization of emissions-intensive sectors like cement will demand large-scale, solid energy solutions—a challenge which today's mine shaft-type GES designs cannot easily meet. Hybrid solutions like those of Qin et al. [74] and Tardy et al. [75] in this case become increasingly pertinent. Qin et al. [74] consider innovative integrated energy management technologies that incorporate storage with peak load management in excavators at mines, and Tardy et al. [75] emphasize the worth of hybridizing wind and pumped hydro storage to achieve enhanced energy resilience for isolated mining operations. Our findings indicate that similar hybrid or multi-technology approaches may be needed to achieve GES in mine shafts as a worthy contributor to the transformation of Poland.

Jiang et al. [76] and Lyu et al. [77] offer additional evidence for the innovative reuse of mining facilities and resources for thermal and mechanical energy storage. Their results validate the conceptual attractiveness of GES as a mine life extension measure—a theme most central to our endeavour. Nonetheless, as our findings show, the particular running and maintenance issues of sizing shaft hoist arrangements to fit gravity-based energy storage, as well as power output finiteness, are overarching concerns. These results are in agreement with those of Yixi et al. [78], who highlight the necessity for optimization and system responsiveness in matching industrial loads with variable renewable energy sources—a goal which the gravity-based storage systems we examined meet in only partial measure.

Our research also corresponds to the issues of Zhao et al. [79] and Liu et al. [80], who refer to the time-dependent degradation and life-cycle cost effect of energy storage in the mining environment. For Polish shaft mines, that means ongoing investment in shaft maintenance, water drainage, and ventilation—expenses which would have the effect of largely stripping gravity-based storage systems of their already slim economic competitiveness. These life-cycle issues are again emphasized further by Li et al. [81] and Zhang et al. [82], whose energy storage optimization for flexible industrial and mining load work substantiates the limitation in using gravity-based storage as a stand-alone-technology solution.

Tang et al. [83] and Nuñez et al. [84] further substantiate this point with better control structures and phase-change material-based storage in the mining context, respectively. Even if these studies discuss other storage media and conditions of operation, they confirm our overall conclusion: that gravity energy storage in mine shafts, as a standalone solution, does not have the operational variability and storage capacity to become a leading balancing mechanism in the energy mix of Poland in the future. In place of that, as Ochmann et al. [85] and Yang et al. [86] observe in their study of post-mining energy systems, a more holistic approach involving various types of storage and numerical optimization techniques is also expected to be more effective.

Schmidt et al. [87] and Liao et al. [88] also corroborate the necessity of a strong geomechanical and technical analysis during retrofitting mining structures to energy storage, supplementing our own evidence towards intensive reinforcement and monitoring in Polish instances. Finally, Sheng et al. [89] present the promise of big data-based management in optimizing energy storage performance, providing a sequence of future studies that would assist in improving the operational performance of gravity-based energy storage systems in spite of the technical and economic constraints of their implementation.

Technological path dependency theory, in which the inertia caused by previous design choices and operational models is emphasized, is a robust interpretative tool to explain these findings [90–93]. As the paper shows, shafts in mines were intended to transport material and people rather than lift and lower huge composite weights in repeated cycles. This historical context has given rise to hoisting technology focusing on speed, security, and dependability for personnel transport but being too weak and energy-inefficient to be appropriate for extensive-scale energy storage purposes [94,95]. Thus, however theoretically desirable the gravity energy potential of deep mining shafts is, underlying technical and design limitations create tall hurdles to utilizing them as feasible repositories.

The calculations and mechanical analysis in this study highlight how this historical path dependency actually carries over into current technical issues. The limitations of existing shaft hoists on carrying load and the structural modifications needed in order to provide for energy storage operations are some points that reflect an evident clash between initial design purpose and current repurposing goals. This is the key argument from technological path dependency theory: that the risks and costs of retro-fitting existing systems generally outweigh the benefits of reuse if new functionality significantly differs from existing functionality [96–98]. In gravity energy storage, the need for precision-controlled raising and lowering of heavy composite blocks through a restricted shaft space is a challenge not readily overcome without widespread and expensive technological upgrading—upgrades, perhaps, that would make the concept economically noncompetitive [99–105].

The general implications of the work extend beyond the particular framework of Polish mine shafts. The research proposes that the process of integrating previous industrial infrastructure into new energy grids in the future should be considering the limitation by technological path dependency right from the planning stage onward. As much as recycling old assets is a central tenet of circular economy principles and consonant with higher goals of sustainability, the pragmatic realization of such projects must also reconcile these visions against deeply ingrained boundaries of incumbent technical regimes. Policymakers and engineers can thus craft more realistic and affordable approaches that respect the original design constraints and functional purpose of existing infrastructure while advancing the transition toward cleaner and more sustainable energy systems. Important conclusions that follow from the technical calculations in Section 4 are the gap between the theoretical cycle-by-cycle energy output of a gravitational fall and the system energy needs of a national grid power supply network. An estimate by the research suggests that a 70 Mg load dropped at 5 m/s under ideal circumstances discharges less than 80 kWh per cycle—i.e., no more than approximately 0.08 MWh of power (Figure 6). Even considering round-the-clock operation, the total energy input of such a system would still be minute compared to that of existing technology like pumped-storage hydropower, which with much better scalability and efficiency can provide hundreds of MWh per cycle [26,52]. For instance, the 2300 MW Turów pumped-hydro scheme, assumed to supply up to 160 GWh, shows the statistical disparity between shaft-based gravity systems and more large-scale, traditional solutions [36]. The numbers emphasize an important consideration: despite how theoretically consistent mine-based GES systems are with low-carbon policy aims, even at the barest minimum levels they are not capable of achieving such that they would be likely viable as primary or secondary grid-scale storage means.

This conclusion is further confirmed by the statistical depth distribution of Polish mine shafts (Figure 7). As ARP SA data illustrate, the majority of the shafts are less than 600 m deep, and less than 10% are both deep enough and not burdened by the hoisting structure, i.e., Koeppe wheels or guiding frames [25,69]. Since stored potential energy is proportional to mass and depth (Formula (1)), these depth limitations alone restrict the practicality of delivering significant energy capacity. This technical limitation is then further exacerbated by the low permissible mass loadings of Polish shafts, normally ranging from 30 to 80 tons, which significantly lower the energy output per cycle. Conversely, innovations such as Gravitricity, which is tested in deeper mines of over 1000 m, are far more promising in their application [43,44] but do not have widespread applicability to the Polish situation. Upgrading the infrastructure for energy storage under existing technological conditions thus does not statistically justify either peak-shaving on a daily basis or long-term grid stabilization, and thus investment claims are questionable.

Lifecycle considerations raise further questions of long-term economic viability for gravity storage in mine shafts. Polish mine shaft hoists were not designed for the cyclical dynamic and stresses of energy storage operations [25,69], implying a greater mechanical wear profile for round-the-clock load-lifting regimes. Extrapolated to a notional 50-year operating lifetime—a common energy transition planning assumption [26]—these systems would likely have gigantic maintenance and retrofitting expenses. This result is consistent with findings from studies by Liu et al. [80] and Sheng et al. [89], which highlight lifecycle deterioration and the economic penalty of time-dependent material degradation in energy storage facilities. Analogously, the requirements of constant dewatering and the ventilation of closed shafts have fixed costs of operation that also offset the anticipated advantages of the reuse of infrastructure [27,37]. As such, despite optimistic assumptions on circular economy advantages, technical and economic factors presented here reduce the statistical accuracy of the long-term cost–benefit analysis commonly applied to justify the installation of GES within heritage mining infrastructure.

The feasibility lessons from this research chart the fine line between the promise and challenge of GES systems in the context of Poland's ongoing energy transition. Technically, the research indicates that although GES presents a novel solution for using discarded coal mine shafts, the significant adaptation of existing shaft hoisting gear—i.e., reinforcing load-carrying capacity and compliance with safety specifications—is a colossal operation and engineering task. Such limitations imply the necessity of thorough technical examination and appraisal studies prior to undertaking any large-scale projects involving mine-based GES systems.

At the policy level, the work contributes to the strategic value of integrating circular economy principles—like recycling old mining infrastructure—into Poland's strategy of decarbonization. Yet it is also a warning that technological path dependencies offered in the present design and operation of old industrial plants can potentially limit the real application of these aims significantly, without major policy and financial incentives. Therefore, the conclusions of this study are of particular interest to energy policymakers, regulators, and business executives, and present an earthed view on the importance of balancing environmental aspirations with technical feasibility and economic imperatives in the quest for climate neutrality and energy security.

6. Conclusions

The principal quantitative findings of the study confirm that the potential energy return from the gravity energy storage (GES) devices on installed Polish mine shafts is tightly constrained by structural and operational problems. In particular, for one 70-ton load descending at 5 m/s, the maximum theoretical power return is around 2.75 MW for about 100 s, delivering less than 0.08 MWh of stored energy per cycle. Since the majority of shafts in Polish hard coal mines are less than 600 m, and acceptable payloads range from 30 to 80 tons, the calculated storage capacity is on the margin. The study also indicates that less than 10% of available shafts do not have hoisting equipment that would render retrofitting, and thus scalability limiting, unfeasible. These values, compared to national grid or industrial energy requirements, are indicative of the technical feasibility of GES only to a very limited extent and under very exceptional circumstances.

This article proves that, theoretically, gravity energy storage in Polish mine shafts is possible within reach of circular economy and sustainability goals, but in practice, it is significantly hindered, directly answering all three research questions posed (RQ1–RQ3). In order to respond to RQ1, the paper presents an exhaustive characterization of economic and technical limitations—i.e., the low rates of hoisting, limited depths, and prohibitively costly retrofitting of the installed infrastructure—combined that constrain energy supply to levels far below desired levels to enable useful grid support. Such remarks suggest justification by quantitative modelling and operating estimates to demonstrate that even on the optimistic assumption, single-cycle energy production is marginal (<0.1 MWh), so mass application is not possible without costly modifications [25,26,69]. For RQ2's scenario, the comparison explicitly situates GES as far less competitive than potential alternatives in the guise of pumped-storage hydropower (PHES) or battery energy storage systems (BESSs), both of which possess higher scalability, flexibility, and energy density [36,59]. That shaft-based GES is not even a competitive player on the terms of these technologies even when under the hypothesis of infrastructure reuse underscores its niche status within Poland's energy storage portfolio. Lastly, in RQ3, while the concept of recycling mine shafts is consilient with national and EU-level policy—i.e., the European Green Deal, Circular Economy Action Plan, and Poland's Energy Policy to 2040 (PEP2040)—pragmatic reality depends on an immense amount of technological and financial intervention. Therefore, as much as GES in mine shafts is appealing to the green transformation and just transition principle, the study concludes that for it to be applied, it has to be extremely discriminatory and based on hybrid solutions derived from the incident and public policy tools to reconcile sustainability aspirations with economic and operational requirements.

The novelty of this paper lies primarily in its focus on the Polish context, which has thus far remained underexplored in the international literature on gravity energy storage (GES). While an increasing number of studies point to the limited scalability of GES in existing mine shafts, this article makes a unique contribution by offering a detailed technical and economic analysis tailored to the geological, infrastructural, and industrial conditions specific to Poland. Its most significant added value is the in-depth quantification of operational parameters such as shaft load capacity, working depth, operation time, and the unit amount of energy that can be stored. This enables a realistic assessment of the deployment potential of GES in a country where mining sector restructuring and the challenge of repurposing post-mining infrastructure represent urgent issues within the broader framework of a just energy transition.

To best address the given weaknesses of GES in Polish mine shafts, subsequent studies will have to design specific hybrid system configurations and simulation regimes aimed at the physical and operational limitations of the facilities. A viable option is the integration of gravity energy storage with high-efficiency battery systems (BESSs), where GES serves for rapid-response, short-duration oscillations and batteries for longer-duration capacity levelling—thus optimizing response time and the depth of discharge. Another potential model is integration with compressed air energy storage (CAES), utilizing the mine shaft for both mass conveyance and as a pressure vessel, leveraging existing voids for dual utilization. Artificial intelligence algorithms and digital twin technology could be employed to dynamically manage the flow of energy, forecast peak demand, and optimize the load cycles of the hoisting machinery in real time. These proposals, based on techno-economic modelling, would give practical blueprints for making GES from a conceptually viable but mediocre solution into a real-world contributor to Poland's energy transition strategy.

The results of this research contrast sharply between the policy ambitions of sustainable energy transitions—specifically, those preferring circular economy approaches and infrastructure repurposing—and the economic and technical realities of gravity energy storage (GES) in Polish mine shafts. Although the strategic fit to EU policy objectives, like the European Green Deal and Clean Planet for All strategy, and national policy objectives like Poland's Energy Policy 2040 (PEP2040) is indubitable, the deployment of GES in this format is faced with stringent feasibility limitations.

The full reuse of retired mining infrastructure for energy storage from a policy perspective sits very well within the principles of resource efficiency and just transition. It also addresses the socio-economic imperative of stretching the life cycle of industrial infrastructure and maintaining jobs in ex-mining areas. Our analysis illustrates, however, that alignment of such is, at a minimum, symbolic. Existing shaft geometry, low lifting capacities, and structural limitations are all such that—despite firm political will—effective storage capacity per installation is at best marginal, and unlikely to fulfil national or regional flexibility ambitions.

Policy rhetoric consequently needs to shift from one of blanket optimism to one of a more cautious and evidence-informed stance. In particular, mine shaft reuse needs to be repositioned not as a scalable energy technology but as a niche or complementary technology. Potential applications for selective use could be remote applications, microgrids, or industrial parks where special geology and infrastructural circumstances render GES feasible on a limited scale. In such applications, shaft-based GES can help with system resiliency and peak-shaving, particularly if combined with other technologies like compressed air energy storage (CAES) or battery energy storage systems (BESSs).

The limitations of the study are related to the availability of empirical evidence on the gravity energy storage systems applied in Polish mine shafts and their impact on the generalizability of the results. The further analysis is mainly based on theoretical calculations and case study comparisons without any real-world demonstration projects confirming the technical and economic viability of such systems.

Future studies must focus on field tests and pilot projects to verify the theoretical models of gravity energy storage in retired mine shafts and to determine the technical parameters for actual application. Research is also warranted to assess the prospects of integrating gravity-based storage with other forms of renewable energy and hybrid storage systems, i.e., coupling gravitational storage with battery or compressed air options. In addition, the study of sophisticated digital monitoring and control technologies can be applied to maximize the operating performance and long-term viability of gravity energy storage systems in mine infrastructure.

Author Contributions: Conceptualization, B.G., R.W. and K.T.-O.; methodology, K.T.-O.; software, R.W. and B.G.; validation, B.G. and R.W.; formal analysis, K.T.-O. and B.G.; investigation, B.G., R.W. and W.G.; resources, R.W. and B.G.; data curation, B.G. and K.T.-O.; writing—original draft preparation, B.G., R.W., J.K. and K.T.-O.; writing—review and editing, R.W., B.G., J.K. and W.G.; visualization, B.G.; supervision, B.G., R.W. and K.T.-O.; project administration, B.G., R.W. and W.G.; funding acquisition, B.G., R.W. and K.T.-O. All authors have read and agreed to the published version of the manuscript.

Funding: This research received no external funding.

Data Availability Statement: Data are contained within the article.

Conflicts of Interest: The authors declare no conflicts of interest.

:eferences

Aneke, M.; Wang, M. Energy storage technologies and real life applications—A state of the art review. *Appl. Energy* **2016**, *179*, 350–377. [CrossRef]

Quaranta, E.; Boes, R.M.; Hunt, J.D.; Szabò, S.; Tattini, J.; Pistocchi, A. Considerations on the existing capacity and future potential for energy storage in the European Union's hydropower reservoirs and pumped-storage hydropower. *J. Energy Storage* **2024**, *104*, 114431. [CrossRef]

Tong, W.; Lu, Z.; Sun, J.; Zhao, G.; Han, M.; Xu, J. Solid gravity energy storage technology: Classification and comparison. *Energy Rep.* **2022**, *8*, 926–934. [CrossRef]

Tong, W.; Lu, Z.; Chen, W.; Han, M.; Zhao, G.; Wang, X.; Deng, Z. Solid gravity energy storage: A review. *J. Energy Storage* **2022**, *53*, 105226. [CrossRef]

Morstyn, T.; Botha, C.D. Gravitational Energy Storage with Weights. Encyclopedia. Available online: www.marketsandmarkets.com/Market-Reports/gravity-energy-storage-market-213578483.html (accessed on 17 March 2025).

Galetakis, M.; Biotakis, G.; Deligiorgis, V.; Varouchakis, E. Transforming Decommissioned Mines to a Gravity Energy Storage System. *Mater. Proc.* **2023**, *15*, 14. [CrossRef]

Li, L.; Wang, B.; Jiao, K.; Ni, M.; Du, Q.; Liu, Y.; Li, B.; Ling, G.; Wang, C. Comparative techno-economic analysis of large-scale renewable energy storage technologies. *Energy AI* **2023**, *14*, 100282. [CrossRef]

Jennings, A. Conveyor Energy Storage: A Battery and Pumped Hydro Alternative, Power-News and Technology for the Global Energy Industry. 2020. Available online: https://www.powermag.com/conveyor-energy-storage-a-battery-and-pumped-hydro-alternative/ (accessed on 7 June 2025).

Hossain, E.; Faruque, H.M.R.; Sunny, M.S.H.; Mohammad, N.; Nawar, N. A comprehensive review on energy storage systems: Types, applications, and challenges. *Energies* **2020**, *13*, 2781. [CrossRef]

). Koolen, D.; De Felice, M.; Busch, S. *Flexibility Requirements and the Role of Storage in Future European Power Systems*; Publications Office of the European Union: Luxembourg, 2023. [CrossRef]

1. Quaranta, E.; Georgakaki, A.; Letout, S.; Mountraki, A.; Ince, E.; Gea Bermudez, J. *Clean Energy Technology Observatory: Hydropower and Pumped-Storage Hydropower in the EUROPEAN Union—2024 Status Report on Technology Development, Trends, Value Chains and Markets*; Publications Office of the European Union: Luxembourg, 2024. Available online: https://publications.jrc.ec.europa.eu/repository/handle/JRC139225 (accessed on 17 March 2025).

12. Overall Strategic Analysis of Clean Energy Technology in the European Union. Strategic Analysis and Outlook. Genera Publications, European Comission,6 February 2025. Available online: https://setis.ec.europa.eu/publications-and-documents clean-energy-technology-observatory/ceto-reports-2024_en (accessed on 20 May 2025).
13. Goal 17: Revitalize the Global Partnership for Sustainable Development. Sustainable Development Goals. United Nations Available online: https://sdgs.un.org/goals/goal17 (accessed on 17 March 2025).
14. Zillman, D.; Godden, L.; Paddock, L.; Roggenkamp, M. Regulating electricity storage in the European Union: How to balance technical and legal innovation. In *Innovation in Energy Law and Technology: Dynamic Solutions for Energy Transitions;* Zillman, D Godden, L., Paddock, L., Roggenkamp, M., Eds.; Oxford University Press: Oxford, UK, 2018; pp. 155–172.
15. Zhou, C.; Richardson-Barlow, C.; Fan, L.; Cai, H.; Zhang, W.; Zhang, Z. Towards organic collaborative governance for a more sustainable environment: Evolutionary game analysis within the policy implementation of China's net-zero emissions goals *J. Environ. Manag.* **2025**, *373*, 123765. [CrossRef]
16. Directive (EU) 2019/944. It Was Amended in 2024 by Directive (EU) 2024/1711: Directive (EU) 2024/1711 of the European Parliament and of the Council of 13 June 2024 Amending Directives (EU) 2018/2001 and (EU) 2019/944 as Regards Improving the Union's Electricity Market Design (Text with EEA Relevance). Available online: https://eur-lex.europa.eu/eli/dir/2024/17 1/oj/eng (accessed on 25 March 2025).
17. European Commission. A Clean Planet for All: A European Strategic Long-Term Vision for a Prosperous, Modern, Competitive and Climate Neutral Economy: Brussels, Belgium, 28 November 2018. Available online: https://eur-lex.europa.eu/legal-content EN/TXT/?uri=CELEX:52018DC0773 (accessed on 25 March 2025).
18. United Nations. Paris Agreement. United Nations Framework Convention on Climate Change. 2015. Available online https://unfccc.int/sites/default/files/english_paris_agreement.pdf (accessed on 20 April 2025).
19. Ozturk, Z.; Demirci, A.; Terkes, M.; Yumurtaci, R. Optimal planning of solar PV-based electric vehicle charging stations empowered by energy storage system: Feasibility and green charge potential. *Renew. Energy* **2025**, *255*, 123715. [CrossRef]
20. Li, C.; Yao, M.; Yao, X.; Li, C. A hybrid processing technology for fabricating lead zirconate-based ceramics with high energy storage density, high efficiency, and low sintering temperature. *J. Mater.* **2025**, *11*, 101077. [CrossRef]
21. Qin, Z.; Ma, J.; Zhu, M.; Khan, T. Advancements in energy storage technologies: Implications for sustainable energy strategy and electricity supply towards sustainable development goals. *Energy Strateg. Rev.* **2025**, *59*, 101710. [CrossRef]
22. Korski, J.; Rak, Z. Magazynowanie energii w szybach kopalnianych—Grawitacyjny magazyn z cięgnem łańcuchowym. In Proceedings of the Komtech, Szczyrk, Poland, 4–6 November 2024. [CrossRef]
23. Matos, C.R.; Carneiro, J.F.; Silva, P.P. Overview of large-scale underground energy storage technologies for integration of renewable energies and criteria for reservoir identification. *J. Energy Storage* **2019**, *21*, 241–258. [CrossRef]
24. Siostrzonek, T. The mine shaft energy storage system—Implementation threats and opportunities. *Energies* **2024**, *16*, 5615 [CrossRef]
25. Kula, J.; Kamiński, P.; Stecuła, K.; Prostański, D.; Matusiak, P.; Kowol, D.; Kopacz, M.; Olczak, P. Technical and Economic Aspects of Electric Energy Storage in a Mine Shaft. *Budryk Case Study* **2021**, *14*, 7337.
26. Hunt, J.D.; Zakeri, B.; Jurasz, J.; Tong, W.; Dąbek, P.B.; Brandão, R.; Patro, E.R.; Đurin, B.; Filho, W.L.; Wada, Y.; et al. Underground Gravity Energy Storage: A Solution for Long-Term Energy Storage. *Energies* **2023**, *16*, 825. [CrossRef]
27. Korski, J. Solid gravity energy storage in mine shafts—Feasibility and functionality. *Min. Mach.* **2024**, *42*, 232–239.
28. Tobór-Osadnik, K.; Gajdzik, B.; Strzelec, G. Configurational Path of Decarbonisation Based on Coal Mine Methane (CMM) An Econometric Model for the Polish Mining Industry. *Sustainability* **2023**, *15*, 9980. [CrossRef]
29. Gajdzik, B.; Tobór-Osadnik, K.; Wolniak, R.; Grebski, W.W. European Climate Policy in the Context of the Problem of Methane Emissions from Coal Mines in Poland. *Energies* **2024**, *17*, 2396. [CrossRef]
30. Wang, X.; Zhou, F.; Ling, Y.; Wei, K.; Kang, J. Overview and Outlook on Utilization Technologies of Low-Concentration Coal Mine Methane. *Energy Fuels* **2021**, *35*, 15398–15423. [CrossRef]
31. Kaczmarek, J. The Balance of Outlays and Effects of Restructuring Hard Coal Mining Companies in Terms of Energy Policy of Poland PEP 2040. *Energies* **2022**, *15*, 1853. [CrossRef]
32. Likwidacja Kopalń na Nowych Zasadach (Title in Polish). Document. Available online: https://www.prawo.pl/biznes/ likwidacja-kopaln-na-nowych-zasadach-jest-projekt-ustawy,530875.html (accessed on 17 May 2025).
33. Energy Policy of Poland (EPP2040) Until 2040, (Title in Polish) Polityka Energetyczna Polski do 2040 r. (PEP2040). Available online: https://www.gov.pl/web/klimat/polityka-energetyczna-polski (accessed on 11 May 2023).
34. INSTRAT: Kopalnie w Polsce—Zaktualizowana Baza na Energy.Instrat.pl. Available online: https://instrat.pl/kopalnie-w-polsce-aktualizacja-2024/#:~:text=Wed%C5%82ug%20stanu%20na%20grudzie%C5%84%202024,oraz%205%20z%C5%82%C3%B3%C5%BC%20w%C4%99gla%20brunatnego (accessed on 20 April 2025).
35. Report (Title in Polish) Polska Ścieżka Transformacji Energetycznej; Polski Komitet Energii Elektrycznej: Warszawa, Poland 2022. Available online: https://pkee.pl/publications/raport-ey-i-pkee-polska-sciezka-transformacji-energetycznej/ (accessed on 27 May 2023).

Rafał, K.; Grabowski, P. Magazynowanie energii. *Energetyka ACADEMIA Mag. Pol. Akad. Nauk.* **2021**, *1*, 34–40. [CrossRef]

Saigustia, C.; Robak, S. Review of Potential Energy Storage in Abandoned Mines in Poland. *Energies* **2021**, *14*, 6272. [CrossRef]

Morstyn, T.; Chilcott, M.; McCulloc, M.D. Gravity energy storage with suspended weights for abandoned mine shafts. *Appl. Energy* **2019**, *239*, 201–206. [CrossRef]

Kawalec, W.; Król, R. Generating of Electric Energy by a Declined Overburden Conveyor in a Continuous Surface Mine. *Energies* **2021**, *14*, 4030. [CrossRef]

Schnepper, T.; Kühn, M.; Kempka, T. Pumped hydropower storage operation in open-pit lignite mines does not compromise the pit lake and groundwater chemistry (no. EGU24-3947). In Proceedings of the Copernicus Meetings, Vienna, Austria, 14–19 April 2024.

Krassakis, P.; Karavias, A.; Zygouri, E.; Roumpos, C.; Louloudis, G.; Pyrgaki, K.; Koukouzas, N.; Kempka, T.; Karapanos, D. GIS-Based Assessment of Hybrid Pumped Hydro Storage as a Potential Solution for the Clean Energy Transition: The Case of the Kardia Lignite Mine, Western Greece. *Sensors* **2023**, *23*, 593. [CrossRef]

Menéndez, J.; Fern'andez-Oro, J.M.; Galdo, M.; Loredo, J. Efficiency analysis of underground pumped storage hydropower plants. *J. Energy Storage* **2020**, *28*, 101234. [CrossRef]

Przekształcanie Opuszczonych Kopalń W Baterie Grawitacyjne. *Magazyn Fotowoltaika News*, 7 February 2023. Available online: https://magazynfotowoltaika.pl/przeksztalcanie-opuszczonych-kopaln-w-baterie-grawitacyjne/ (accessed on 25 April 2025).

First Commercial Gravity Storage for Energy Planned in Finnish Mine. Available online: https://www.mining.com/first-commercial-gravity-storage-for-energy-planned-in-finnish-mine/ (accessed on 7 June 2025).

Bentham, J. Europe's mines look to gravity energy storage for green future. *Global Mining Review*, 13 May 2024. Available online: https://www.globalminingreview.com/mining/13052024/europes-mines-look-to-gravity-energy-storage-for-green-future/ (accessed on 25 March 2025).

World's First 26MW Gravity Energy Storage Project Nears Completion in Rudong. 2024. Available online: https://en.nantong.gov.cn/2024-06/21/c_998345.htm (accessed on 20 April 2025).

Mercer, P. (Snowy 2.0) Australia's divisive plan for a vast underground 'battery' Snowy 2.0 project. *BBC*, 17 October 2019. Available online: https://www.bbc.com/news/world-australia-49202175 (accessed on 20 March 2025).

Morgan, S. Decommissioned Mines Converted to Energy Storage. News 7 March 2023, I.M3, Institute of Materials, Minerals Mining, News. Available online: https://www.iom3.org/resource/decommissioned-mines-converted-to-energy-storage.html (accessed on 7 June 2025).

Bluszcz, A.; Tobór-Osadnik, K.; Tomiczek, K.; Mansor, N.S.; Awang, H. The use of geomatics tools in critical infrastructure management. *Inżynieria Miner.* **2023**, *1*, 169–174. [CrossRef]

Bluszcz, A.; Tobór-Osadnik, K.; Manowska, A.; Smetana, M. Functionalities of geoinformatic systems in crisis events caused by climate change. In Proceedings of the 24th International Multidisciplinary Scientific GeoConference SGEM 2024, Albena, Bulgaria, 1–7 July 2024. [CrossRef]

Cahyono, M. Hybrid Models for Solving the Colebrook–White Equation Using Artificial Neural Networks. *Fluids* **2022**, *7*, 211. [CrossRef]

Ostraszewski, M.; Jurasz, J.; Kaźmierczak, B. Transforming Coal Pits into Renewable Energy Sources: The Potential of Pumped Hydro Energy Storage in the Bełchatów Lignite Mine. In Proceedings of the EGU General Assembly Conference Abstracts, Vienna, Austria, 14–19 April 2024; p. 6474.

Álvarez, H.; Domínguez, G.; Ordóñez, A.; Menéndez, J.; Álvarez, R.; Loredo, J. Mine Water for the Generation and Storage of Renewable Energy: A Hybrid Hydro–Wind System. *Int. J. Environ. Res. Public Health* **2021**, *18*, 6758. [CrossRef]

Gao, R.; Wu, F.; Zou, Q.; Chen, J. Optimal dispatching of wind-PV-mine pumped storage power station: A case study in Lingxin Coal Mine in Ningxia Province, China. *Energy* **2022**, *243*, 123061. [CrossRef]

Schmidt, F.; Menéndez, J.; Konietzky, H.; Pascual-Muñoz, P.; Castro, J.; Loredo, J.; Sánchez, A.B. Converting closed mines into giant batteries: Effects of cyclic loading on the geomechanical performance of underground compressed air energy storage systems. *J. Energy Storage* **2020**, *32*, 101882. [CrossRef]

Prado, L.Á.D.; Menéndez, J.; Bernardo-Sánchez, A.; Galdo, M.; Loredo, J.; Fernández-Oro, J.M. Thermodynamic analysis of compressed air energy storage (CAES) reservoirs in abandoned mines using different sealing layers. *Appl. Sci.* **2021**, *11*, 2573. [CrossRef]

Bartela, Ł. A hybrid energy storage system using compressed air and hydrogen as the energy carrier. *Energy* **2020**, *196*, 117088. [CrossRef]

Yuan, L.; Tang, W.; Thomas, R.A.; Soles, J. Experimental study on suppression of lithium iron phosphate battery fires. *Min. Metall. Explor.* **2024**, *41*, 637–645. [CrossRef] [PubMed]

Wu, T.; Zhou, Y.; Niu, J. An energy storage system for smart coal mine emergency power supply. *Energy Storage Sci. Technol.* **2015**, *4*, 104. [CrossRef]

60. Baigorri, J.; Zaversky, F.; Astrain, D. Massive grid-scale energy storage for next-generation concentrated solar power: A review o the potential emerging concepts. *Renew. Sustain. Energy Rev.* **2023**, *185*, 113633. [CrossRef]
61. Tan, K.M.; Babu, T.S.; Ramachandaramurthy, V.K.; Kasinathan, P.; Solanki, S.G.; Raveendran, S.K. Empowering smart grid: A comprehensive review of energy storage technology and application with renewable energy integration. *J. Energy Storage* **2021**, *39*, 102591. [CrossRef]
62. Zhou, K.; Yang, S.; Chen, Z.; Ding, S. Optimal load distribution model of microgrid in the smart grid environment. *Renew. Sustain. Energy Rev.* **2014**, *35*, 304–310. [CrossRef]
63. Bąk, P.; Sukiennik, M.; Kowal, B. The Main Drivers of the Raw Materials and ICT Sectors in Poland Using PESTEL Analysis. *Energies* **2025**, *18*, 1987. [CrossRef]
64. Götz, M.; Lefebvre, J.; Mörs, F.; Koch, A.M.; Graf, F.; Bajohr, S.; Kolb, T. Renewable Power-to-Gas: A technological and economic review. *Renew. Energy* **2016**, *85*, 1371–1390. [CrossRef]
65. Menéndez, J.; Ordóñez, A.; Álvarez, R.; Loredo, J. Energy from closed mines: Underground energy storage and geothermal applications. *Renew. Sustain. Energy Rev.* **2019**, *108*, 498–512. [CrossRef]
66. Andrusikiewicz, W. Możliwości magazynowania gazu ziemnego i energii w utworach solnych na terenie Polski (Potential for natural gas and energy storage in salt formations in Poland). *Inżynieria Miner.* **2024**, *2*, 143–150.
67. Kiliński, A. *Przemysłowe Procesy Realizacji: Podstawy Teorii*; WNT: Warszawa, Poland, 1976.
68. Halliday, D.; Resnick, R. *Fundamentals of Physics*, 12th ed.; John Wiley & Sons: Hoboken, NJ, USA, 2023.
69. Zmysłowski, T. *Górnicze Maszyny Wyciągowe: Część mechaniczna*; Śląsk Sp. z o.o.: Warszawa, Poland, 2004.
70. Gajdzik, B.; Wolniak, R.; Grebski, W. Electricity and heat demand in steel industry technological processes in Industry 4.0 conditions. *Energies* **2023**, *16*, 787. [CrossRef]
71. Gajdzik, B.; Wolniak, R.; Grebski, W. An econometric model of the operation of the steel industry in Poland in the context of process heat and energy consumption. *Energies* **2022**, *15*, 7909. [CrossRef]
72. Stecuła, K.; Wolniak, R.; Grebski, W.W. AI-Driven Urban Energy Solutions—From Individuals to Society: A Review. *Energies* **2023**, *16*, 7988. [CrossRef]
73. Gajdzik, B.; Wolniak, R.; Grebski, W. An Econometric Analysis of CO_2 Emission Intensity in Poland's Blast Furnace–Basic Oxygen Furnace Steelmaking Process. *Sustainability* **2025**, *17*, 4045. [CrossRef]
74. Qin, T.; Li, Y.; Quang, L.; Yao, Y.; Yang, L. A Novel Integrated Energy Management Strategy of Energy Storage System for a Pure Electric-Driven Mining Hydraulic Excavator. *Chin. J. Mech. Eng. (Engl. Ed.)* **2025**, *38*, 58. [CrossRef]
75. Tardy, A.; Rousse, D.R.; Mungyeko Bisulandu, B.-J.R.; Ilinca, A. Enhancing Energy Sustainability in Remote Mining Operations Through Wind and Pumped-Hydro Storage; Application to Raglan Mine, Canada. *Energies* **2025**, *18*, 2184. [CrossRef]
76. Jiang, J.; Cao, Y.; Li, G.; Zhao, J.; Liu, C. Using Industrial Mining Solid Waste through Conversion to Phase-Change Materials for Thermal Energy Storage. *ChemPlusChem* **2025**, *90*, e202400519. [CrossRef]
77. Lyu, J.; Han, W.; Zhang, X.; Pan, Y.; Dai, L. Evolution Law of the Activation and Energy Storage of Fault with Hard Rock Bridge under Mining Influence. *Meitan Xuebao/J. China Coal Soc.* **2025**, *50*, 901–916.
78. Yixi, C.; Liu, X.; Li, Y.; Danzeng, B.; Wang, L. Optimal Configuration Strategy of Energy Storage Considering Flexible Response of High Energy-Consuming Industrial and Mining Loads in Independent Microgrid. *Front. Energy Res.* **2025**, *13*, 1536668. [CrossRef]
79. Zhao, P.; Chang, Z.; Li, S.; Shao, Q.; Pei, W. Study on Time-Varying Mechanism of Energy Dissipation Structure in Depressed Pressure Gas Transport and Storage Area of Inclined Thick Coal Seam Mining. *Zhongguo Kuangye Daxue Xuebao/J. China Univ. Min. Technol.* **2025**, *54*, 202–214.
80. Liu, Q.; Tong, Y.; Liu, B.; Yang, J.; Feng, Y. Life Cycle Cost-Oriented Optimization of Hybrid Energy Storage System for Mining Haul Truck. In *Proceedings of the 19th Annual Conference of China Electrotechnical Society*; Lecture Notes in Electrical Engineering; Springer: Berlin/Heidelberg, Germany, 2025; Volume 1406, pp. 112–120.
81. Li, Y.; Wang, L.; Liu, X.; Danzeng, B.; Xia, Q. Coordinated Optimization Method of Energy Storage and Flexible Demand Response for Industrial and Mining Loads. *J. Web Eng.* **2025**, *44*, 167–192. [CrossRef]
82. Zhang, P.-F.; Gu, X.-B.; Guo, W.-Y.; Zhu, Z.-Q.; Guo, L. Bending Energy Storage Mechanical Model of Layered Composite Roof Structure in Coal Mining. *Sci. Rep.* **2024**, *14*, 30976. [CrossRef]
83. Tang, Q.; Hu, M.; Bian, Y.; Peng, X.; Li, K. Optimal Energy Efficiency Control Framework for Distributed Drive Mining Truck Power System with Hybrid Energy Storage: A Vehicle-Cloud Integration Approach. *Appl. Energy* **2024**, *374*, 123989. [CrossRef]
84. Nuñez, S.M.; Trujillo Preuss, F.E.; Masip Macía, Y. Numerical Study for the Design of a Thermal Energy Storage System with Multiple Tunnels Based on Phase Change Material: Case Study Mining in Chile (Thermal Storage in Off-Grid Industrial Applications). *Appl. Sci.* **2024**, *14*, 3690. [CrossRef]
85. Ochmann, J.; Jurczyk, M.; Rusin, K.; Bartela, Ł.; Uchman, W. Solution for Post-Mining Sites: Thermo-Economic Analysis of a Large-Scale Integrated Energy Storage System. *Energies* **2024**, *17*, 1970. [CrossRef]

76. Yang, Y.; Cao, M.; Zhao, W.; Zhong, Y.; Lyu, P. Optimal Operation Model of Complex Integrated Energy System for Mining Areas with Coalbed Methane and Compressed Air Energy Storage. *Zhongguo Dianji Gongcheng Xuebao/Proc. Chin. Soc. Electr. Eng.* **2024**, *44*, 2257–2269.
77. Schmidt, F.; Menéndez, J.; Konietzky, H.; Álvarez, L.; Bernardo-Sánchez, A. Technical Feasibility of Lined Mining Tunnels in Closed Coal Mines as Underground Reservoirs of Compressed Air Energy Storage Systems. *J. Energy Storage* **2024**, *78*, 110055. [CrossRef]
78. Liao, Y.; Wang, T.; Ren, Z.; Li, J.; Zou, X. Multi-Well Combined Solution Mining for Salt Cavern Energy Storages and Its Displacement Optimization. *Energy* **2024**, *288*, 129792. [CrossRef]
79. Sheng, X.; Yang, Q.; Ye, R.; Tao, L.; Zhang, Z. Big Data Feature Mining Method for Energy Storage System of Photovoltaic Power Station in Typical Scenarios. In Proceedings of the 2024 IEEE 4th International Conference on Information Technology, Big Data and Artificial Intelligence (ICIBA 2024), Beijing, China, 6 December 2024; pp. 569–574.
80. Parker, M.O.; Masrani, S. Overcoming Technological Path Dependency: Role of Dynamic Capabilities, Deliberate and Emergent Actions. *Strategy Leadersh.* **2025**, *53*, 393–414. [CrossRef]
81. Meijer, A.; Thaens, M. Path Dependency of Smart Cities: How Technological and Social Legacies Condition Smart City Development. In *Smart Cities and Smart Governance*; Public Administration and Information Technology; Springer: Berlin/Heidelberg, Germany, 2021; Volume 37, pp. 31–42.
82. Vadén, T.; Majava, A.; Toivanen, T.; Hakala, E.; Eronen, J.T. To Continue to Burn Something? Technological, Economic and Political Path Dependencies in District Heating in Helsinki, Finland. *Energy Res. Soc. Sci.* **2019**, *58*, 101270. [CrossRef]
83. Onufrey, K.; Bergek, A. Self-Reinforcing Mechanisms in a Multi-Technology Industry: Understanding Sustained Technological Variety in a Context of Path Dependency. *Ind. Innov.* **2015**, *22*, 523–551. [CrossRef]
84. Ntostoglou, E.; Martin, V.; Khatiwada, D.; Urban, F. Path-Dependencies in the Transition to Sustainable Biowaste Valorization: Lessons from a Socio-Technical Analysis of Sweden and Greece. *Waste Manag.* **2025**, *192*, 47–57. [CrossRef]
85. Kim, J.; Yoon, J.; Lee, J.-D. Technological Speciation: Navigating New Needs Through Trial and Error—A Rifle Case Study. *J. Evol. Econ.* **2024**, *34*, 927–952. [CrossRef]
86. Streb, J. The Cliometric Study of Innovations. In *Handbook of Cliometrics*, 3rd ed.; Springer: Berlin/Heidelberg, Germany, 2024; pp. 447–468.
87. Chiodo, M.; Müller, D.; Sienknecht, M. Educating AI Developers to Prevent Harmful Path Dependency in AI Resort-to-Force Decision Making. *Aust. J. Int. Aff.* **2024**, *78*, 210–219. [CrossRef]
88. Nye, D.E. A Model for Heterogeneous Energy Transitions. *Archimedes* **2023**, *67*, 21–50.
89. Taberna, A.; Filatova, T.; Roventini, A.; Lamperti, F. Coping with Increasing Tides: Evolving Agglomeration Dynamics and Technological Change Under Exacerbating Hazards. *Ecol. Econ.* **2022**, *202*, 107588. [CrossRef]
90. Kuo, C.-I. The Adoption Speed of Scientific Knowledge: The Moderating Role of Path Dependency on Scientific Knowledge. In Proceedings of the IEEE International Conference on Industrial Engineering and Engineering Management, Kuala Lumpur, Malaysia, 7–10 December 2022; pp. 1321–1326.
91. Castillo, A.A. Digital Transformation and the Public Sector Workforce: An Exploration and Research Agenda. In Proceedings of the 14th International Conference on Theory and Practice of Electronic Governance, Athens, Greece, 6–8 October 2021; pp. 471–475.
92. de Percy, M.; Batainah, H. Identifying Historical Policy Regimes in the Canadian and Australian Communications Industries Using a Model of Path Dependent, Punctuated Equilibrium. *Policy Stud.* **2021**, *42*, 42–59. [CrossRef]
93. Bondarenko, A.V.; Islamov, S.R.; Ignatyev, K.V.; Mardashov, D.V. Laboratory Studies of Polymer Compositions for Well-Kill under Increased Fracturing. *Perm J. Pet. Min. Eng.* **2020**, *20*, 37–48. [CrossRef]
94. Cavalheiro, E.A.; Oliveira, I.R.; Leandro, D.; Kontz, L.B. Governance, development, and environment: Pathways to a sustainable future. *Sustain. Futures* **2025**, *10*, 100813. [CrossRef]
95. Belousov, A.; Lushpeev, V.; Sokolov, A.; Sultanbekov, R.; Tyan, Y.; Ovchinnikov, E.; Shvets, A.; Bushuev, V.; Islamov, S. Experimental Research of the Possibility of Applying the Hartmann–Sprenger Effect to Regulate the Pressure of Natural Gas in Non-Stationary Conditions. *Processes* **2025**, *13*, 1189. [CrossRef]

energies

MDPI

Article

Regional Disparities, Spatial Effects, and the Dynamic Evolution of Distorted Energy Prices in China

Zhiyuan Gao [1,2], Ziying Jia [1] and Yu Hao [3,4,5,*]

1 School of Economics and Management, Beijing Institute of Petrochemical Technology, Beijing 102617, China; gaozhiyuan@bipt.edu.cn (Z.G.); 2024540089@bipt.edu.cn (Z.J.)
2 Development Research Centre of Beijing New Modern Industrial Area, Beijing 102617, China
3 School of Economics and Management, North University of China, Taiyuan 030051, China
4 School of Economics, Beijing Institute of Technology, Beijing 100081, China
5 Faculty of Economics, Shenzhen MSU-BIT University, Shenzhen 518172, China
* Correspondence: haoyuking@gmail.com

Abstract

The distortion of energy prices has become an important obstacle to the high-quality development of China's economy. Moreover, energy price distortions are not merely a domestic issue. They may trigger carbon leakage by diverting emissions-intensive production to countries with cheaper energy. Although the existing literature has extensively examined the effects of energy price distortions, two significant research gaps remain. First, most studies treat energy price distortions merely as an influencing factor, lacking a systematic analysis that places it at the core. Second, the spatial correlation characteristics of energy price distortions are often overlooked. This study measures the degree of energy price distortions across Chinese provinces from 2000 to 2022 and employs methods such as the Global Moran's I, Local Moran's I, and kernel density estimation to systematically analyze the spatial correlation, spatial distribution of coordination indices, and dynamic evolution patterns of these distortions. The results reveal that: (1) the overall degree of energy price distortions in China exhibited a trend of rising first and then declining, with significant regional disparities; (2) the regional gap followed an "expansion-contraction" trajectory; (3) there is notable spatial autocorrelation, with high-distortion areas concentrated in Northeast China, the middle reaches of the Yellow River, and Northwest China; and (4) the dynamic evolution suggests that distortion levels in high- and medium-value regions may continue to decline, while those in low-value regions may increase. This study fills a critical gap in the systematic spatial analysis of energy price distortions and provides new empirical evidence and policy insights for advancing market-oriented reforms in energy markets.

Keywords: energy price distortion; Theil index; spatial effect; kernel density function; regional disparity

1. Introduction

As a global economic phenomenon, energy price distortion is widespread in developing countries and transitional economies. International experience indicates that government intervention in energy markets, lagging factor market reforms, and market monopolies often result in energy prices that fail to reflect the true scarcity of resources and the environmental externalities they generate [1,2]. Such distortions not only lead to inefficiencies in resource allocation but also transmit through price signals to influence industrial

https://doi.org/10.3390/en18133465

structure, energy consumption patterns, and carbon emission intensity, becoming a key obstacle to green economic transformation. Although energy price distortions are common, their manifestations and underlying mechanisms vary significantly across economies. As the world's largest energy consumer and carbon emitter, China presents a particularly distinctive case. Price liberalization marked the decisive turning point in China's shift from a command economy to a market-oriented one. Under central planning, resources were misallocated, goods were perennially scarce, and the entire economic engine was in need of overhaul. Reformers, therefore, had to find a breakthrough powerful enough to reset incentives yet subtle enough to avoid social upheaval. After limited experiments in tax and fiscal policy, they zeroed in on prices, the economy's most fundamental signal. A transparent, flexible price-formation mechanism could unlock efficient resource allocation, spur production, and guide consumption far more effectively than any decree. Nowhere is this signal more critical than in energy. When energy prices are distorted, consumption habits warp, investment skews, and carbon footprints balloon. China's ascent to the leading global user of energy and source of carbon emissions is therefore at odds with its new blueprint for green, productivity-led growth, and it magnifies vulnerabilities in national energy security. Correcting these price signals is no longer just an economic tweak; it is a strategic imperative for sustainable development.

As China's market reforms deepen, price signals now govern the allocation of most final goods, yet factor-market reforms remain comparatively lagging. The central government continues to set prices for seven categories—including transmission and distribution tariffs, as well as oil and gas pipeline transportation—thereby preserving multiple pricing regimes within a single economy and inevitably generating energy-price distortions. The coexistence of planned- and market-based mechanisms produces two distinct forms of distortion: policy-induced distortions, rooted in market imperfections, and market-induced distortions, stemming from sub-optimal policy design. Although existing research has examined these distortions extensively, most studies treat them chiefly as explanatory variables that influence other outcomes. The largest body of work explores how distorted energy prices affect improvements in energy efficiency and factor substitution. For instance, Xu et al. [3] developed an extended vintage-capital model under imperfect factor markets and showed that distorted factor prices reshape the accumulation of high-efficiency capital vintages, thereby altering energy demand and CO_2 emissions. Wang et al. [4] argued that, amid China's rapid development, distorted energy prices have become a principal obstacle to energy conservation and emission reduction. Sun et al. [5] introduced a joint-frontier parametric linear programming method based on the Shephard energy-distance function to measure total-factor energy efficiency in manufacturing and to derive shadow energy prices through cost minimization.

In light of the above, this study focuses on 30 Chinese provinces, and, based on a scientifically constructed measure of energy price distortions, employs multiple quantitative methods to clarify regional disparities and the dynamic evolution of their distribution. Compared with the existing literature, this paper makes several novel contributions. First, it provides a rigorous measurement of energy price distortion. While previous studies have primarily conducted qualitative analyses or relied on single-indicator case studies, this research draws on the fundamental connotations and characteristics of energy price distortion and utilizes the Cobb-Douglas (C-D) production function to comprehensively estimate provincial-level distortions across China. Second, it accurately identifies regional disparities in distortion levels. Although current research acknowledges the imbalance and inadequacy of energy price distortion, little attention has been paid to the extent and sources of these disparities. To address this gap, this study introduces the Theil index and its decomposition method to clarify the magnitude and sources of regional variation, offering valuable insights for reforming China's energy pricing system. Third, it investigates the dynamic evolution of energy price distortion from a multidimensional perspective. Moving beyond the limitations of static analysis, the study incorporates both temporal and spatial dimensions by applying stochastic kernel density estimation and Markov chain analysis to capture the dynamic shifts in relative positions and state transition probabilities of distortion levels, aiming to provide empirical evidence and systematic understanding to support energy pricing reform.

The remainder of this paper is organized as follows. Section 2 reviews the relevant literature; Section 3 details the research methodology and data; Section 4 reports the empirical results; and Section 5 offers conclusions and policy recommendations.

2. Literature Review and Theoretical Framework

2.1. Literature Review

(1) Measurement Methods of Energy Price Distortion

The production function approach, shadow pricing, and frontier analysis are the three primary methodologies used to estimate energy price distortion. Compared to the other two, the production function method is widely adopted due to its ease of data acquisition, simplicity in model construction, and its ability to directly reveal the gap between marginal prices and marginal productivity. Sun et al. [5] developed a joint frontier model based on the Shephard energy distance function to measure total-factor energy efficiency in China's manufacturing sector, deriving shadow energy prices through cost minimization. Wu et al. [6] focused on China's heavy industry and used the price gap method to estimate absolute price distortions across multiple energy types—including coal, oil, and natural gas—over the period 2003–2019. Ju et al. [2] applied a path analysis framework to construct three distortion indicators across five energy types. He found that although China's energy pricing system has undergone partial marketization, significant distortions remain. Relative and dynamic distortions were found to promote economic growth, while absolute distortions constrained it. Moreover, achieving carbon reduction goals depends on reducing such distortions. Ouyang and Sun [1] employed a shadow pricing model combined with stochastic frontier analysis (SFA) to compare actual and shadow prices in China's industrial sector, identifying allocative inefficiencies and potential energy savings. Ju et al. [7] and Sha et al. [8] proposed a theoretical energy pricing method based on marginal opportunity cost, decomposing the theoretical price into the marginal costs of production, utilization, and externalities. They used this framework to estimate distortions in coal, oil, and other energy types, establishing a broader framework for analyzing energy market distortions. Qiao et al. [9] adopted a ratio-based method to estimate capital and labor price distortions at the firm level. Capital distortion was estimated by comparing actual capital costs with the theoretical interest rate, while labor distortion was measured as the ratio of per capita labor cost to the industry average wage. Based on observed market prices, this approach reflects deviations from ideal resource allocation and is a typical method for measuring relative price distortion. Sun et al. [10] incorporated ecological efficiency, estimated using a bootstrap-DEA method, as an indicator of China's environmental performance, and used a translog production function to estimate factor price distortions. Zhang et al. [11] proposed a production function model with time-varying elasticity and used a factor income share approach based on national accounts to estimate capital and labor price distortions. They argue that this method captures temporal variation more effectively than the C-D or translog models. Peng, Zhang, and Xu [12] used a non-radial directional distance function to investigate the effect of energy price distortion on energy efficiency in China, introducing a spatial correlation perspective and confirming that distortion significantly affects energy efficiency. Wang et al. [13] developed a translog production function combined with a stochastic frontier analysis model (SFA), using cities as the unit of analysis to evaluate total-factor energy efficiency and energy price distortion in China's high-energy-consuming industries during 2003–2019.

(2) Economic Impacts of Energy Price Distortion

Energy price distortion is more than a market anomaly as it profoundly affects macroeconomic outcomes. This section reviews its economic impacts, focusing on efficiency, resource allocation, and industrial structure, and explores the mechanisms through which these effects occur. Most existing studies focus on specific dimensions of energy price distortion, particularly its effects on efficiency and resource allocation. Xu and Tan [14] examined the implications of eliminating energy misallocation in China, concluding that correcting energy price distortions could significantly increase economic output and contribute to global energy savings. Wang et al. [13] employed a translog production function and stochastic frontier analysis (SFA) to investigate the impact of energy price distortion on total-factor energy efficiency (TFEE) in energy-intensive industries across Chinese cities. Their results showed that energy price distortion significantly suppresses efficiency improvements. Zamani et al. [15], using translog cost function estimation based on Iranian manufacturing data from 2006–2018, found that rising energy prices reduce energy intensity and enhance energy and labor efficiency, while having only minor effects on capital and material inputs. These studies indirectly support the view that market-oriented pricing and the reduction of distortions enhance both resource allocation efficiency and energy utilization. In addition, some research examines the broader systemic and macroeconomic effects of energy price distortion. Kirikkaleli and Darbaz [16], using Toda-Yamamoto causality, Fourier Toda, and Spectral BC tests, identified a bidirectional causality between energy prices and the food price index across different frequency domains. Punzi [17] employed a dynamic stochastic general equilibrium (DSGE) model of a small open economy to assess how energy price uncertainty affects macroeconomic variables. Croonenbroeck and Hüttel [18] proposed a simplified method to quantify the efficiency loss resulting from forecast errors in renewable energy prices. Their findings suggest that price forecast errors significantly impair resource allocation efficiency, particularly in volatile energy types such as solar power, while the effects for wind power were negligible. Furthermore, they observed time-dependent patterns in efficiency loss, underscoring the importance of improving forecast accuracy to reduce deadweight losses in energy allocation.

(3) Environmental Impacts of Energy Price Distortion

With the growing global awareness of environmental protection, the environmental impacts of energy price distortion have increasingly become a focus of academic research. This section reviews how energy price distortion affects the environment directly or indirectly by influencing energy consumption patterns, carbon emission intensity, and related pathways. Scholars have explored the structural implications of energy price distortion. At the industrial sector level, accelerating the marketization of factor prices, particularly energy, is considered essential for advancing green industrial transformation and achieving China's "dual carbon" goals. Xu et al. [3], Li Ke et al. [19], and Gao et al. [20] demonstrated that correcting price distortions not only improves energy efficiency and emission performance but also unlocks substantial growth potential for high-quality industrial development. Xu et al. estimated that eliminating distortions in energy and other factor prices could reduce industrial coal consumption by approximately 225 million tonnes and CO_2 emissions by 793 million tonnes annually, along with a reduction in air pollution. Li Ke, Xu Chang, and Tang Liwei found that mitigating distortions—especially in competitive energy markets—substantially increases industrial green total-factor productivity (GTFP), with the most urgent needs observed in eastern provinces and those under economic or fiscal pressure. Gao et al. further estimated that fully eliminating capital and energy market distortions could yield a 10% annual increase in green industrial development potential, with greater gains in central and western China. At the sectoral level, Gao et al. [21] showed that energy price distortion contributes to inefficient industrial structures, affecting both energy consumption intensity and carbon emission intensity. Wang, Bai, and Xie [22] estimated the extent of distortion in China's oil pricing between 2004 and 2016, identifying positive price distortion in the transport sector. Their findings revealed significant substitution effects between oil, coal, and electricity, moderated by factor substitutability.

A systematic review of existing studies reveals several key research gaps that remain to be addressed. First, there are limitations related to data availability. Most current research relies on provincial-level or industry-level data [12], lacking long-term, high-frequency, and interprovincially comparable datasets. Some studies [10] estimate energy prices using proxy variables, which may introduce measurement bias. Second, there is a noticeable geographic coverage bias. Existing findings are predominantly concentrated in economically developed eastern regions [13], with insufficient attention to energy-rich areas in central and western China, making it difficult to reveal spatial heterogeneity across the entire territory. Third, there is a mismatch in the time dimension. Many studies are limited to data before 2019 [6] and fail to capture the structural shifts in the energy pricing mechanism that have occurred following the implementation of the "dual carbon" policy. To address these issues, this study focuses on 30 provinces in China. Based on a scientifically constructed measure of energy price distortion, the study investigates the imbalance and inadequacy of energy price distortions across regions. It introduces the Theil index and its decomposition method to clarify the extent and sources of regional disparities, and examines the dynamic evolution of the distribution of distortion levels from multiple dimensions. The aim is to provide empirical evidence and generalizable insights to support energy pricing reform.

(4) Spatial Analysis of Energy Price Distortion

In recent years, spatial econometric approaches have been widely applied in the study of energy pricing and resource allocation. Methodologically, spatial econometrics is grounded in solid theoretical foundations. As noted by Anselin [23], spatial econometric models are effective in identifying spatial dependence and heterogeneity among variables. Following Chen et al. [24], we first employed Moran's I test to confirm the spatial autocorrelation of energy price distortion, which aligns with the five fundamental principles of spatial econometrics proposed by Paelinck et al. [25]. For model selection, we adopted the spatial modeling framework suggested by Baltagi et al. [26] and ultimately employed the Spatial Lag Model (SLM) to capture the spatial spillover effects of energy price distortion.

Regarding the spatial analysis of energy price distortion, Elhorst [27] provided a systematic overview of spatial panel data models, offering a methodological foundation for studying spatial effects in energy pricing. In cross-country research, Wu et al. [28] employed a Spatial Durbin Model to reveal the suppressive effects of energy prices on both local and neighboring countries' carbon emissions in the EU. Burnett et al. [29] applied a spatial panel data model to examine spatial dependence in U.S. energy markets and found significant negative spatial spillover effects of energy prices across states. At the methodological level, the seminal work by LeSage and Pace [30] provides theoretical guidance for constructing spatial weight matrices in energy price research. The Generalized Spatial Two-Stage Least Squares (GS2SLS) method, developed by Kelejian et al. [31], is widely used to address endogeneity in studies of energy markets. For instance, Hua et al. [32] applied this method to examine the spatial impact of technology market development on energy efficiency. In the Chinese context, Zhang et al. [33] adopted the Dagum Gini coefficient decomposition and spatial Markov chain methods to explore the dynamic spatial distribution of energy price distortions in the industrial sector. Their findings showed that energy distortions were more pronounced than labor distortions, with east-west disparities and spatial proximity identified as key drivers of this evolution. Peng et al. [12] used a spatial panel model to assess the impact of energy price distortions on energy efficiency, finding that distortions not only significantly reduced local energy efficiency but also generated spatial spillover and signaling effects in neighboring provinces, demonstrating strong spatial dependence. Additionally, Li et al. [34], using Geographically Weighted Regression (GWR), demonstrated that spatial heterogeneity implies that energy pricing policies should be tailored to the industrial structure and technological foundation of each region to maximize both economic and environmental benefits.

2.2. Theoretical Framework

The formation of energy misallocation can be examined from both the supply and demand sides. On the supply side, low energy efficiency in production results in a marginal output value of energy that is significantly lower than its total cost of use. On the demand side, factors such as price controls may cause energy prices to diverge from their true value, leading to overconsumption [3]. During China's industrialization, many enterprises operated inefficiently due to technological limitations characteristic of specific stages of development. To maintain economic growth, government authorities implemented relatively loose environmental regulations during particular historical periods. This policy stance effectively encouraged an extensive growth model. Under such conditions, enterprises only bore the explicit costs of energy, while negative externalities—such as environmental pollution—were shifted onto society. As a result, the marginal output of energy remained far below its full social cost over the long term, ultimately causing distortions in resource allocation [4].

The evolution of environmental regulation policy in China has exhibited a clear pattern of institutional innovation. Rather than merely resorting to administrative price increases, policymakers sought to internalize environmental externalities, thereby restructuring the cost framework faced by enterprises. This policy design forced firms to make one of three choices: optimize their energy consumption patterns, implement energy-saving technological upgrades, or exit the market altogether. As a result, the institutional arrangement achieved a dual policy objective: maintaining relative stability in nominal energy prices while significantly improving energy use efficiency, thereby enhancing the overall efficiency of resource allocation [5].

From a regional perspective, the central government has taken a cautious stance on energy price marketization reforms out of concern for public welfare and livelihood protection. This cautious approach has led to a nationwide trend of energy price convergence. In theory, under perfect competition, firms should make production decisions such that the marginal output of energy equals its cost of use, in which case regional differences in pricing should not exist [35]. However, due to the heterogeneous environmental carrying capacities across regions and in light of limited autonomy over price regulation, local governments have adjusted enterprise energy costs by applying varying intensities of environmental regulation. This policy practice has resulted in significant regional differentiation in the implicit cost of energy use. To adapt to these institutional changes, firms at the micro level have had to improve energy productivity through technological innovation, aligning their operations with region-specific energy cost structures. At the macro level, this has ultimately manifested as regional disparities in energy allocation efficiency [36].

Based on the preceding literature review and theoretical framework, this study proposes the following hypotheses:

H1. *Energy price distortions in China display significant regional disparities, following a spatial pattern of higher distortion in the east and lower distortion in the west.*

H2. *Energy price distortions exhibit significant spatial dependence, characterized by high-high and low-low clustering.*

H3. *The spatial effects of energy price distortions reflect a pattern of club convergence, whereby regions with similar levels of distortion tend to cluster together.*

3. Research Methodology and Data Processing

3.1. Research Methodology

This study follows a "measurement-analysis-validation" research framework and consists of the following steps: (1) Measurement of energy price distortion: A provincial-level distortion index is constructed based on the C-D production function. (2) Decomposition of regional disparities: The Theil index is employed to examine spatial and temporal differences. (3) Test of spatial effects: Global and local Moran's I indices are used to identify spatial clustering patterns. (4) Analysis of dynamic evolution: Kernel density estimation and Markov chain models are combined to forecast evolving trends. (5) Robustness Checks: Robustness tests on the convergence mechanism to verify the stability and consistency of the findings.

(1) C-D Production Function

To capture the economic essence of factor-market distortions and to trace their temporal dynamics, this study adopts the production-function approach to evaluate provincial energy-price distortions. Grounded in the classical marginal principle, the method compares a factor's observed price with its marginal revenue product to infer the degree of distortion. The C-D production function is widely used in long-term analyses of factor distortions due to its transparency, decomposability, and the ease of marginal product calculation [37,38]. In this study, the assumption of constant returns to scale is relaxed to improve the model's adaptability to regional heterogeneity in China. The choice of the C-D function over the translog specification is based on the following considerations. First, the use of fixed elasticity coefficients in the C-D function simplifies parameter estimation and helps avoid the multicollinearity often introduced by interaction terms in translog models, making it particularly suitable for long-term panel data at the provincial level. Second, while the translog function provides greater flexibility in capturing time-varying elasticities of factor substitution, the primary objective of this study is to measure the relative degree of energy price distortion, rather than to track changes in elasticity over time. Third, the translog function requires higher data quality and a larger sample size, while provincial-level energy price data in China are subject to frequent adjustments in statistical definitions and classification standards. These inconsistencies undermine data comparability and reliability, thereby limiting the applicability of the translog specification in this context. The procedure begins by assuming that each province operates under a C-D technology, such that its production function is specified as:

$$Y_{it} = A_{it} \cdot K_{it}^{\alpha} L_{it}^{\beta} E_{it}^{\gamma} \tag{1}$$

Taking the natural logarithm of both sides of Equation (1) produces the linearized specification:

$$\ln Y_{it} = \ln A_{it} + \alpha \ln K_{it} + \beta \ln L_{it} + \gamma \ln E_{it} + \mu_{it} \tag{2}$$

Drawing on the shadow price model proposed by Ouyang and Sun [1] and integrating the price distortion measurement framework developed by Ju et al. [7], this study measures the degree of energy price distortion by comparing actual energy prices with their marginal revenue product. Given the marked heterogeneity in provincial development stages, the constant-returns-to-scale assumption is relaxed, permitting $\alpha + \beta + \gamma \neq 1$. Building on Equations (1) and (2), the marginal revenue product of energy for each province in year t can be further expressed as:

$$MRP_{E_{it}} = \gamma \cdot \frac{Y_{it}}{E_{it}} \tag{3}$$

In Equation (3), i indexes provinces and t denotes years. Comparing the observed energy price $P_{E_{it}}$ with its marginal revenue product $MRP_{E_{it}}$ yields the annual energy-price distortion for each province, expressed as:

$$Dist_{E_{it}} = \frac{P_{E_{it}} - MRP_{E_{it}}}{MRP_{E_{it}}} \tag{4}$$

The energy-price distortion index $Dist_{E_{it}}$ captures the deviation of a province's actual energy price from its market-equilibrium level in year t. When $Dist_{E_{it}} > 0$, meaning the observed price exceeds the marginal revenue product of energy, production units typically curb energy demand, substitute toward capital or labor, or enhance energy-use efficiency to conserve energy. Conversely, $Dist_{E_{it}} < 0$ indicates that the observed price lies below the equilibrium benchmark, suggesting scope for profitable expansion of energy input. Nonetheless, as energy consumption rises, the marginal return to additional energy diminishes, and its utilization efficiency correspondingly declines.

(2) Theil T Index (GE(1))

This study employs the Theil T index, a member of the Generalized Entropy Index family, to measure energy distortion. Originally introduced by Theil [39], this method has been widely adopted in the analysis of regional disparities [40]. The Theil index is a measure of regional disparities, with larger values indicating greater differences. It is decomposable and can distinguish between within-region and between-region disparities. Energy distortion typically appears when certain regions or industries draw disproportionately on high-efficiency or low-cost energy, undermining economy-wide allocation. To capture this phenomenon, we apply the generalized entropy index GE(1). Unlike many alternative inequality measures, GE1 is especially sensitive to areas with extreme distortion, sharply illuminating the "resource-misallocation" pattern of excessive energy use. Further, its additivity and decomposability allow us to separate total distortion into between-region and within-region components, revealing the precise origins and structural contours of the problem. GE(1) strikes a balance between the sensitivity of GE(0) to low distortions and the tendency of GE(2) to overweight extremely high distortions, making it well-suited for capturing resource misallocation, where excessive energy possession by certain regions is a key concern. The additive decomposability of GE(1) provides clear, policy-relevant insights, whereas GE(2)'s use of squared terms may exaggerate outliers, and GE(0) may underrepresent areas with high levels of distortion. GE(1) is widely used in regional disparity studies, ensuring consistency and comparability with existing literature on China's spatial heterogeneity. The corresponding computational formula is presented below.

$$T_t = \frac{1}{k}\sum_{i=1}^{k}\left(\frac{E_{it}}{\overline{E_t}} \times \ln \frac{E_{it}}{\overline{E_t}}\right) \tag{5}$$

In Equation (5), T_t takes values in the interval [0, 1] and rises monotonically with the severity of energy-price distortion. Values approaching 0 indicate convergence in provincial distortion levels, whereas values nearing 1 signal pronounced inter-provincial disparities.

Here, n denotes the number of provinces, E_{it} denotes the energy-price distortion of province i, and $\overline{E}$ refers to the national mean.

$$T_{jt} = \frac{1}{n_j}\sum_{i=1}^{n_j}\left(\frac{E_{ijt}}{\overline{E_{jt}}} \times \ln\frac{E_{ijt}}{\overline{E_{jt}}}\right) \tag{6}$$

In Equation (6), T_{jt} denotes the overall Theil index for region j; n_j is the number of provinces in region j; E_{ijt} represents the energy-price distortion level of province i within region j; and $\overline{E_{jt}}$ is the mean distortion level for region j.

$$T_t = T_{wt} + T_{bt} = \sum_{j=1}^{8}\left(\frac{n_j}{n} \times \frac{\overline{E}_{jt}}{\overline{E}_t} \times T_{jt}\right) + \sum_{j=1}^{8}\left(\frac{n_j}{n} \times \frac{\overline{E}_{jt}}{\overline{E}_t} \times \ln\frac{\overline{E}_{jt}}{\overline{E}_t}\right) \tag{7}$$

In Equation (7), the overall Theil index for digital-rural development is decomposed into a within-region component T_{wt} and a between-region component T_{bt}. The terms $\frac{T_{wt}}{T_t}$ and $\frac{T_{bt}}{T_t}$ denote the respective contribution shares of these two components to the national disparity, while $\frac{E_{jt}}{E_t} \times \frac{T_{jt}}{T_t}$ captures the share of each region in the aggregated within-region inequality. The expression E_{jt} is the sum of energy-price distortions for all provinces in region j, and E_t is the corresponding national total.

3.2. Data Processing

The calculation of the marginal revenue product of energy relies primarily on data from standard statistical yearbooks. Provincial-level GDP, labor input, and capital stock are sourced from successive editions of the China Statistical Yearbook, while energy consumption figures are sourced from the China Energy Statistical Yearbook. Given the lack of officially published energy price data, researchers typically estimate them by inflating benchmark prices from 2003, as reported for 36 large and medium-sized cities in the China Price Yearbook 2004. However, this approach has notable limitations: the range of energy types is incomplete, and the selected cities do not adequately represent all 30 provinces. To enhance accuracy, this study adopts an alternative estimation method based on data from the Compilation of Data from the Third National Industrial Census. Compared with existing studies, our method covers more provinces and energy types, thereby enhancing representativeness. The process involves three main steps. First, the census provides both energy consumption volumes and corresponding prices for each province in 2000; multiplying these yields the total energy cost per province. Second, energy consumption is converted into standard coal equivalents, enabling the derivation of uniform provincial energy prices for 2000. Third, annual provincial energy prices for 2001–2022 are estimated by adjusting the 2000 baseline using the yearly Fuel and Power Price Index. To ensure the validity of the model specification, we conducted panel data diagnostics using the LLC, IPS, and Fisher-ADF tests to examine variable stationarity. The results show that the null hypothesis of a unit root for energy price distortion is rejected at the 1% significance level, indicating that the variable is stationary.

4. Empirical Analysis

4.1. Baseline Measurements

Table 1 reports provincial energy-price distortion indices for 2000–2022 and reveals three main patterns. First, at the national level, distortion has fallen sharply, dropping from 2.030 in 2000 to 0.115 in 2022, causing a 94.33% decline. Second, all eight economic regions exhibit a similar downward trajectory, yet their distortion levels differ markedly. Ranked from highest to lowest, the ordering is: East Coastal, North Coastal, Northeast

Comprehensive, South Coastal, Middle Yangtze, Southwest, Middle Yellow River, and the Greater Northwest. The East and North Coastal regions display notably higher distortion than all others, while the Middle Yellow River and Greater Northwest show the lowest levels. Third, at the provincial scale, Shanghai, Jiangsu, Guangdong, Tianjin, and Zhejiang lead the distortion ranking, clustering within the East, North, and South Coastal zones. This pattern suggests that economically advanced or policy-favored areas more readily obtain low-cost energy through administrative intervention or market power, thereby exacerbating price distortion. These regions often benefit from subsidized electricity tariffs, preferential energy supply, and large firms that secure discounted prices, preventing energy prices from reflecting true supply-demand conditions and resource scarcity. Consequently, energy becomes over-allocated to energy-intensive, not necessarily efficient industries, hindering clean-energy uptake and efficiency gains and imposing a clear allocative-efficiency loss.

Table 1. Regional Energy Price Distortion Indices.

Area	2000	2008	2016	2022	Mean	Area	2000	2008	2016	2022	Mean
China	2.030	0.552	0.232	0.115	0.620	Henan	1.944	0.558	0.283	0.194	0.636
Liaoning	2.233	0.833	0.414	0.162	0.855	Shaanxi	1.910	0.501	0.173	0.088	0.541
Jilin	2.497	0.729	0.444	0.223	0.842	Middle Yellow River	5.894	1.568	0.652	0.372	1.739
Heilongjiang	1.664	0.864	0.410	0.222	0.778	Anhui	1.420	0.505	0.190	0.091	0.506
Northeast Comprehensive	6.394	2.427	1.268	0.607	2.474	Jiangxi	1.552	0.494	0.187	0.083	0.509
Beijing	2.056	0.643	0.314	0.157	0.707	Hubei	2.153	0.609	0.296	0.144	0.700
Tianjin	3.324	0.832	0.268	0.140	0.952	Hunan	1.599	0.341	0.170	0.084	0.432
Hebei	1.395	0.460	0.173	0.112	0.482	Middle Yangtze	6.724	1.948	0.843	0.401	2.147
Shandong	2.560	0.562	0.245	0.133	0.713	Guangxi	1.713	0.486	0.193	0.086	0.538
Northern Coast	9.335	2.497	1.001	0.543	2.854	Chongqing	1.618	0.320	0.147	0.066	0.422
Shanghai	4.422	1.136	0.554	0.270	1.352	Sichuan	1.419	0.351	0.170	0.080	0.414
Jiangsu	4.324	0.908	0.368	0.178	1.182	Guizhou	0.643	0.266	0.112	0.053	0.246
Zhejiang	3.288	0.750	0.317	0.141	0.904	Yunnan	1.335	0.375	0.142	0.061	0.405
Eastern Coast	12.034	2.793	1.240	0.589	3.439	Southwest	6.728	1.799	0.764	0.345	2.025
Fujian	2.793	0.547	0.212	0.095	0.743	Gansu	1.243	0.520	0.199	0.121	0.499
Guangdong	3.983	0.761	0.329	0.164	1.018	Qinghai	1.190	0.313	0.093	0.066	0.343
Hainan	1.926	0.396	0.122	0.048	0.490	Ningxia	0.993	0.252	0.082	0.026	0.235
Southern Coast	8.702	1.703	0.663	0.307	2.252	Xinjiang	1.673	0.750	0.168	0.077	0.608
Shanxi	0.788	0.308	0.110	0.061	0.276	Greater Northwest	5.099	1.835	0.542	0.290	1.685
Inner Mongolia	1.253	0.201	0.086	0.029	0.287						

Note: Means in the table are arithmetic averages over the period 2000–2022.

Figure 1 illustrates a clear decline in nationwide energy price distortions and a convergence of regional disparities between 2000 and 2022. Two notable trends emerge. First, the spatial concentration of high-distortion "hot spots" diminished significantly: in 2000, the most severe distortions (shown in dark green) were concentrated along the eastern coast as well as in Xinjiang, Gansu, and Liaoning. By 2022, however, all provinces had shifted into medium- or low-distortion categories, with top-tier distortion levels nearly disappearing. Second, the historical "east-high/west-low" gradient narrowed considerably. Initially, distortions were most pronounced in the East Coast and Northeast, while the major inland energy-producing provinces remained relatively stable. In the later years, distortions declined most rapidly in the coastal regions, while those in the northwest also receded, significantly reducing the east-west disparity. These trends suggest that market-oriented reforms—such as the liberalization of electricity, natural gas, and refined oil prices, the development of a unified national power market, and the implementation of carbon-trading pilot programs—have effectively curtailed distortionary administrative subsidies.

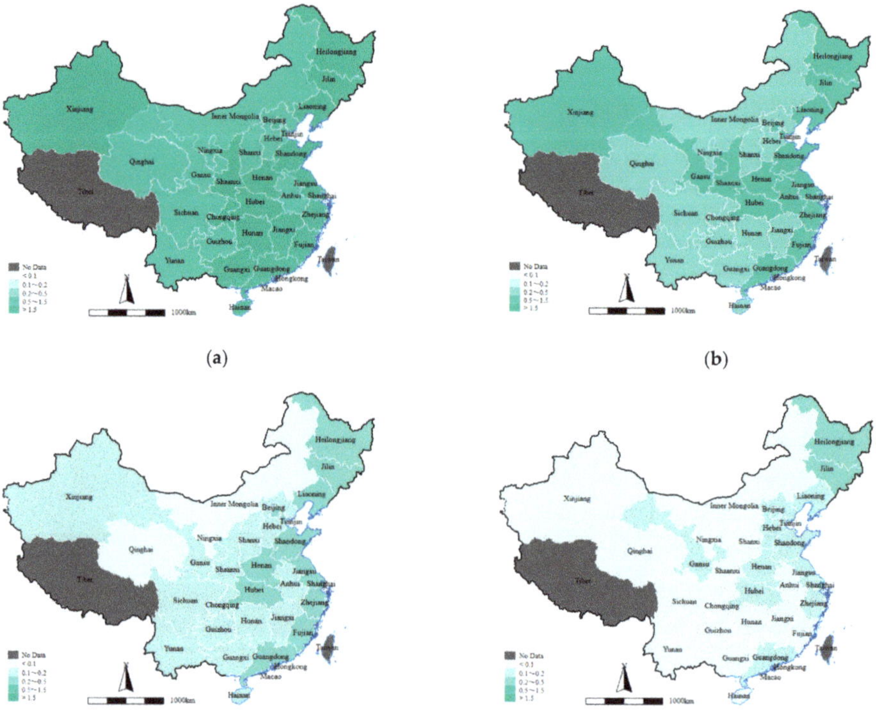

(a) (b)

(c) (d)

Figure 1. Trends in Energy-Price Distortion Levels in China. (**a**) Distribution of Energy-Price Distortion in 2000. (**b**) Distribution of Energy-Price Distortion in 2008. (**c**) Distribution of Energy-Price Distortion in 2016. (**d**) Distribution of Energy-Price Distortion in 2022.

4.2. Regional Disparities

Using the Theil index, we quantify nationwide, within-region, and between-region differences in energy price distortions and compute each region's contribution to the aggregate disparity, thereby revealing the spatial pattern and temporal dynamics of distortion.

Figure 2 indicates a "narrowing-widening" trajectory for regional gaps in distortion from 2000 to 2022. Specifically, both absolute distortion levels and their regional dispersion declined during 2000–2008, surged sharply over 2008–2016, and then leveled off during 2016–2022. Over the entire sample period, the amplitude of fluctuation in the between-region component is a modest 1.03%, whereas the within-region component swings by 90.11%. Hence, the inter-regional gap remains comparatively stable—albeit with a slight upward drift—while intra-regional inequality is far more volatile; even so, the magnitude of the between-region disparity consistently exceeds that of the within-region component. Ranking the volatility of provincial distortion by region yields the following order (high to low): Greater Northwest, East Coast, Middle Yellow River, South Coast, Middle Yangtze, Northeast Comprehensive, Southwest, and North Coast.

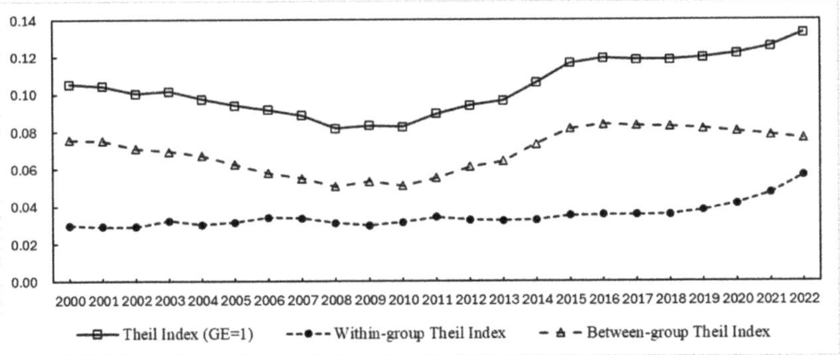

Figure 2. Temporal Evolution of Regional Theil Index for Energy-Price Distortion.

As shown in Table 2, the contributions of between-region and within-region disparities to energy-price distortion exhibit an inverse relationship, with the between-region contribution consistently exceeding the within-region component throughout the observation period.

Table 2. Regional Theil Index and Contribution Rates of Energy-Price Distortion.

Year		2000	2008	2016	2022	Change
Between-Region Gap and Contribution (%)		0.0757 (71.76)	0.0508 (62.08)	0.0839 (70.23)	0.0765 (57.45)	0.0008 (−14.31)
Within-Region Gap and Contribution (%)	Northeast Comprehensive	0.0137 (1.36)	0.0026 (0.47)	0.0007 (0.10)	0.0102 (1.34)	−0.0035 (−0.02)
	Northern Coast	0.0465 (6.76)	0.0234 (4.31)	0.0216 (2.60)	0.0071 (0.83)	−0.0394 (−5.93)
	Eastern Coast	0.0085 (1.58)	0.0144 (2.97)	0.0294 (4.38)	0.0371 (4.75)	0.0286 (3.17)
	Southern Coast	0.0423 (5.74)	0.0347 (4.36)	0.0752 (5.98)	0.1101 (7.35)	0.0678 (1.61)
	Middle Yellow River	0.0572 (5.25)	0.0709 (6.59)	0.105 (8.22)	0.2105 (17.05)	0.1533 (11.80)
	Middle Yangtze	0.0133 (1.39)	0.0201 (2.89)	0.026 (2.63)	0.0299 (2.61)	0.0166 (1.22)
	Southwest	0.0449 (4.70)	0.02 (2.65)	0.016 (1.47)	0.0156 (1.17)	−0.0293 (−3.53)
	Greater Northwest	0.0183 (1.45)	0.0892 (12.07)	0.0676 (4.40)	0.1178 (7.43)	0.0995 (5.98)
	Within-Region Total	0.0298 (28.24)	0.031 (37.92)	0.0356 (29.77)	0.0566 (42.55)	0.0268 (14.31)
Overall Gap		0.1055	0.0818	0.1195	0.1331	0.0166

Note: Numbers in parentheses indicate contribution rates (%).

Specifically, ranking regions by their within-region contribution in descending order yields: Middle Yellow River, Greater Northwest, South Coast, East Coast, Middle Yangtze, Northeast Comprehensive, Southwest, and North Coast. Within-region inequality continues to widen in the East Coast, South Coast, Middle Yellow River, and Middle Yangtze regions; in contrast, it has declined in the Northeast Comprehensive, North Coast, and Southwest. In the Greater Northwest, within-region disparity initially widened before narrowing, but remains slightly above its initial level.

Meanwhile, the between-region gap has displayed "convergence" or "club convergence" since its peak. As national reforms in energy pricing and market liberalization have advanced, the earlier spikes in interregional distortion have been suppressed, and regions have begun to align more closely in both pricing mechanisms and allocative efficiency—hallmarks of club convergence. Moreover, the divergent trajectories of within-region disparities reflect heterogeneity in reform progress and market structures: differences in resource endowments, industrial composition, and policy enforcement have allowed internal gaps to expand in certain areas. These patterns underscore the need for targeted, region-specific policy interventions to foster coordinated development within regions.

This table highlights three distinct features of regional disparities in energy price distortion across China from 2000 to 2022. The first feature is a reversal in the relative contributions of interregional and intraregional disparities. Specifically, the contribution of interregional differences declined from 71.76% to 57.45%, while that of intraregional differences rose from 28.24% to 42.55%. This shift indicates that disparities among provinces, rather than between broader regions, have increasingly become the primary source of variation. The second feature is pronounced regional heterogeneity. For example, the contribution of internal disparities within the middle reaches of the Yellow River jumped from 5.25% to 17.05%, reflecting a widening policy implementation gap among major energy-producing provinces such as Shanxi and Shaanxi. This growing divergence can be attributed to three main factors. First, the lock-in effect of policy bias—with over 80% of energy enterprises being state-owned and receiving substantial subsidies—has weakened price signals and market responsiveness. Second, structural constraints in resource endowment play a role. Provinces such as Shanxi and Xinjiang, which serve as major energy suppliers, remain affected by legacy mechanisms of planned pricing, leading to persistently higher distortion levels compared to energy-consuming provinces. Third, uneven market integration across regions has resulted in mismatches between supply and demand, further exacerbating spatial disparities in energy pricing. The third feature is the stage-specific nature of distortion trends. Since 2016, the overall Theil index has shown signs of stabilization (rising modestly from 0.1195 to 0.1331), suggesting that efforts to develop a unified national electricity market during the 13th Five-Year Plan have yielded some success. Nevertheless, interprovincial disparities remain above 2008 levels, indicating that ongoing reforms have not yet fully addressed the historical legacy of price distortions.

4.3. Spatial Distribution

To reveal the geographic-spatial distribution of China's energy-price distortion, this study takes a spatial-econometric perspective and employs the global Moran's I statistic and Moran scatterplots to examine spatial autocorrelation and clustering. Moran's I ranges from -1 to 1 and measures spatial autocorrelation, with positive values indicating the clustering of similar values and negative values indicating the clustering of dissimilar values. To analyze the spatial effects of energy price distortion, this study first constructs a spatial weight matrix W. Considering that geographically adjacent regions are more likely to exhibit spatial interactions, we define the spatial weights using a binary contiguity matrix.

Table 3 shows that Moran's I for energy-price distortion is statistically positively significant at the 1% level throughout 2000–2022, implying strong spatial autocorrelation and a broad clustering tendency across regions. The global Moran's I remains positive and relatively stable over this period, yet exhibits a downward trend overall. This decline suggests that, although spatial clustering persists, its intensity is weakening. The spatial dependence of energy-price distortion has diminished over time, though a certain degree of spatial linkage still remains. Such a trajectory indicates that the spatial distribution of distortion is gradually optimizing, interregional differences are narrowing, and the pattern is moving toward a more balanced state. This evolution not only reflects progress toward integrated energy markets but also demonstrates a coordinated improvement in distortion levels across regions.

Table 3. Global Moran's I for Energy-Price Distortion in China.

Year	Moran's I	*p*-Value	Year	Moran's I	*p*-Value
2000	0.401 ***	0.000	2012	0.271 ***	0.006
2001	0.427 ***	0.000	2013	0.251 ***	0.010
2002	0.388 ***	0.000	2014	0.271 ***	0.006
2003	0.378 ***	0.000	2015	0.277 ***	0.005
2004	0.389 ***	0.000	2016	0.277 ***	0.005
2005	0.330 ***	0.001	2017	0.133 *	0.078
2006	0.296 ***	0.003	2018	0.269 ***	0.006
2007	0.282 ***	0.005	2019	0.260 ***	0.008
2008	0.260 ***	0.008	2020	0.248 ***	0.010
2009	0.282 ***	0.005	2021	0.230 **	0.015
2010	0.258 ***	0.009	2022	0.203 **	0.026
2011	0.253 ***	0.009			

Notes: *, **, and *** indicate statistical significance at the 10%, 5%, and 1% levels, respectively.

While the global Moran's I captures the overall spatial autocorrelation of energy-price distortion, it may obscure inter-regional clustering patterns. To address this limitation, we employ the local Moran's I to directly visualize each region's distortion value and its spatial lag. In Figure 3, the horizontal axis represents the local distortion, while the vertical axis reflects its spatially lagged value. The four quadrants correspond to distinct spatial clustering patterns: HH (high-high) and LL (low-low) indicate positive spatial correlation (clustering), whereas LH (low-high) and HL (high-low) suggest negative spatial correlation (spatial heterogeneity).

The observed decline in Moran's I index can be attributed to three main factors. First, the progressive weakening of local governments' administrative intervention in energy pricing—due to the electricity market reform initiated after 2003, the launch of carbon trading pilots in 2015, and the construction of a unified national electricity market starting in 2017—has reduced the spatial correlation of price distortions stemming from policy discrepancies across regions. Second, the development of ultra-high-voltage (UHV) transmission networks and cross-regional energy delivery initiatives such as the "West-to-East Power Transmission" project has facilitated the broader optimization of energy resource allocation, thereby diminishing the homogeneity of distortions among geographically adjacent areas. Third, the gradual removal of interprovincial barriers to the mobility of energy resources, along with the initial operation of the national carbon market, has led to increasingly market-oriented price signals and the gradual emergence of "club convergence" patterns in regional price distortions.

Figure 3 reveals that, in all four benchmark years (2000, 2008, 2016, and 2022), most provinces fall into the HH and LL quadrants, indicating a persistent pattern of positive spatial autocorrelation in energy-price distortion.

Table 4 details the spatial distribution of provinces across the four quadrants of the Moran scatterplot for energy-price distortion during 2000–2022. The third quadrant (LL) consistently contains the largest share of provinces, and its spatial footprint remains relatively stable. Specifically, 18 provinces fell into LL in 2000; this number declined to 11 in 2008, rose to 15 in 2016, and stood at 12 in 2022. These provinces are mainly located in the Middle Yellow River, Southwest, Northwest, and Middle Yangtze regions—areas that are relatively underdeveloped or resource-dependent—and exhibit strong positive spatial autocorrelation. In contrast, economically advanced regions such as the East Coast, South Coast, North Coast, and Northeast predominantly lie in the other quadrants of the Moran scatterplot. In less developed and resource-based areas, energy supply structures tend to be homogeneous and subject to tighter policy controls, producing more concentrated and stable distortion patterns. Conversely, in developed regions where market mechanisms are more mature, energy-price distortions display a more diversified and dynamic spatial configuration.

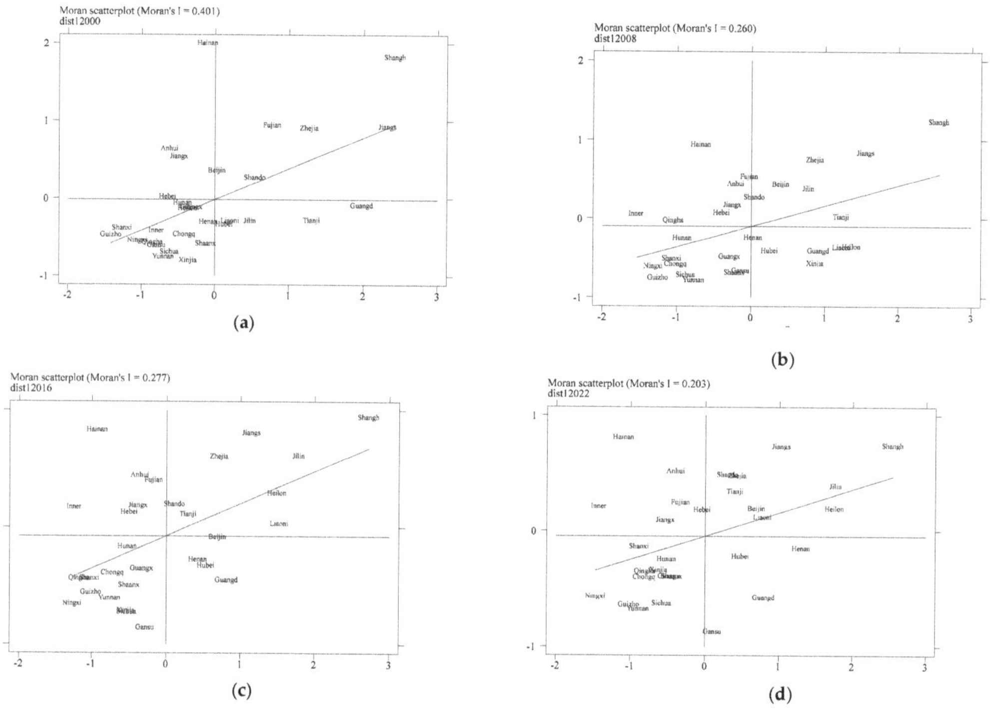

Figure 3. Local Moran's I Scatterplot for Energy-Price Distortion Index. (**a**) Local Moran's I cluster map in 2000. (**b**) Local Moran's I cluster map in 2008. (**c**) Local Moran's I cluster map in 2016. (**d**) Local Moran's I cluster map in 2022.

Table 4. Provincial Distribution across Moran Scatterplot Quadrants for Energy-Price Distortion.

Quadrant	I	II	III	IV
2000	Shanghai, Jiangsu, Zhejiang, Fujian, Shandong, Beijing	Hainan, Anhui, Jiangxi, Hebei	Shanxi, Guizhou, Inner Mongolia, Ningxia, Qinghai, Yunnan, Sichuan, Xinjiang, Chongqing, Shaanxi, Gansu, Henan, Heilongjiang, Guangxi, Hunan	Liaoning, Hubei, Tianjin, Guangdong, Jilin
2008	Shanghai, Jiangsu, Zhejiang, Jilin, Beijing, Tianjin, Shandong	Inner Mongolia, Qinghai, Hebei, Jiangxi, Fujian, Anhui, Hainan	Shanxi, Chongqing, Ningxia, Guizhou, Sichuan, Yunnan, Hunan, Guangxi, Shaanxi, Gansu	Liaoning, Guangdong, Heilongjiang, Xinjiang, Hubei, Henan
2016	Shanghai, Jiangsu, Zhejiang, Jilin, Heilongjiang, Liaoning, Shandong, Tianjin	Hainan, Anhui, Fujian, Jiangxi, Inner Mongolia, Hebei	Qinghai, Chongqing, Guizhou, Yunnan, Ningxia, Gansu, Sichuan, Shaanxi, Guangxi, Xinjiang, Shanxi, Hunan	Beijing, Henan, Hubei, Guangdong
2022	Shanghai, Jiangsu, Jilin, Heilongjiang, Liaoning, Beijing, Tianjin, Zhejiang, Shandong	Hainan, Anhui, Fujian, Jiangxi, Inner Mongolia, Hebei	Ningxia, Guizhou, Yunnan, Qinghai, Xinjiang, Chongqing, Shaanxi, Shanxi, Hunan, Guangxi, Sichuan	Henan, Hubei, Guangdong, Gansu

Table 4 discusses the evolution of quadrant distributions, which demonstrates a pronounced pattern of spatial polarization. Eastern provinces such as Shanghai and Jiangsu consistently remained in the first quadrant (HH), with spatial lag coefficients exceeding 0.7 ($p < 0.05$), forming a "high-distortion club." The number of provinces in the third quadrant (LL) decreased from 18 to 12, with western provinces such as Xinjiang and Qinghai experiencing a decline in Moran's I values by between 0.08 and 0.22. This indicates that the Western Development Strategy has weakened the spatial clustering of low-distortion areas. In 2022, provinces such as Guangdong and Gansu appeared in the fourth quadrant (HL), suggesting that some economically developed regions have taken the lead in reducing distortions through market-oriented reforms.

4.4. Dynamic Evolution

The above analysis demonstrates that China's energy-price distortions exhibit significant spatial heterogeneity. To explore their temporal trajectories and variation characteristics, this study applies a Gaussian kernel estimation to uncover dynamic trends in regional distortion, providing a basis for future forecasting. In light of China's socioeconomic development and regional characteristics, the 30 provincial-level units are categorized into Eastern, Central, and Western regions to better reflect differences in economic structure, resource endowment, policy environment, and energy-consumption habits. This classification facilitates the identification of interregional heterogeneity and clustering in distortion and enables targeted analysis of each region's dynamic evolution. We select cross-sectional data for 2000, 2008, 2016, and 2022, and use their kernel density curves to examine how distortion evolves in each region and its influence on the overall pattern.

Figure 4a–d presents three-dimensional kernel density plots of energy-price distortion for China as a whole and its three major regions.

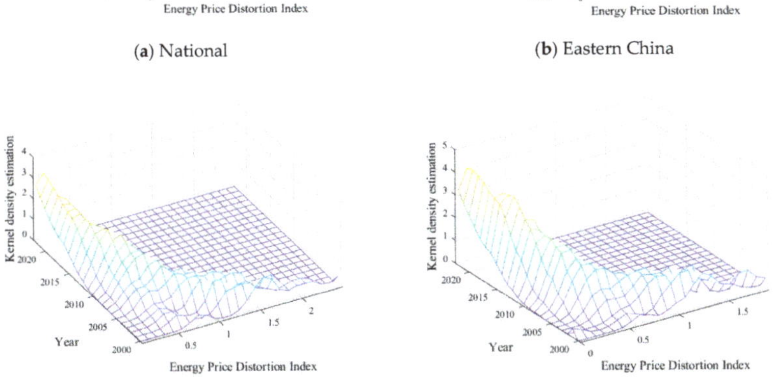

(a) National

(b) Eastern China

(c) Central China

(d) Western China

Figure 4. Evolutionary Trends of Regional Energy-Price Distortion in China. Different colors are used solely for visualization purposes and do not indicate any specific grouping.

(1) Location Shift: Over time, the distribution curves for both the national aggregate and each region consistently shift leftward, indicating a steady decline in energy-price distortion. This trend suggests that distortions are easing and gradually approaching zero, reflecting the improved functioning of energy markets.

(2) Dispersion Pattern: The absolute dispersion of distortion levels varies across regions and over time. Nationally, the peak of the density curve rises while its width narrows, indicating a convergence of distortion levels across provinces. In the Western region, the peak height increases with some fluctuations, suggesting that although distortion is declining, regional coordination is strengthening and distortion levels are becoming more concentrated. The Eastern region exhibits the widest distribution, reflecting substantial interprovincial variation and greater volatility.

(3) Tail Behavior: The distribution curves exhibit right-skewed tails at both national and regional levels, indicating that certain provinces continue to experience relatively high levels of distortion. This long-tail phenomenon is most evident in the Eastern region, where pockets of high distortion remain prominent.

(4) Peak Structure: The Central region transitions from a bimodal to a unimodal distribution, signaling that the initial polarization in distortion levels has eased and values are becoming more balanced. In the Eastern region, a multimodal distribution—indicative of distinct tiers of distortion—gradually evolves into a single peak, suggesting increasing concentration and stability in the region's distortion profile.

The underlying drivers of the "east-high, west-low" gradient in energy price distortion can be summarized in three main dimensions. First, eastern provinces initially attracted energy-intensive industries by offering tax incentives and electricity subsidies, which led to a persistent deviation of energy prices from marginal costs. In contrast, resource-rich western provinces such as Shaanxi and Xinjiang, which rely heavily on fiscal revenues from resource-based taxes, tend to maintain lower energy prices to sustain local industrial development. Second, although the eastern region is more economically advanced, its energy sector is often characterized by stronger administrative monopolies. In comparison, western provinces have progressively promoted market-based pricing through mechanisms such as the "West-to-East Electricity Transmission" program, thereby enhancing interregional energy integration and pricing flexibility. Third, the eastern region is dominated by export-oriented manufacturing, where energy demand is relatively inelastic and highly sensitive to price changes, making it more susceptible to policy interventions. Conversely, the West enjoys an abundance of coal, wind, and solar resources, resulting in lower energy supply costs and fewer institutional barriers to implementing market-oriented reforms.

To further investigate the transition dynamics across different levels of energy-price distortion, this study employs Markov-based transition probabilities, detailed in Table 5. All three distortion categories demonstrate strong persistence: self-transition probabilities are 99.44% for Type I (low distortion), 85.92% for Type II (moderate distortion), and 88.77% for Type III (high distortion). Notably, downward transitions are more likely than upward ones, suggesting a general trend toward stabilization and improvement over time.

A more detailed analysis reveals that the distortion level of surrounding areas significantly influences local transitions: Low-distortion neighborhoods (I): Local low-distortion regions exhibit exceptional stability, with self-transition probabilities above 99.44%. Moderate-distortion areas in these neighborhoods have a 32.43% chance of improving to low distortion—more than double the baseline probability (14.08%)—indicating that favorable surroundings promote improvement. Moderate-distortion neighborhoods (II): Low-distortion areas remain stable, while 28.89% of high-distortion areas shift downward to moderate distortion, compared to a baseline of 10.87%. This suggests that a moderate environment can effectively suppress more severe distortion. High-distortion neighborhoods

(III): A striking 33.33% of low-distortion areas transition upward to high distortion, while high-distortion areas show a self-transition probability of 92.21%, indicating increased persistence and a strong assimilation effect. These findings underscore the critical role of spatial context: low-distortion areas remain resilient in favorable environments, while high-distortion neighborhoods exert a contagious "escalation" effect on their surroundings, reinforcing localized patterns of inefficiency.

Table 5. Traditional and Spatial Markov Transition Probability Matrices.

	Neighborhood Context	t/(t + 1)	I	II	III	Observations
Traditional		I	0.9944	0.0000	0.0056	178
		II	0.1408	0.8592	0.0000	206
		III	0.0036	0.1087	0.8877	276
Spatial	I	I	1.0000	0.0000	0.0000	126
		II	0.3243	0.6757	0.0000	37
		III	0.0000	0.0000	0.0000	0
	II	I	1.0000	0.0000	0.0000	49
		II	0.1181	0.8819	0.0000	127
		III	0.0000	0.2889	0.7111	45
	III	I	0.6667	0.0000	0.3333	3
		II	0.0476	0.9524	0.0000	42
		III	0.0043	0.0736	0.9221	231

Table 5 analyzes the transition probability matrix and reveals three key patterns in the evolution of energy price distortions. First, the self-transition probability for the high-distortion state reaches 88.77%, which is higher than that for the low-distortion state, suggesting the existence of a "high-distortion trap," though weaker than expected. Second, provinces neighboring high-distortion regions face a 33.33% probability of worsening, whereas only 14.08% of high-distortion provinces improve when surrounded by low-distortion neighbors. This indicates that negative spillover effects outweigh positive diffusion, as reflected in the differences in coefficients. Third, provinces in a medium-distortion state have a 32.43% probability of improving when located near low-distortion areas, significantly higher than in neutral environments (14.08%). This highlights the potential of regional coordinated reform as a breakthrough strategy and underscores the importance of implementing preventive measures in areas where energy prices have not yet experienced serious distortions. Meanwhile, provinces already in a low-distortion state should maintain their existing policy frameworks while exploring further opportunities to optimize energy pricing mechanisms. For provinces still in high or medium-distortion states, policymakers should carefully consider the long-term impacts and sustainability of interventions, avoiding short-term actions or excessive administrative interference.

4.5. Robustness Checks

Club convergence refers to the phenomenon where economic variables (such as income or productivity) exhibit convergence trends within specific subgroups or "clubs." However, this convergence is limited to these subgroups and does not extend across the entire sample. In other words, certain regions with similar economic or social backgrounds tend to follow comparable growth patterns. To verify the robustness of our findings, we adopt the grouped α-convergence test method proposed by Barro and Sala-i-Martin [41] and Quah [42,43]. This method groups the sample based on distortion values, revealing whether convergence exists within each subgroup and analyzing the dynamic convergence patterns between groups. In our analysis, we use the same classification approach as in the Markov chain grouping and conduct convergence tests within each subgroup. Unlike the preliminary analysis based on the Markov chain transition matrix, the grouped α-convergence test directly quantifies the strength and robustness of convergence and statistically validates the significance of convergence trends within each group. When combined with Markov chain analysis, the grouped α-convergence test provides a more systematic and robust analytical framework.

Table 6 reports the empirical results, showing that the high-, medium-, and low-distortion groups all exhibit significant convergence. Specifically, the convergence coefficients are −0.033 for the high-distortion group, −0.017 for the medium-distortion group, and −0.015 for the low-distortion group, with p-values less than 0.01 for all groups. This confirms convergence within each group at the 1% significance level. These findings support the club convergence hypothesis, demonstrating that economic variables within each distortion group follow consistent convergence patterns. The grouped α-convergence test further confirms the presence of club convergence within each subgroup, enhancing the robustness of our conclusions and providing strong empirical support for related policy formulation.

Table 6. Convergence Trends in Different Distortion Groups.

Group	Coefficient	t-Value	p-Value	Convergence Pattern
High Distortion	−0.033 ***	−7.79	0.000	Convergence
Medium Distortion	−0.017 ***	−10.16	0.000	Convergence
Low Distortion	−0.015 ***	−10.90	0.000	Convergence

Notes: *** indicate statistical significance at the 1% level.

5. Conclusions and Recommendations

Unlike previous studies that mainly focus on the determinants of energy price distortions, this study investigates their spatiotemporal dynamics, providing new evidence for understanding the marketization process of China's energy sector. It reveals a "gradient lock-in" effect, where the negative spillovers from high-distortion provinces exert significantly stronger influences on surrounding regions than the positive diffusion effects from low-distortion areas. This supports the existence of a "policy-driven convergence" pattern, which is distinctive to transitional economies and differs from the natural convergence paths observed in developed markets. Additionally, by utilizing the decomposability of the Theil index, this study quantifies the dominant role of interregional disparities in contributing to overall distortion, offering a more refined structural explanation than traditional descriptive analysis. Based on estimates of energy-price distortion across 30 Chinese provinces from 2000 to 2022, this study employs the Theil index and spatial econometric methods to analyze regional disparities and spatial effects. In addition, kernel density estimation and a Markov chain model are used to assess the dynamic evolution of energy-price distortion in China. The main conclusions are as follows:

Energy price distortion in China has declined substantially, exhibiting a distinct "high in the east, low in the west" spatial pattern. From 2000 to 2022, the national distortion index decreased by 94.33%, indicating a significant retreat in price distortions and a steady progression toward more rational, market-based pricing mechanisms. Although the east-west divergence has narrowed over time, regional heterogeneity remains pronounced. The East and North Coastal regions continue to exhibit the highest distortion levels, largely due to policy biases and market concentration in economically developed areas. In contrast, regions such as the Middle Yellow River and the Greater Northwest report lower distortion levels, reflecting differences in economic development and the efficiency of energy resource allocation.

The evolution of regional disparities follows a "narrowing-widening-stabilizing" trajectory. Specifically, the regional gap narrowed between 2000 and 2008, widened rapidly from 2008 to 2016, and stabilized thereafter. The contribution of interregional disparities to overall distortion consistently exceeds that of intraregional differences. However, volatility within regions remains substantial, reflecting uneven progress in market reforms and divergent institutional conditions. In some cases, internal disparities in distortion have even expanded.

Energy price distortion displays significant positive spatial autocorrelation and clustering. Although the strength of these spatial effects has declined over time, indicating a gradual movement toward spatial equilibrium, a moderate degree of spatial dependence still persists. Local Moran's I statistics further indicate the presence of stable high-high and low-low clusters in resource-dependent and less-developed regions, while economically advanced areas exhibit more heterogeneous and dynamic spatial patterns.

Energy price distortion has continued to decline across the country and in all major regions, accompanied by a gradual narrowing of regional disparities. In the Central region, the previous pattern of polarization has diminished, resulting in a more balanced, unimodal distribution. In the Western region, distortion has become more concentrated, whereas the Eastern region exhibits the highest volatility, along with a pronounced right-skewed tail, indicating that severe distortion persists in some provinces. Markov analysis confirms that distortion states are highly stable, with downward transitions occurring more frequently than upward shifts, suggesting a steady improvement under stable conditions. Moreover, spatial neighborhood effects remain significant, exhibiting a clear pattern of club convergence.

Based on the findings above, several targeted policy recommendations are proposed:

Further advance market-oriented energy pricing reforms, with a focus on mitigating distortions in the Eastern Coastal region. Reform efforts should prioritize accelerating the transition toward market-based pricing by phasing out policy subsidies and administrative controls, strengthening market competition, and preventing large enterprises from leveraging monopolistic advantages to obtain artificially low energy prices. Prices should more accurately reflect supply-demand dynamics and the scarcity of energy resources.

Implement regionally differentiated reform strategies to promote balanced intra-regional development. In regions with pronounced internal disparities—such as the Middle Yellow River, Greater Northwest, and South Coast—reform measures should be tailored to local resource endowments, industrial structures, and policy implementation capacities. Differentiated pricing strategies can enhance allocative efficiency and reduce internal distortion gaps. For the eastern coastal regions, it is recommended to gradually phase out electricity subsidies for energy-intensive industries, implement market-based pricing mechanisms, and strengthen antitrust regulation. For the central and western regions, it is advisable to establish regional energy trading centers, improve grid interconnectivity, and provide financial support to facilitate the energy transition.

Enhance spatially coordinated governance to optimize regional clustering effects. Leveraging insights from spatial economic analysis, interregional cooperation, and policy alignment should be strengthened to contain distortion spillovers caused by spatial clustering. Cross-regional regulatory mechanisms and information-sharing platforms can support greater policy consistency and promote the integration of national energy markets. To address potential resistance from local protectionism, it is recommended that energy marketization be incorporated into the performance evaluation systems of local governments.

Support structural transformation and energy diversification in resource-based and less-developed regions. These areas should receive targeted support to broaden their energy supply mix, promote clean energy deployment, and foster innovation. Reducing dependence on single energy sources will improve market flexibility and support the rationalization of pricing mechanisms.

Establish a dynamic monitoring and evaluation framework. By integrating spatiotemporal data, kernel density estimation, and spatial diagnostics, authorities can regularly track the evolution of energy-price distortion across regions, identify high-risk areas early, and adjust policy responses accordingly. This will ensure the sustained and adaptive advancement of energy pricing reform. These recommendations take into account regional disparities and the gradual nature of reform. For instance, electricity pricing reforms can be piloted in free trade zones before being gradually expanded nationwide.

Develop a tiered and progressive regional coordination mechanism to overcome administrative boundary constraints through a three-level framework of "intra-provincial reform, regional cooperation, and cross-regional coordination." The plan includes establishing a unified regional electricity market with a "contract + spot" trading model and reserving 15% of capacity for inter-provincial transactions. It also calls for an interconnected pipeline network based on the "beneficiary pays" principle to support cross-provincial infrastructure development. A federated learning–based data platform will enable real-time load sharing at 15-min intervals. Supporting measures include legal revisions, incentive mechanisms, and a dynamic evaluation system. The implementation follows a three-phase roadmap, aiming to establish nationally unified market rules by 2030 and to achieve policy synergy through institutional innovation rather than geographic restructuring.

Author Contributions: Conceptualization, Z.G. and Y.H.; Methodology, Z.G. and Y.H.; Software, Z.J.; Formal analysis, Z.J.; Resources, Z.G.; Writing—original draft, Z.G. and Z.J.; Writing—review & editing, Z.G. and Y.H.; Visualization, Z.J.; Supervision, Z.G. and Y.H.; Funding acquisition, Z.G. and Y.H. All authors have read and agreed to the published version of the manuscript.

Funding: This research was funded by [the National Social Science Fund Project] grant number [24FJYB049] and Beijing Institute of Petrochemical Technology 2025 Graduate Education Reform and Practice Project: Development of a Dual-Carbon-Focused Talent Cultivation System for Economics and Management Master's Programs.

Data Availability Statement: The data presented in this study are available on request from the corresponding author due to restrictions imposed by the data provider.

Conflicts of Interest: The authors declare no conflict of interest.

References

1. Ouyang, X.; Sun, C. Energy savings potential in China's industrial sector: From the perspectives of factor price distortion and allocative inefficiency. *Energy Econ.* **2015**, *48*, 117–126. [CrossRef]
2. Ju, K.; Su, B.; Zhou, D.; Wu, J. Does energy-price regulation benefit China's economy and environment? Evidence from energy-price distortions. *Energy Policy* **2017**, *105*, 108–119. [CrossRef]
3. Xu, J.; Wang, E. The path to energy savings and CO_2 emission reductions in China's industrial sector from the perspective of factor price distortions correction–Based on an extended capital vintage model. *Energy* **2024**, *313*, 134004. [CrossRef]
4. Wang, X.; Liao, B.; Cheng, L.; Li, J.; Yang, X.; Wang, X. Energy Price Distortions and Urban Carbon Emission Efficiency: Evidence from China's Energy-Intensive Sectors. *Sustainability* **2024**, *16*, 9351. [CrossRef]
5. Sun, Y.; Li, H.; Zhu, B. Factor market distortion, total factor energy efficiency and energy shadow price: A case of Chinese manufacturing industry. *Energy* **2024**, *307*, 132739. [CrossRef]
6. Wu, L.; Xu, C.; Zhu, Q.; Zhou, D. Multiple energy price distortions and improvement of potential energy consumption structure in the energy transition. *Appl. Energy* **2024**, *362*, 122992. [CrossRef]
7. Ju, K.; Wang, Q.; Liu, L.; Zhou, D. Measurement of the price distortion degree for exhaustible energy resources in China: A discount rate perspective. *Emerg. Mark. Financ. Trade* **2019**, *55*, 2718–2737. [CrossRef]

Sha, R.; Li, J.; Ge, T. How do price distortions of fossil energy sources affect China's green economic efficiency? *Energy* **2021**, *232*, 121017. [CrossRef]

Qiao, S.; Zhao, D.H.; Guo, Z.X.; Tao, Z. Factor price distortions, environmental regulation and innovation efficiency: An empirical study on China's power enterprises. *Energy Policy* **2022**, *164*, 112887. [CrossRef]

Sun, X.; Loh, L.; Chen, Z.; Zhou, X. Factor price distortion and ecological efficiency: The role of institutional quality. *Environ. Sci. Pollut. Res.* **2020**, *27*, 5293–5304. [CrossRef]

Zhang, S.; Chen, C.; Huang, D.; Hu, L. Measurement of factor price distortion: A new production function method with time-varying elasticity. *Technol. Forecast. Soc.* **2022**, *175*, 121363. [CrossRef]

Peng, C.; Zhang, J.; Xu, Z. Does Price Distortion Affect Energy Efficiency? Evidence from Dynamic Spatial Analytics of China. *Energies* **2022**, *15*, 9576. [CrossRef]

Wang, X.; Wang, H.; Liang, S.; Xu, S. The Influence of Energy Price Distortion on Region Energy Efficiency in China's Energy-Intensive Industries from the Perspectives of Urban Heterogeneity. *Sustainability* **2021**, *14*, 88. [CrossRef]

Xu, M.; Tan, R. Removing energy allocation distortion to increase economic output and energy efficiency in China. *Energy Policy* **2021**, *150*, 112110. [CrossRef]

Zamani, R.; Tash, M.N.S.; Norouzi, A.; Hosseinzadeh, R. Impact of energy price reform on energy intensity in Iranian manufacturing industries. *Energy Effic.* **2024**, *17*, 71. [CrossRef]

Kirikkaleli, D.; Darbaz, I. The causal linkage between energy price and food price. *Energies* **2021**, *14*, 4182. [CrossRef]

Punzi, M.T. The impact of energy price uncertainty on macroeconomic variables. *Energy Policy* **2019**, *129*, 1306–1319. [CrossRef]

Croonenbroeck, C.; Hüttel, S. Quantifying the economic efficiency impact of inaccurate renewable energy price forecasts. *Energy* **2017**, *134*, 767–774. [CrossRef]

Li, K.; Xu, C.; Tang, L. The important of eliminating energy market distortions: The perspective of industrial green productivity. *Energy Econ.* **2024**, *132*, 107472. [CrossRef]

Gao, K.; Yuan, Y. Does market-oriented reform make the industrial sector "Greener" in China? Fresh evidence from the perspective of capital-labor-energy market distortions. *Energy* **2022**, *254*, 124183. [CrossRef]

Gao, B.; Huang, Z.; Zhang, T.; Sun, X.; Song, M. Exploring the impact of industrial land price distortion on carbon emission intensity: Evidence from China. *Land* **2022**, *12*, 92. [CrossRef]

Wang, X.; Bai, M.; Xie, C. Investigating CO_2 mitigation potentials and the impact of oil price distortion in China's transport sector. *Energy Policy* **2019**, *130*, 320–327. [CrossRef]

Anselin, L. A local indicator of multivariate spatial association: Extending Geary's C. *Geogr. Anal.* **2019**, *51*, 133–150. [CrossRef]

Chen, Y. Spatial autocorrelation equation based on Moran's index. *Sci. Rep.* **2023**, *13*, 19296. [CrossRef]

Paelinck, J.H.; Klaassen, L.H.; Ancot, J.; Verster, A.; Wagenaar, S. *Spatial Econometrics*; Saxon House: Wendouree, Australia, 1979.

Baltagi, B.H.; Pace, R.K.; LeSage, J.P.; Liesenfeld, R.; Richard, J.F.; Vogler, J.; Craioveanu, M.; Terrell, D.; Egger, P.H.; Kesina, M.; et al. *Spatial Econometrics: Qualitative and Limited Dependent Variables*; Emerald Group Publishing: Bingley, UK, 2016.

Elhorst, J.P. *Spatial Econometrics: From Cross-Sectional Data to Spatial Panels*; Springer: Heidelberg, Germany, 2014; Volume 479.

Wu, J.; Abban, O.J.; Boadi, A.D.; Charles, O. The effects of energy price, spatial spillover of CO_2 emissions, and economic freedom on CO_2 emissions in Europe: A spatial econometrics approach. *Environ. Sci. Pollut. Res.* **2022**, *29*, 63782–63798. [CrossRef]

Burnett, J.W.; Bergstrom, J.C.; Dorfman, J.H. A spatial panel data approach to estimating US state-level energy emissions. *Energy Econ.* **2013**, *40*, 396–404. [CrossRef]

LeSage, J.; Pace, R.K. *Introduction to Spatial Econometrics*; Chapman and Hall/CRC: New York, NY, USA, 2009.

Kelejian, H.H.; Prucha, I.R. Specification and estimation of spatial autoregressive models with autoregressive and heteroskedastic disturbances. *J. Econom.* **2010**, *157*, 53–67. [CrossRef]

Hua, Q.; Liu, G.; Sun, D.; Zhu, L. Spatial effects of technology market development on energy efficiency: Heterogeneity analysis based on the characteristics of technology market. *Technol. Forecast Soc.* **2022**, *185*, 122008. [CrossRef]

Zhang, Z.; Wang, Z.; Ji, Y.; Liang, S. Dynamic evolution of spatial distribution of energy factor allocation efficiency: Industrial sector in China. *Environ. Dev. Sustain.* **2024**, *27*, 13883–13901. [CrossRef]

Li, W.; Sun, W.; Li, G.; Jin, B.; Wu, W.; Cui, P.; Zhao, G. Transmission mechanism between energy prices and carbon emissions using geographically weighted regression. *Energy Policy* **2018**, *115*, 434–442. [CrossRef]

Wang, J.; Enilov, M.; Kizys, R. Does M&A activity spin the cycle of energy prices? *Energy Econ.* **2024**, *137*, 107781.

Sha, R.; Qian, J.; Li, C.; Ge, T. A win-win opportunity for economic growth and carbon emissions reduction in China: The perspective from correcting energy price distortions. *Energy Strateg. Rev.* **2024**, *53*, 101406. [CrossRef]

Xu, S.; Song, Y.; Yang, Y. Embedded carbon emissions in the inter-industrial sectors of China: A new analysis of total factor productivity and capacity utilization. *J. Environ. Manag.* **2025**, *388*, 125970. [CrossRef]

Li, C. Climate change impacts on rice production in Japan: A Cobb-Douglas and panel data analysis. *Ecol. Indic.* **2023**, *147*, 110008. [CrossRef]

Theil, H. *Economics and Information Theory*; North-Holland Publishing Company: Amsterdam, The Netherlands, 1967.

40. Shorrocks, A.F. The class of additively decomposable inequality measures. *Econom. J. Econom. Soc.* **1980**, *48*, 613–625.
41. Barro, R.J.; Sala-i-Martin, X. Convergence. *J. Political Econ.* **1992**, *100*, 223–251. [CrossRef]
42. Quah, D.T. Convergence empirics across economies with (some) capital mobility. *J. Econ. Growth* **1996**, *1*, 95–124. [CrossRef]
43. Quah, D.T. Twin peaks: Growth and convergence in models of distribution dynamics. *Econ. J.* **1996**, *106*, 1045–1055. [CrossRef]

energies

MDPI

rticle

conomic Growth and Energy Consumption in Thailand: vidence from the Energy Kuznets Curve Using rovincial-Level Data

hanakhom Srisaringkarn and Kentaka Aruga *

Graduate School of Humanities and Social Sciences, Saitama University, 255 Shimo-Okubo, Sakura-ku, Saitama 338-8570, Japan; srisaringkarn.t.447@ms.saitama-u.ac.jp
* Correspondence: aruga@mail.saitama-u.ac.jp

Abstract

This study investigates the relationship between economic growth and energy consumption using the Energy Kuznets Curve (EKC) framework. Spatial econometric models, including the Spatial Panel Lag Model and the Spatial Dynamic Panel Lag IV Model, are employed to capture both spatial and dynamic effects. The results indicate that energy consumption in Thailand is spatially clustered, with energy use tending to spill over into neighboring provinces and concentrating in specific regions. Key factors that positively influence energy consumption include gross provincial product (GPP) per capita, population density, and road density. Regions characterized by favorable climates, sufficient infrastructure, and high levels of economic activity exhibit higher per capita energy consumption. The EKC analysis reveals a U-shape relationship between GPP per capita and energy consumption in the BKK&VIC, CE, EA, WE, and NE regions. As many regions continue to experience rising energy consumption, the findings underscore the importance of Thailand adopting more efficient energy usage strategies in tandem with its economic development.

Keywords: energy consumption; Energy Kuznets Curve; spatial dynamic model; remote sensing and GIS; provincial-level data; Thailand

1. Introduction

Thailand is a developing country whose manufacturing sector relies heavily on energy. From Figures 1–3 it can be observed that within the Association of Southeast Asian Nations (ASEAN), Thailand has the highest electricity consumption per capita and the highest electricity production from oil sources among all ASEAN countries, excluding Malaysia, Brunei, and Singapore, which have larger economies than Thailand. In contrast, Thailand has the lowest share of combustible renewables and waste compared to its neighboring developing countries with smaller economies, such as Myanmar, Cambodia, and Lao PDR. Thailand's economic expansion has been driven by industrialization, informatization, urbanization, and agricultural modernization, all of which rely heavily on energy consumption. According to the Office of the National Economic and Social Development Board, Thailand's electricity consumption grew at an average annual rate of 3.5% between 1990 and 2019, while the economy expanded at an average annual rate of 3.7% over the same period. This suggests a potential correlation between electricity consumption and economic development in Thailand.

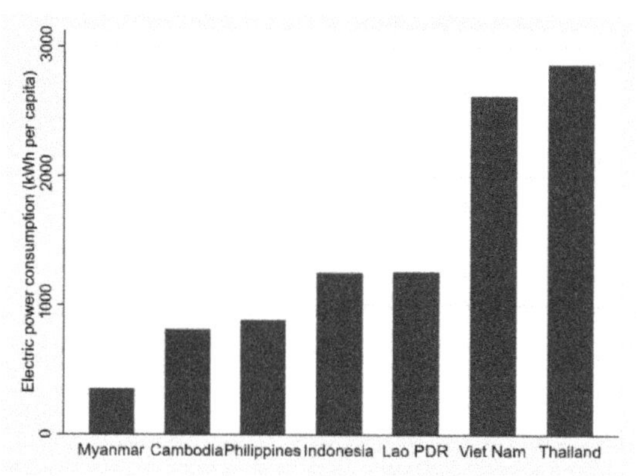

Figure 1. Electric power consumption (kWh per capita) in 2022. Source: World Bank [1].

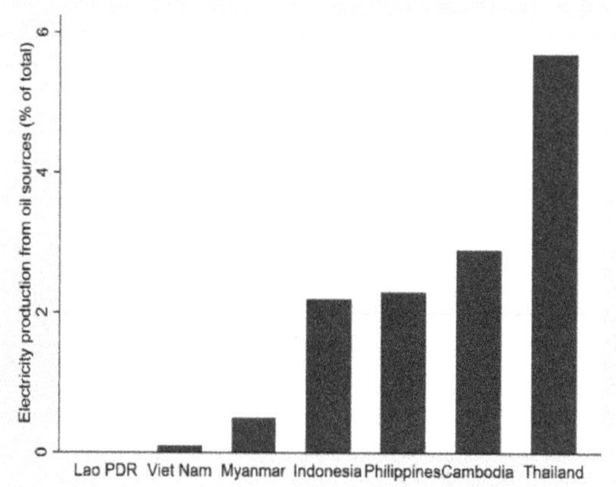

Figure 2. Electricity production from oil sources (% of total) in 2022. Source: World Bank [2].

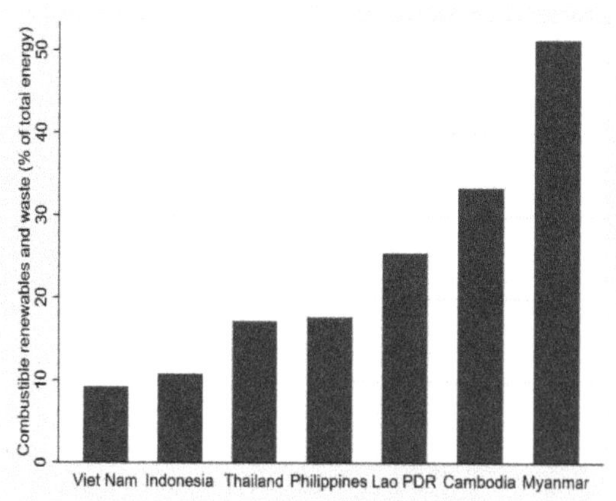

Figure 3. Combustible renewables and waste (% of total energy) in 2022. Source: World Bank [3].

Domestic electricity demand in Thailand, which fluctuates in line with economic activity (Figure 4), exhibits distinct patterns across various economic sectors. Over the past

10 years (average from 2013 to 2023), electricity consumption was distributed as follows: the industrial sector accounted for 46.7%, the household sector for 25.1%, the business sector for 24.0%, and other sectors for 4.2% of the country's total electricity consumption. Within the industrial sector, the highest electricity-consuming industries include food processing, steel and basic metals, rubber and rubber products, electronics, plastics, and textiles. In the business sector, the largest electricity users are primarily tourism-related services, including hotels, restaurants, nightclubs, apartments, guesthouses, department stores, and retail and wholesale establishments [4].

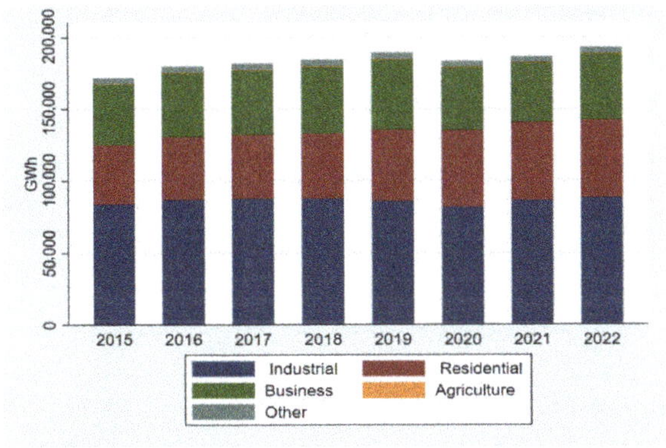

Figure 4. Electricity consumption by sector for 2015–2022. Source: Energy Policy and Planning Office (EPPO [5]).

As illustrated in Figure 5, the trends in energy consumption and economic growth from 2015 to 2022 were closely correlated. Electricity can be generated from both renewable and non-renewable energy sources (Figure 6a,b). When electricity is generated from renewable sources, it can significantly reduce greenhouse gas emissions. Conversely, electricity generation based on non-renewable sources has been found to have a negative impact on environmental quality, particularly when consumption increases [6]. Figure 6a,b show that more than 85% of electricity generation in 2015 and 71% in 2022 came from non-renewable sources. This indicates that electricity generation in Thailand still largely relies on traditional energy sources, which are associated with high emissions during the production process and are gradually being depleted. This highlights the unsustainability of Thailand's current electricity consumption patterns.

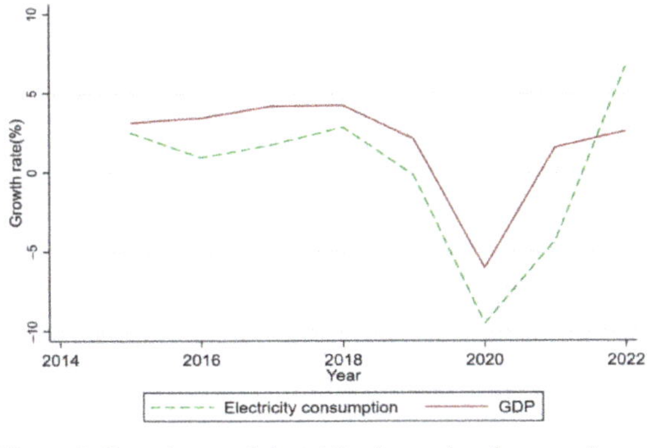

Figure 5. Growth rate of electricity demand and economic growth for 2015–2022. Source: Energy Policy and Planning Office (EPPO [7]).

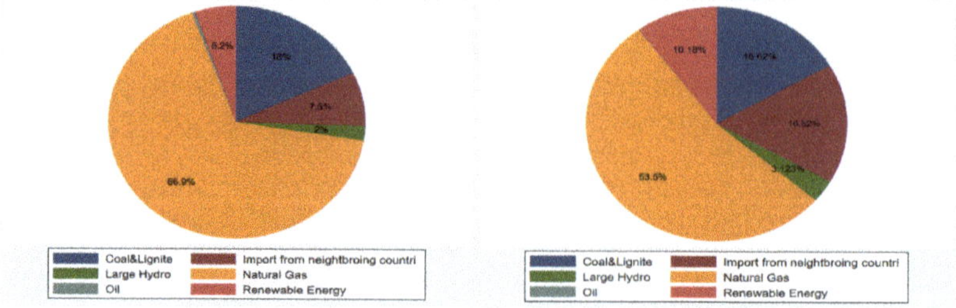

(**a**) (**b**)

Figure 6. (**a**) Power generation (GWh) by fuel in 2015; (**b**) power generation (GWh) by fuel in 2022. Source: Energy Policy and Planning Office (EPPO [8]).

Thailand consumes a high amount of energy, with relatively limited domestic energy resources compared to its neighboring countries. Thailand is the largest energy consumer in ASEAN and imports significant amounts of natural gas for electricity generation, ranking second only to Singapore, which has a larger economy. As a result, Thailand is the most reliant on imported natural gas among the ASEAN countries [9]. The Electricity Generating Authority of Thailand (EGAT), the state-owned electricity producer, accounts for 40% of the country's total electricity generation capacity. To reduce dependency and spread the risk associated with relying heavily on natural gas, particularly from neighboring countries, the Thai government has developed a plan to diversify the energy mix. This includes reducing the current 70% reliance on natural gas by increasing the share of coal in electricity production while incorporating clean coal technologies to mitigate environmental impacts [10]. It can be seen that if energy demand continues to increase in the future, Thailand may face challenges related to energy security and become increasingly dependent on energy imports from neighboring countries. Moreover, the current electricity generation technologies used in Thailand significantly contribute to environmental pollution and are unsustainable in the long term [10]. Therefore, this study applies the Energy Kuznets Curve (EKC) theory to examine the relationship between economic growth and energy consumption, to understand how energy consumption in Thailand may evolve as the economy grows. The goal is to support effective energy planning that aligns with the country's energy consumption trends. In addition, the study aims to provide policy recommendations that align with the Thai government's target of increasing the share of renewable energy, such as solar, wind, and biomass, to 15–20% by 2036 [10].

Gross regional product (GRP), GRP per capita, energy consumption, and energy consumption per capita by region in Thailand are presented in Table 1. Figure 7 shows the boundaries of Thailand's 77 provinces. The region with the highest GRP and average energy consumption per capita is Bangkok and vicinity (BKK&VIC), followed by the northeastern region (NE) and the southern region (SO). The remaining regions, such as the northern region (NO), exhibit the lowest levels of economic activity [11]. However, Bangkok and vicinity (BKK&VIC), the eastern region (EA), the central region (CE), and the western region (WE) are combined in this study, as they are classified as high-economic-growth pole regions in central Thailand. In this study, we use provincial-level data from all 77 provinces of Thailand as representative data points and categorize them into the four areas mentioned above to analyze the impact of economic growth on energy consumption based on the EKC theory. This regional approach aims to provide more comprehensive and tailored policy recommendations compared to a whole-region analysis alone, taking into account variations in economic structure, geography, and environmental conditions across different regions.

Table 1. Gross regional product (GRP), GRP per capita, energy consumption, and energy consumption per capita by region (USD).

Regions	2022 GRP	2022 GRP per Capita	2022 Average Energy Consumption (GWh)	2022 Average Energy Consumption (per Capita) (kWh/person)
Northeastern (NE)	USD 498 billion	USD 2724	1,162,050,000	1101
Northern (NO)	USD 373 billion	USD 3329	990,185,072	1436
Southern (SO)	USD 394 billion	USD 4053	1,207,050,000	1688
Bangkok and Vicinity (BKK&VIC)	USD 2.2 trillion	USD 13,210	7,554,832,704	3861
Integrated Regions				
BKK&VIC, CE, EA, WE	USD 3.6 trillion	USD 40,857	5,212,023,913	3970

Source: Office of the National Economic and Social Development Council (NESDC [12]), Energy Policy and Planning Office (EPPO [13]).

Figure 7. Thailand map. Source: Authors.

Several studies have investigated the impact of economic growth on energy consumption and air pollution using various methodologies, such as Autoregressive Distributed Lag (ARDL) cointegration analysis, Fully Modified Ordinary Least Squares (FMOLS), and Dynamic Ordinary Least Squares (DOLS), predominantly relying on provincial-level, ground-based data in Thailand [6,14–18]. Additionally, Maneejuk et al. [19] proposed the Simultaneous Smooth Transition Kink Equations (SKE) model, in which each regression parameter can exist in two different regimes, lower and upper, to examine the presence of the Kuznets Curve in Thailand's economy over the past two decades. A research gap is identified through the review of previous studies, presenting an opportunity to extend the existing literature. This study aims to fill that gap by utilizing provincial-level data in Thailand, combined with environmental data from satellite observations, to offer a more localized and specific analysis of the impacts occurring within the country. Moreover, this study employs the Spatial Dynamic Panel IV Model (SDPD IV), which captures dynamic effects over time and accounts for spatial spillover effects across regions in Thailand.

Therefore, this study examines how energy consumption and economic growth in Thailand and its regions are likely to evolve in the short and long term under continued economic growth. It also seeks to assess whether energy consumption will become more sustainable. Additionally, the study investigates the characteristics of energy consumption concentration in Thailand and analyzes the relationship patterns between energy consumption and its influencing factors. The ultimate goal is to enhance the understanding of energy consumption patterns associated with economic growth in Thailand and to support the formulation of appropriate, region-specific energy policies in the future.

The following section presents the theoretical framework of the study. The second section outlines the analytical methods used, while the third section presents the empirical results. The fourth section offers a discussion of the findings. Finally, the last section provides the conclusions and policy implications.

1.1. Energy Kuznets Curve Theory

The Kuznets Curve, introduced by Kuznets [20], was developed to explore the relationship between economic growth and income inequality. Over the past five decades, Grossman and Krueger [21] extended this concept to develop the Environmental Kuznets Curve. This theory hypothesizes a relationship between various indicators of environmental degradation and per capita income. In the early stages of economic growth, pollution emissions increase, and environmental quality deteriorates. However, beyond a certain level of per capita income (which varies depending on the indicator), this trend reverses, and further economic growth leads to environmental improvement. This implies that environmental impacts or emissions per capita follow an inverted U-shape relationship with per capita income [21–24]. This study applies the Energy Kuznets Curve (EKC) theory, which is conceptually similar to the Environmental Kuznets Curve. However, the EKC focuses specifically on the relationship between economic growth and energy consumption, as discussed by Suri and Chapman [25]. Understanding the relationship between economic growth and energy consumption, as outlined in the EKC theory, will help to identify the direction of energy consumption as the economy continues to expand. This insight can guide the planning and management of sustainable and efficient energy use in the future. Although various empirical studies have confirmed the EKC hypothesis in some instances, it is not universally applicable across all types of environmental degradation, income groups, or national contexts [26]. Throughout this paper, the EKC theory generally refers to the Energy Kuznets Curve.

Table 2 summarizes findings from various studies examining the impact of economic growth on energy consumption and air pollution. Some studies have identified different EKC and Environmental Kuznets Curve patterns, including monotonic increases, inverted U-shape, U-shape, and N-shape. This study incorporates variables based on the research findings summarized in Table 2 to analyze the relationship between economic growth and energy consumption in each region of Thailand. The aim is to identify EKC patterns that can inform the development of important policy recommendations for the country. This study adopts an alternative approach, utilizing provincial-level and satellite data to analyze the EKC pattern in each region, in contrast to previous studies that relied on country-level data and ground-based measurements. As a result, this study provides more localized insights into Thailand's energy consumption and offers more precise implications for national energy policy.

Table 2. A review of the relationship between economic growth and energy consumption.

EKC Patterns	Authors	Dependent Variables	Independent Variables
Monotonically rising curve	Dasgupta, et al. [27], Ang [28], Halicioglu [29], Chandran and Tang [30], Al-Mulali, et al. [31], Pablo-Romero and De Jesús [32]	Energy use, annual emissions of CO_2	Gross value added per capita (GVApc), share of agriculture employment, foreign direct investment (FDI), transport energy consumption, labor force, exports and imports

Table 2. *Cont.*

EKC Patterns	Authors	Dependent Variables	Independent Variables
Inverted U-shape	Suri and Chapman [25], Kurniawan and Managi [33], Nguyen and Kakinaka [34], Hien [35], Kibria, et al. [36], Maneejuk, et al. [17], Shahzad and Aruga [37]	Energy per capita, coal consumption, renewable energy consumption, non-renewable energy consumption, electricity consumption, CO_2 emission	Real GDP, manufacturing value added (MVA), industrial value added (IVA), import, export, urbanization, share of secondary industry value added, trade openness, real oil price in country, fossil fuel share, population density
U-shape	Ozcan [38], Chandran and Tang [30], Wang, et al. [39], Dyrstad, et al. [40], Deichmann, et al. [41], Borozan [42], Aruga [43], Srisaringkarn and Aruga [44]	Electricity production, energy intensity, energy consumption in households per capita, annual emissions of CO_2, SO_2, suspended particulate matter	GDP, non-fossil energy, fossil energy, percentage of total energy consumption used in industry, transport, residential, services, agriculture and nonenergy use, energy taxes, energy prices, tertiary education, risk of poverty, climate conditions, population density
N-shape	Grossman and Krueger [21], Aslanidis and Xepapadeas [45], Sinha Babu and Datta [46], Mahmood, et al. [47]	Primary energy, oil, natural gas, coal consumption, hydroelectricity consumption, air quality index	GRP per capita, GRP, GRP square, GRP cubic, temperature, import shares, share of the tertiary industry, environmental degradation index and population

1.2. Socioeconomic Factors Affecting Energy Consumption

In addition to the main variables commonly used to explore the EKC theory, several other studies have examined the relationship between energy consumption and environmental impact. For example, Zhang et al. [48] tested the EKC hypothesis for Pakistan by analyzing the significance of renewable and non-renewable energy consumption. Their results strongly support the EKC hypothesis in Pakistan, showing that renewable energy plays a dominant role in reducing carbon dioxide emissions. In contrast, non-renewable energy consumption is the primary contributor to increased emissions. Marques et al. [49] analyzed the relationship between economic growth and CO_2 emissions in Australia, using annual data from 1965 to 2016, with a focus on fossil fuel (oil and coal) and renewable energy consumption. Their results support the EKC hypothesis, suggesting that Australia is experiencing increased relative decoupling, which indicates that economic growth continues to drive CO_2 emissions and environmental degradation. Mahmood [50] examined the asymmetric environmental effects of different energy sources and tested the EKC hypothesis in the Gulf Cooperation Council (GCC) region from 1975 to 2019. The findings show that CO_2 emissions resulting from increased natural gas consumption are less severe than those from oil consumption. As a result, the study recommends that GCC countries shift from oil to natural gas to reduce overall CO_2 emissions. Pablo-Romero and De Jesús [32] investigated the relationship between economic growth and energy consumption using the hypothesis postulated for the EKC, which assumes an inverted-U shape relationship between income and energy consumption, for 22 Latin American and Caribbean countries for the period 1990–2011. The hypothesis postulated for the EKC is not supported for the region. Aruga [43] investigated the EKC hypothesis among the 19 countries in the Asia–Pacific region. The study also tests the EKC hypothesis for low-, middle-, and high-income groups within the region. The test results of both models suggest that the EKC hypothesis holds for the whole Asia–Pacific region. However, the test performed on the three different income groups revealed that the hypothesis only holds for the high-income group. The hypothesis was not apparent for the low- and middle-income groups.

Based on the literature review, this study utilizes variables such as energy consumption per capita, GPP per capita, and population density, which are commonly used to examine the relationship between economic growth and energy consumption. However, unlike previous studies, this research introduces additional control variables, such as road density, and instrumental variables, including nighttime land surface temperature, rainfall, and wind speed, all derived from satellite data at the provincial level in Thailand. Satellite data offer high spatial resolution and low-cost accessibility, making them a valuable alternative for developing countries that lack comprehensive environmental data collection across all provinces.

2. Materials and Methods

The overview of the research workflow for this study includes all key steps, beginning with the background and a literature review based on the Environmental Kuznets Curve theory, as discussed by Grossman and Krueger [21] and Suri and Chapman [25]. This theory serves as the main framework for this study. Next, the data and methodology are established by identifying stationary variables, followed by conducting spatial analysis tests to determine the most appropriate spatial econometric model.

In this context, we use spatial econometrics to address spatial interaction (spatial autocorrelation) and spatial structure (spatial heterogeneity) in regression models for panel data. This method incorporates both spatial (geographical) and time dimensions. The focus on location and spatial interaction has recently gained prominence not only in applied but also in theoretical econometrics, particularly in models such as the Spatial Lag Model (SLM), the Spatial Error Model (SEM), and the Spatial Dynamic Panel Model (SDPD) [51]. After identifying stationary variables, serial correlation, and multicollinearity problems, we conduct a Spatial Correlation Test (LM-test) to determine whether the SLM or SEM model is more appropriate. Additionally, the Hausman test is used to decide whether a fixed-effects or random-effects model should be employed. Once the appropriate model type (SLM or SEM) is identified, it is integrated into the Spatial Dynamic Panel Model (SDPD) framework for estimating the variables. This study adopts the Spatial Dynamic Panel Instrumental Variables Model (SDPD IV), which provides consistent and accurate estimates, particularly when the sample size is large. The selection of instrumental variables is validated using the two-stage least squares (2SLS) method. This approach ultimately aims to identify the most appropriate model for estimating the relationships among variables. Furthermore, a Generalized Additive Model (GAM) analysis is employed as a supplementary method to enhance policy-relevant insights, similar to the role of hot spot analysis. Finally, all analytical results are synthesized to provide policy recommendations. These steps are illustrated in Figure 8.

Based on previous studies [17,35–37,40,41,43], Equation (1) illustrates the functional relationship between energy consumption per capita (*Energycons_pc*) and the variables grounded in the EKC theory, as identified through the literature review summarized in Table 2. In addition, this study incorporates a road density variable derived from satellite-based data to generate meaningful insights.

2.1. Model

$$Energycons_{pc} = \left(GPPpc,\ GPPpc^2, Pop\_dens,\ Roaddens\right) \quad (1)$$

All the variables used in the analysis are presented in Table 3. In this study, energy consumption per capita (*Energycons_pc*) is the dependent variable. The independent variables include gross provincial product per capita (*GPPpc*), population density (*Pop_dens*), and road density (*Roaddens*). Land surface temperature at night (*Landtemp_night*), rainfall (*Rainfall*), and wind speed (*Windspeed*) are used as instrumental variables. Gross provincial product per capita squared ($GPPpc^2$) is included to capture the nonlinear relationship between $GPPpc^2$ and energy consumption per capita, which may take the form of a turning point, typically represented by an inverted U-shape or U-shape curve. Three underlying effects explain the inverted U-shape: the scale effect (as the economy grows, energy consumption increases, leading to more pollution), the composition effect (a structural shift from low-tech to high-tech industries as the country develops), and the technical effect (benefits from high-tech and cleaner technologies following industrial transformation, resulting in reduced pollution during later stages of economic growth). Economies that exhibit all three effects are expected to follow the EKC pattern, which is characterized by an inverted U-shape [26].

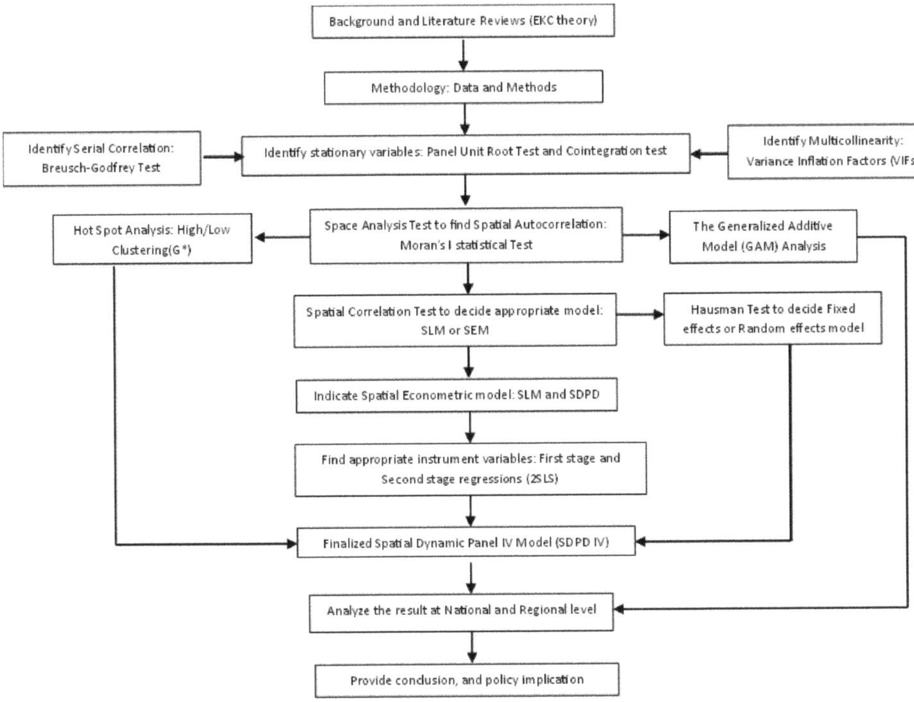

Figure 8. Overview of the steps of the methods. Source: Authors.

Table 3. Data description.

Types	Variables	Descriptions	Years	Sources	Expected Sign	References
Dependent Variable	*Energy consumption per capita (Energycons_pc)*	A variable that represents a province's economic activity using electricity consumption per capita as an indicator (kWh/person)	2015–2022	Metropolitan Electricity Authority (MEA), Provincial Electricity Authority (PEA), and Electricity Generating Authority of Thailand (EGAT)		

Table 3. *Cont.*

Types	Variables	Descriptions	Years	Sources	Expected Sign	References
Independent Variables	*Gross Provincial Product per capita (GPPpc)*	A variable that represents economic growth per capita in each province; reflects the level of economic output per person in that province (baht/person)	2015–2022	Office of the National Economic and Social Development Council (NESDC)	+/−	[30,38,43,44]
	Population density (Pop_ dens)	A variable that represents a province's population density; indicates how many people live per unit of area in that province (unit/km^2)	2015–2022	Office of the National Economic and Social Development Council (NESDC)	+	[33,36,41]
	Road density (Roaddens)	A variable that represents the road density of a province in a given year (length)	2015–2022	Geofabrik	+	[30,52,53]
Instrumental Variables	*Land surface temperature at Night (Landtemp_ night)*	A variable that represents a province's land surface temperature at night; indicates how high the surface temperature is during nighttime in that province (Celcius)	2015–2022	MOD11A1.061 Terra Land Surface Temperature and Emissivity Daily Global 1km; bands: LST_Night_1km		[54–64]
	Rainfall (Rainfall)	A variable that shows the average annual rainfall in a province (mm/y) (millimeters/year)	2015–2022	CHIRPS Daily: Climate Hazards Center InfraRed Precipitation With Station Data (Version 2.0 Final); bands: precipitation		[54–64]
	Windspeed (Windspeed)	A variable that shows the average annual wind speed and direction in a province (m/s) (meters per second)	2015–2022	GLDAS-2.1: Global Land Data Assimilation System; bands: Wind_f_inst		[54–56,58,60,63,64]

The results of the panel unit root tests used to determine whether the variables in the model are stationary are shown in Appendix A, Table A1. If the variables are not stationary, it can lead to spurious regression results. The Levin, Lin, and Chu (LLC) test and the Im, Pesaran, and Shin (IPS) test were applied. The results indicate that most variables are stationary at the level, i.e., I(0). However, some variables, such as *Pop_dens* of the northeastern (NE) region and *Landtemp_night* of the southern (SO) region, are stationary at the first difference, i.e., I(1), after differencing. Additionally, the Breusch–Godfrey test was conducted to identify the presence of serial correlation, and variance inflation factors (VIFs) were calculated to detect multicollinearity issues (Appendix B, Table A2).

2.2. *Moran's I Statistical Test*

Developed by Moran [65], Moran's I test is a statistical tool used to assess spatial autocorrelation in regression residuals by measuring their correlation with neighboring observations. It is used to confirm the presence of spatial correlation in a dataset. If the null hypothesis of no spatial autocorrelation is rejected, the OLS estimates are considered misspecified due to the omission of spatial relationships. Equation (2) presents the mathematical formulation of Moran's I test.

$$I = \frac{\left[\sum_i \sum_j w_{i,j}(x_i - \overline{x})(x_j - \overline{x})\right]}{\sum_i \sum_j w_{i,j}\left(x_i - \overline{x}\right)^2} \tag{2}$$

Here, Wi,j is the spatial weight matrix, and x_i and x_j are the values of the variable of interest in the respective areas. A statistically significant Moran's I result confirms the presence of spatial effects.

As shown in Table 4, the K-nearest neighbors (K-NN) method was used to construct the spatial weight matrix as K-NN can generate a weight matrix regardless of absolute distance. This approach is particularly suitable for provinces like Phuket, which, being an island, does not share borders with any other province in Thailand. The Moran's I values are close to 1 and statistically significant at the 1% level (p-value < 0.001), indicating strong spatial clustering of energy consumption. Given this spatial dependence, the study applies spatial econometric methods, specifically the Spatial Panel Lag Model (SLM) and the Spatial Dynamic Panel Lag Model (SDPD), to analyze the spatial diffusion and effects of energy consumption across regions.

Table 4. Moran's I statistical test for the year 2022.

Variable: Energy Consumption per Capita (Energycons_pc)	Whole Region	BKK&VIC, CE, EA, WE	NE	NO	SO
Moran's I	0.658	0.932	0.961	0.884	0.994
E(I)	−0.013	−0.013	−0.013	−0.013	−0.013
SE(I)	0.088	0.091	0.094	0.094	0.089
Z(I)	7.690	10.419	10.363	9.507	11.269
p-value	0.001	0.001	0.001	0.001	0.001

2.3. *High/Low Clustering (G*) (Hot Spot Analysis)*

High/low clustering (Getis-Ord General G or G* test), also known as "hot spot analysis", was developed by Getis and Ord [66]. This test is a tool for measuring the value of some variables that cluster in certain areas. The null hypothesis states that there is no spatial clustering of feature values. This method can identify spatial cluster patterns in each region, thereby improving the interpretability of the results and facilitating policy implications. Equation (3) represents the mathematical representation of high/low clustering:

$$G = \frac{\sum_{i=1}^{n} \sum_{j=1}^{n} w_{i,j} x_i x_j}{\sum_{i=1}^{n} \sum_{j=1}^{n} x_i x_j}, \ \forall j \neq i \tag{3}$$

where x_i and x_j are the attribute values of variables in the interest area i and j, respectively, and $w_{i,j}$ is the spatial weight matrix. When the test shows that the null hypothesis is rejected, it confirms the concentration of value and clustering.

2.4. *LM Test for Spatial Dependence (LM-Lag, LM-Error), LM Test for Random Effects, LM Test for Serial Correlation*

We conducted the LM test for spatial dependence to determine the most appropriate model for the dataset and to assess whether a lag or error model is more suitable. Based on the results, we can then select the specific model type to be used throughout the study.

Table 5 shows that Moran's I statistic is 2.554, with a corresponding p-value of 0.01, which is less than 0.05. This indicates that the sample data exhibits spatial correlation. Although, the LM-Error and LM-Lag tests are insignificant. However, the results of Moran's I statistical test remain statistically significant, indicating that it is appropriate to apply a spatial econometric model to capture the spillover effects of variables influencing neighboring provinces. The choice of a spatial econometric model depends on the form of spatial dependence between units and the type of spatial interaction effect that is present. Although taking the Dynamic Spatial Durbin model (DSDM) as a benchmark model is theoretically reasonable, further tests should be conducted to confirm this. We initially estimated several spatial econometric models to identify the most suitable one for this study, based on Equation (5), using the AIC (Akaike information criterion) and BIC (Bayesian information criterion) values as primary criteria for model comparison. The results show that the Spatial Lag Model (SLM) has the lowest AIC and BIC values (AIC: −1840.59 and BIC: −1796.36), while the Spatial Error Model (SEM) has an AIC of −1823.54 and BIC of −1779.31, and the Spatial Durbin Model (SDM) has an AIC of −1830.87 and BIC of −1751.26. Therefore, the SLM method is employed to estimate the results and address the research questions of this study. The Spatial Lag Model (SLM) captures substantive spatial dependence by directly incorporating spatial interaction into the model rather than merely correcting for the potentially biasing effects of spatial autocorrelation that may arise from the use of spatial data or nonlinear parametric restrictions. This makes the SLM approach more suitable than the SEM or other spatial econometric models (e.g., SDM and SDEM) [51,67]. Additionally, the Hausman test rejects the null hypothesis of the random-effects model, indicating that the fixed-effects model is more appropriate for these datasets.

Table 5. Spatial correlation test results.

Tests	Statistic	df	*p*-Value
Moran's I	2.554	1	0.011
Spatial Error			
Lagrange multiplier	1.406	1	0.236
Robust Lagrange multiplier	1.212	1	0.271
Spatial Lag			
Lagrange multiplier	2.154	1	0.142
Robust Lagrange multiplier	1.961	1	0.161

2.5. Spatial Panel Lag Model (SLM)

After conducting the LM test for spatial dependence, we apply the Spatial Panel Lag Model (SLM) to verify the presence of spatial dependence or the similarity of observations in neighboring areas. This model captures the spatial effects of surrounding regions by including a spatially lagged dependent variable. Moreover, the SLM serves as the baseline model for comparison with the Spatial Dynamic Panel Lag Model (SDPD IV) to determine which provides the most accurate estimation results. The conclusion will be drawn based on the model that yields the most reliable outcomes. The mathematical expression of the model is as follows:

$$y_i = \rho W_{i,j} y_j + x_i \beta + u_i \quad (4)$$

where $\rho W_{i,j} y_j$ is the vector of the dependent variable y_j multiplied by the spatial weight matrix $W_{i,j}$, which represents another spatially lagged dependent variable; x_i is the vector of explanatory variables; and β is a vector of parameters. Using this model our EKC model is analyzed using Equation (5):

$$LnEnergycons\_pc_{i,t} = a + \rho W LnEnergycons\_pc_{i,t} + \beta_1 LnGPPpc_{i,t} + \beta_2 LnGPPpc^2{}_{i,t} + \beta_3 LnPop\_dens_{i,t} + \beta_4 LnLandtemp\_night_{i,t} + \beta_5 LnRainfall_{i,t} + \beta_6 LnWindspeed_{i,t} + \beta_7 LnRaoddens_{i,t} + \beta_8 COVID_{i,t} + \mu_i + \lambda_i + \varepsilon_{i,t} \quad (5)$$

In this equation, ρ is the spatial autoregressive coefficient indicating the spatial spillover effect of energy consumption per capita ($Energycons\_pc$), and $WLnEnergycons\_pc_{i,t}$ is the spatially lagged dependent variable of energy consumption per capita ($Energycons\_pc$).

We aim to utilize the full SLM model as a benchmark for comparison with the SDPD IV model, determining which provides more accurate estimation results based on the lowest root mean square error (RMSE) and mean absolute error (MAE) values.

2.6. Spatial Dynamic Panel Lag Instrumental Variables Fixed-Effects Model (SDPD IV)

Regional energy consumption per capita is influenced not only by local factors but also by the region's past energy consumption and by the energy consumption per capita in other regions during both the current and previous periods. This study adopts the Spatial Dynamic Panel Lag IV Model [68,69] to analyze the impact of energy consumption per capita on economic growth in Thailand, expressed as follows:

$$y_{i,t} = \varphi \sum_{j=1}^{n} w_{i,j} y_{j,t} + \rho y_{i,t-1} + x_{i,t}\beta + \gamma_{y,i} f_{y,t} + \varepsilon_{i,t} \quad (6)$$

where $y_{i,t}$ denotes the observation on the dependent variable for individual unit i at time period t, and $x_{i,t}$ is a k × 1 vector of covariates with slope coefficients β. The lagged dependent variable $y_{i,t-1}$ captures dynamic or temporal effects due to state dependence. The error term of the model is composite: $f_{y,t}$ and $\gamma_{y,i}$ denote ry × 1 vectors of latent factors and factor loadings, respectively, and $\varepsilon_{i,t}$ is an idiosyncratic error. The "spatial-lag" variable $w_{i,j}y_{j,t}$ is a weighted average of the outcome variable in the neighboring locations for individual i. $w_{i,j}$ denotes the (i, j)-th element of the N × N spatial weights matrix WN. The spatial-lag coefficient φ is bounded by 1 as well, and its magnitude can be interpreted similarly to the time-lag coefficient ρ [70,71].

This model is used as a benchmark for comparison with the Spatial Lag Model (SLM). For the Spatial Dynamic Panel Lag Model, the asymptotic properties require that both the time and spatial dimensions be large. This implies that, with a sufficiently large sample size, the SDPD IV model is expected to yield more accurate results [51,67]. The rationale behind this approach is to identify the model that provides the most accurate estimates, one that provides precise coefficient estimates in terms of both sign and magnitude, thereby allowing for reliable conclusions to be drawn and aligning with the study's objectives.

Since including the time-lagged dependent variable in the equation may lead to inconsistent estimates, instrumental variable estimators are required [72]. The Hansen J statistic can be used to test whether, for example, the covariates are strictly exogenous with respect to $\varepsilon_{i,t}$ or whether the model parameters are indeed homogeneous across i. Violating either of these assumptions invalidates the model's moment conditions [73].

The present study uses a two-stage least squares (2SLS) model to overcome the endogeneity issue and estimate the effect of energy consumption per capita. The 2SLS model is a widely used econometric method for estimating causal relationships in panel data [74–78]. In the next step, this method assesses the validity of the instrument variables included in the Spatial Dynamic Panel Model because *Landtemp_night*, *Rainfall*, and *Windspeed* might have direct effects on *GPPpc*, $GPPpc^2$, *Pop_dens*, and *Roaddens* (these factors affect human settlement, e.g., [57,59,61,62], but have little direct effect on energy consumption per capita (*Energycons_pc*)). This study utilizes the spxtivdfreg package in Stata MP version 16.1 to estimate the Spatial Dynamic Panel Lag Instrumental Variables (SDPD IV) model.

Meteorological factors can have either positive or negative effects on electricity consumption [64]. However, some studies suggest that these factors, which can reflect climate change, may lead people to migrate to areas with more favorable climatic conditions [54]. Additionally, meteorological variables influence economic growth, labor productivity, and residential location choices in various countries [55,56,58,60,63]. Furthermore, geographic studies in Thailand have found that meteorological factors affect residential settlement patterns and commercial activities among the Thai population [57,59,61,62]. Most studies have found that meteorological factors often have uncertain or varying effects across different geographic areas. Therefore, this study uses these variables as instrumental variables that indirectly influence economic growth, labor productivity, and population settlement. Incorporating them in this way helps improve the accuracy of the SDPD IV model estimates and reduces the volatility associated with meteorological conditions. This approach also contributes to the academic literature by offering conclusions from a more diverse range of perspectives.

In the first stage our main explanatory variables are estimated as follows:

$$\hat{X}_{j,i,,t} = \delta_1 LnLandtemp\_night_{i,t} + \delta_2 LnRainfall_{i,t} + \delta_3 LnWindspeed_{i,t} + \varnothing_j X_{j,i,t-1} + \varepsilon_{i,t} \quad (7)$$

where $X_{j,i,t}$ are the logarithms of GPP per capita ($LnGPPpc$), GPP per capita squared ($LnGPPpc^2$), population density ($LnPop\_dens$), and road density ($LnRoaddens$) where j is variable lists, i is individual unit and t is time period. After completing the estimation in the first-stage equation (stage 1), the estimated coefficients are further analyzed in the second-stage equation (stage 2) to obtain the most accurate estimated equation. The statistical test results of the first-stage regression can be found in Appendix F, Table A6.

The specific form of the second-stage Spatial Dynamic Panel Lag IV Fixed-Effects Model for the whole region, BKK&VIC, CE, EA, WE, NE, NO, and SO is as follows:

$$LnEnergycons\_pc_{i,t} = a + \varphi \sum_{j=1}^{p} w_{i,j} LnEnergycons\_pc_{j,t} + \rho LnEnergycons\_pc_{i,t-1} + \sum_{j=0}^{q} \beta_{1,j} LnGP\hat{P}pc_{i,t-j} + \sum_{j=0}^{q} \beta_{2,j} LnGP\hat{P}pc^2_{i,t-j} + \sum_{j=0}^{q} \beta_{3,j} LnPo\hat{p}\_dens_{i,t-j} + \sum_{j=0}^{q} \beta_{4,j} LnRo\hat{a}ddens_{i,t-j} + \gamma_{y,i} f_{y,t} + \varepsilon_{i,t} \quad (8)$$

2.7. The Generalized Additive Model (GAM)

The spatial dynamic panel econometric model cannot fully capture the interactive effects of meteorological and socioeconomic factors on energy consumption per capita across all regions. Therefore, the Generalized Additive Model (GAM) is employed to evaluate the combined effects of these influential factors. This analytical method enhances the interpretability of the SDPD IV results, provides deeper insights, leads to a more comprehensive conclusion, and offers greater value for policy implications than relying solely on the SDPD IV model. The detailed algorithm of the GAM is presented below [79].

$$g(\mu) = a + f_1(X_1) + f_2(X_2) + \cdots + f_p(X_p) \quad (9)$$

where $\mu = E(Y/X_1, X_2, X_3, \cdots X_p)$; $g(\mu)$ is the contiguous function. f_p is treated as a smooth function explaining the dependent variable. X_p represents the independent variables. In the GAM, the distribution of the response variable is the Gauss–Markov distribution. The GAM was programmed in R. The mgcv was used to improve the GAM's computing performance.

3. Results

3.1. Hot Spot Analysis Results

Hot spot analysis is used to find the spatial cluster patterns within a region. The results of this analysis, along with spatial regression analyses, provide meaningful conclusions.

Figure 9 illustrates the energy consumption per capita in each region of Thailand from 2015 to 2022. Observing the annual changes, it is evident that energy consumption per capita is highly concentrated in the BKK&VIC, CE, EA, and WE regions (indicated in red), which are densely populated areas with diverse economic activities, including shopping malls, business districts, and industrial estates. The economies of these regions are primarily driven by the service and industrial sectors [11]. In contrast, the blue areas, representing the NE and SO regions, exhibit lower energy consumption per capita. This is because the agricultural and service sectors mainly support the economies of these regions. Most residents work as farmers, gardeners, or providers of tourism-related services, resulting in lower overall energy usage compared to the more industrialized and urbanized regions of BKK&VIC, CE, EA, and WE [11,80,81].

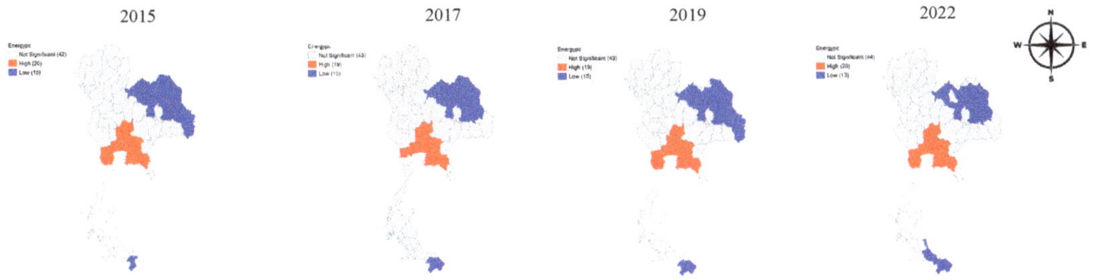

Figure 9. Hot spot analysis. Source: Authors.

Next, we conduct spatial econometric analyses using both the SLM and SDPD IV models with the whole-region dataset (Table 6) to obtain the most accurate estimation results. The best-performing model is then used as a benchmark for modeling other regions (BKK&VIC, CE, EA, WE, NE, NO, and SO).

Table 6 presents the estimation results from different model specifications. Model 1 is estimated using a Static Spatial Panel Lag Fixed-Effects model and serves as the benchmark for comparison. Models 2, 3, and 4 are estimated using the Spatial Dynamic Panel Lag IV Fixed-Effects model for the whole region. Models 2 and 3 separate the variables $LnGPPpc$ and $LnGPPpc^2$ to mitigate the effects of autocorrelation. However, the estimation results of models 2 and 3 do not differ significantly from those of model 4, which includes all variables simultaneously. To compare the accuracy of model 1 and model 4, we refer to the root mean square error (RMSE) and mean absolute error (MAE) values. Model 4 has the lowest RMSE and MAE among all models, indicating that it provides the most accurate estimation. The use of instrumental variables (IVs) in models 2, 3, and 4 is supported by the Hansen J-test p-values, which are greater than 0.05 but less than 0.90. This suggests that the IVs used do not lead to over-identification or overfitting of the model. Therefore, the SDPD IV model (model 4) is selected as the reference analytical method for the study and will be applied to estimate the econometric models for other regions (e.g., BKK&VIC, CE, EA, WE, NE, NO, and SO) as it provides the most accurate estimation results.

From model 4, it can be observed that the coefficients of $LnGPPpc$ and $LnGPPpc^2$ contradict the EKC theory. The model indicates a U-shape relationship between economic growth and energy consumption per capita (*LnEnergycons_pc*), implying that energy consumption per capita increases rapidly as the economy grows beyond a certain point. In other words, in the long run, continuous economic growth in Thailand does not lead to more sustainable energy use; instead, it results in increased energy consumption in an inefficient manner. Population density (*LnPop_dens*) is statistically significant and has a positive effect on energy consumption per capita, indicating that as population density increases, energy consumption also rises.

In addition, the spatial rho (ρ) coefficient in model 4 indicates that energy consumption in one province affects energy consumption in neighboring provinces. This suggests a spillover effect, meaning that energy consumption is not isolated but spatially dependent. This finding is consistent with the results of the hot spot analysis, which show that energy consumption is concentrated in specific regions of Thailand. In other words, when a province exhibits high energy consumption, its neighboring provinces are also likely to experience high energy consumption, reflecting spatial clusters of economic growth in certain areas.

Moreover, the COVID-19 outbreak occurred from 2020 to early 2022. A dummy variable is included in the model to assess its impact on energy consumption per capita. However, the estimation results show that this variable is insignificant. Similarly, the road density variable (*LnRoaddens*) is also insignificant. Nevertheless, the direction of the coefficient remains consistent with our initial hypotheses.

After obtaining the benchmark model from Table 6, we apply this method to estimate the results for the other regional datasets, as shown in Table 7.

Table 6. Regression results of the Spatial Panel Lag and Spatial Dynamic Panel Lag IV Fixed-Effects Models.

Variables: LnEnergycons_pc	Static Spatial Panel Lag Model (Whole Region)		Spatial Dynamic Panel Lag IV Model (Whole Region)					
	Model 1		Model 2		Model 3		Model 4	
LnEnergycons_pc(−1)	na		−0.074 (−1.13)		−0.087 (−1.36)		−0.081 (−1.36)	
LnGPPpc	2.060 (4.75)	***	0.372 (5.53)	***			−1.979 (−2.04)	**
$LnGPPpc^2$	−0.070 (−3.92)	***			0.016 (6.07)	***	0.096 (2.44)	**
LnPop_dens	−0.753 (−8.42)	***	13.117 (1.62)		9.108 (1.69)	*	6.336 (2.06)	**
LnLandtemp_night	−2.572 (−2.10)	**	na		na		na	
LnRainfall	0.013 (1.01)		na		na		na	
LnWindspeed	0.012 (0.46)		na		na		na	
LnRoaddens	0.023 (3.86)	***	0.023 (1.50)		0.023 (1.52)		0.017 (1.09)	
COVID	−0.001 (−0.06)		0.011 (1.75)	*	0.011 (1.78)	*	0.006 (0.84)	
Spatial rho (ρ)	0.336 (7.39)	***	0.372 (2.48)	**	0.347 (2.34)	**	0.327 (2.02)	**
Obs	616		462		462		462	
Instruments nb.			18		18		18	
Hansen J			14.147		13.768		10.450	
(Hansen J p-value)			(0.291)		(0.316)		(0.490)	
AIC	−1840.591							
BIC	−1796.359							
RMSE	20.889		12.703		8.857		6.225	
MAE	20.862		9.302		6.678		4.860	

Note: z-statistics are reported in parentheses. * $p < 0.10$, ** $p < 0.05$, *** $p < 0.01$.

Table 7 presents the estimation results of the Spatial Dynamic Panel Lag IV Fixed-Effects Models, separated by region: BKK&VIC, CE, EA, WE, NE, NO, and SO. The regions BKK&VIC, CE, EA, and WE were grouped together because they exhibit a high concentration of energy consumption per capita, as indicated by the hot spot analysis. Additionally, since the number of provinces in each of these regions is relatively small, they were combined to increase the number of observations and enhance the policy relevance of the analysis.

According to the estimation results, the coefficients of *LnGPPpc* and $LnGPPpc^2$ in the BKK&VIC, CE, EA, WE, and NE regions suggest that the relationship between economic growth and energy consumption per capita contradicts the EKC theory. These regions exhibit a U-shape relationship, suggesting that energy consumption per capita increases as economic growth continues beyond a certain threshold. This implies that long-term economic growth may lead to less efficient energy use in Thailand. In contrast, an insignificant relationship is found between economic growth and energy consumption per capita in the NO and SO regions.

Table 7. Regression results of the Spatial Dynamic Panel Lag IV Fixed-Effects Model (regions).

Variables: LnEnergycons_pc	**BKK&VIC, CE, EA, WE**			**Variables**	**NE**		
	Coef.		**Z-Stat**		**Coef.**		**Z-Stat**
Intercept	48.986	***	3.33	Intercept	58.134	***	2.79
LnEnergycons_pc(−1)	0.495	***	4.43	LnEnergycons_pc(−1)	0.151	***	3.38
LnGPPpc	−7.788	***	−5.53	LnGPPpc	−10.977	***	−2.79
$LnGPPpc^2$	0.319	***	5.63	$LnGPPpc^2$	0.481	***	2.76
LnPop_dens	−0.353		−0.91	LnPop_dens	1.144		0.86
LnRoaddens	0.022	*	1.72	LnRoaddens	0.010	***	3.44
Spatial rho (ρ)	0.631	***	2.61	Spatial rho (ρ)	0.787	***	7.21
Obs		156		Obs		120	
Instruments nb.		18		Instruments nb.		18	
Hansen J		10.063		Hansen J		10.634	
(Hansen J p-value)		0.611		(Hansen J p-value)		0.561	
Variables	**NO**			**Variables**	**SO**		
	Coef.		**Z-stat**		**Coef.**		**Z-stat**
Intercept	−29.319		−0.67	Intercept	−3.996		−0.65
LnEnergycons_pc(−1)	0.297	***	3.77	LnEnergycons_pc(−1)	0.051		0.37
LnGPPpc	4.889		1.23	LnGPPpc	0.531		0.34
$LnGPPpc^2$	−0.211		−1.24	$LnGPPpc^2$	−0.011		−0.17
LnPop_dens	0.251		1.11	LnPop_dens	0.653		0.71
LnRoaddens	0.025	**	2.06	LnRoaddens	0.034	***	3.92
Spatial rho (ρ)	0.731	***	5.87	Spatial rho (ρ)	0.411	***	4.46
Obs		102		Obs		84	
Instruments nb.		17		Instruments nb.		14	
Hansen J		8.201		Hansen J		7.327	
(Hansen J p-value)		0.695		(Hansen J p-value)		0.502	

Note: * $p < 0.10$, ** $p < 0.05$, *** $p < 0.01$.

Other variables, such as road density (*LnRoaddens*), show a positive relationship with energy consumption per capita. This relationship is statistically significant in almost all regions (BKK&VIC, CE, EA, WE, NE, NO, and SO), indicating that as road density increases, energy consumption also increases.

Including the spatial rho (ρ) variable in Table 7 indicates that energy consumption in one province influences energy consumption in neighboring provinces, particularly in the BKK&VIC, CE, EA, WE, NE, NO, and SO regions. This suggests the presence of a spillover effect, which aligns with the results of the hot spot analysis, showing that energy consumption per capita is concentrated in specific areas of Thailand.

3.2. Spatial Dynamic Panel Econometric Model Results (SLM, SDPD IV)

After obtaining the estimation results for each region, we conduct an effect analysis using the Spatial Dynamic Panel Lag IV Fixed-Effects Model. These results provide a more comprehensive understanding of the cumulative effects over time and across regions, as presented in Table 8.

Table 8 presents the effect analysis for the whole region. The results show that the total effect of economic growth on energy consumption per capita is statistically significant in both the short and long runs. Notably, the coefficient indicates that the short-run impact of economic growth on energy consumption per capita is larger than the long-run impact. This suggests that, over time, the influence of economic growth on energy consumption per capita tends to diminish; in other words, the effect becomes regressive in the long term. Population density (*LnPop_dens*) is also statistically significant in both the short- and long-run analyses. However, the cumulative effects of this variable are more pronounced in the short run and tend to decrease over time. This implies that population density has a more substantial influence on energy consumption per capita in the short term. However, its impact gradually diminishes over time, suggesting that increased energy efficiency may occur as population density rises in the future.

Table 8. Effect analysis of the Spatial Dynamic Panel Lag IV Fixed-Effects Model.

Whole Region (n = 462)							
	Variables	**Short-Run Effect**		**Z-Stat**	**Long-Run Effect**		**Z-Stat**
Direct	LnGPPpc	−2.012	**	−2.04	−1.856	**	−1.98
	LnGPPpc²	0.097	**	2.44	0.090	**	2.35
	LnPop_dens	6.444	**	2.08	5.943	**	2.10
	LnRoaddens	0.017		1.09	0.016		1.11
	COVID	0.006		0.84	0.005		0.84
Indirect	LnGPPpc	−0.701		−1.15	−0.582		−1.11
	LnGPPpc²	0.034		1.23	0.028		1.19
	LnPop_dens	2.246		1.37	1.864		1.35
	LnRoaddens	0.006		0.89	0.005		0.87
	COVID	0.002		0.66	0.002		0.67
Total	LnGPPpc	−2.713	*	−1.87	−2.438	*	−1.82
	LnGPPpc²	0.131	**	2.19	0.118	**	2.11
	LnPop_dens	8.690	**	2.12	7.801	**	2.13
	LnRoaddens	0.023		1.06	0.021		1.09
	COVID	0.008		0.80	0.007		0.81

Note: * $p < 0.10$, ** $p < 0.05$.

Additionally, when examining the direct effects (within-region effects), which occur within the provinces themselves, we find that the impact of economic growth on energy consumption per capita is statistically significant in both the short and long runs. The coefficients indicate that the effect is more substantial in the short run but diminishes over time. This suggests that, in the long run, energy consumption increases in a regressive manner as the influence of economic growth weakens. Population density (*LnPop_dens*) is also statistically significant in both the short-run and long-run analyses. However, the cumulative effects of this variable are more pronounced in the short run and tend to decrease in the long run. These findings are consistent with the results observed for the total effects.

In contrast, the indirect effects (spillover effects to neighboring provinces) are insignificant for all variables in both the short-run and long-run analyses.

For the BKK&VIC, CE, EA, WE, NE, NO, and SO regions, economic growth have no cumulative effect on energy consumption per capita, and the other variables are also insignificant.

3.3. GAM Analysis Results

As mentioned earlier, the GAM analysis provides deeper insights than spatial econometric analysis alone. It enhances interpretability and offers a novel approach for deriving policy implications.

Table 9 suggests that most interaction (cross-) terms exhibit a nonlinear relationship with energy consumption per capita. The *p*-values for all cross-terms are significantly below 0.01, indicating a strong association between the influential factors and energy consumption per capita. In Figure 10, panel (a) shows that when both GPP per capita (*GPPpc*) and land surface temperature at night (*Landtemp_night*) are high, energy consumption per capita is also high. Panel (b) illustrates that provinces with high *GPPpc* and high *Rainfall* tend to have higher energy consumption per capita. Panel (c) demonstrates that when *GPPpc* and wind speed (*Windspeed*) are high, energy consumption per capita is also high. Similarly, panel (d) shows that provinces with high *GPPpc* and high road density (*Roaddens*) tend to have higher energy consumption per capita. Areas with good wind circulation, well-connected roads, and easy access are more likely to attract human settlement and economic activity, which leads to increased energy use [57,59,61]. This pattern is consistent with the relationship observed with *Rainfall*. Panel (e) indicates that provinces with high land surface temperature at night (*Landtemp_night*) but low wind speed (*Windspeed*) tend to have high energy consumption per capita. This could be due to poor air circulation and elevated temperatures, which lead to greater use of air conditioners and energy consumption. Such conditions are often found in large urban areas with high levels of economic activity. Finally, panel (f) shows that provinces with both high land surface temperature at night (*Landtemp_night*) and high road density (*Roaddens*) also exhibit high energy consumption per capita. This may reflect highly urbanized areas having increased infrastructure and population density, leading to elevated energy usage.

Table 9. The GAM hypothesis test for energy consumption per capita and the explanatory variables.

	Estimated Degrees of Freedom	F-Statistic	*p*-Value
GPPpc-Landtemp_night	56.4	39.6	<0.01
GPPpc-Rainfall	65.5	35.3	<0.01
GPPpc-Windspeed	76.0	10.6	<0.01
GPPpc-Roaddens	71.5	247.5	<0.01
Landtemp_night-Windspeed	11.7	4.9	<0.01
Landtemp_night-Roaddens	5.1	8.8	<0.01

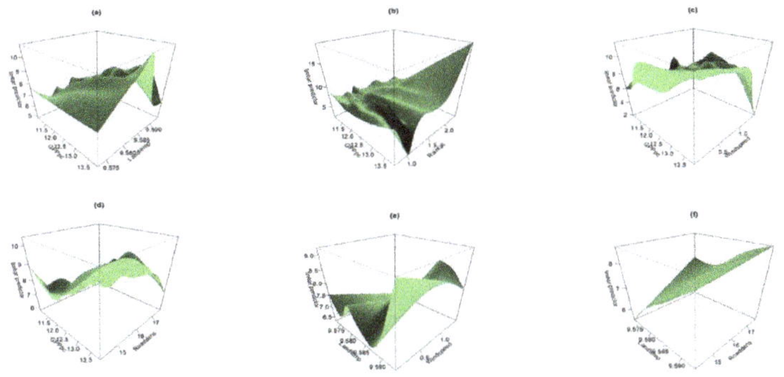

Figure 10. GAM analysis: Three-dimensional effect graph of the driving factors of the variation in energy consumption per capita; (**a**) GPPpc and Landtemp_night; (**b**) GPPpc and Rainfall; (**c**) GPPpc and Windspeed; (**d**) GPPpc and Roaddens; (**e**) Landtemp_night and Windspeed; (**f**) Landtemp_night and Roaddens. Source: Authors.

In conclusion, provinces with high GPP per capita (*GPPpc*), land surface temperature at night (*Landtemp_night*), rainfall (*Rainfall*), wind speed (*Windspeed*), and road density (*Roaddens*) tend to exhibit higher energy consumption per capita. These indicators reflect levels of urbanization and human habitation, which are often associated with the presence of business districts, shopping malls, markets, and industrial estates [57,59,61,62,82,83], typically concentrated in only a few provinces.

4. Discussion

The results from the Moran's I statistical test and hot spot analysis indicate that energy consumption exhibits a spatial clustering pattern. This means that energy consumption is highly concentrated in regions with intense economic activity and established economic centers, such as BKK&VIC, CE, EA, and WE. In contrast, energy consumption is less concentrated in areas farther from these economic hubs, particularly in regions primarily driven by the agricultural sector, such as NE and SO. Furthermore, high energy consumption in one area tends to lead to increased energy consumption in surrounding areas due to spillover effects. Based on these findings, this study applies spatial econometric methods to analyze the spatial distribution of energy consumption and its impact on neighboring provinces.

From model 4 (Spatial Dynamic Panel Lag IV Fixed-Effects Model), it is evident that the relationship between GPP per capita and energy consumption contradicts the EKC theory, which typically suggests an inverted U-shape relationship between economic growth and environmental impact [40–43]. Instead, the results indicate a U-shape relationship between economic growth and energy consumption per capita, suggesting that after the economy reaches a certain level of growth, energy consumption increases rapidly. In other words, continued economic growth leads to rising energy consumption and inefficient energy use. Additionally, population density is statistically significant and positively correlates with energy consumption per capita, indicating that higher population density is associated with increased energy use. This is because areas with high population density tend to have large economic activities and are highly urbanized [59,61,62]. Moreover, the spatial autocorrelation coefficients (rho) from both the Spatial Lag Model (SLM) and the Spatial Dynamic Panel Lag IV Model (SDPD IV) are statistically significant, indicating a strong spillover effect. This means that an increase in energy consumption in one province also leads to increased energy consumption in neighboring provinces.

In addition, the analysis of GPP per capita and its effect on energy consumption in regions such as BKK&VIC, CE, EA, WE, and NE reveals that the relationship between economic growth and energy consumption per capita contradicts the EKC theory. These regions exhibit a U-shape relationship, indicating that energy consumption per capita increases as economic growth surpasses a certain threshold. In contrast, no significant relationship was found between economic growth and energy consumption per capita in the NO and SO regions. Using the Spatial Dynamic Panel Lag IV Model (SDPD IV), we also estimated the effects of other variables on energy consumption per capita across different regions of Thailand. The results show that road density has a positive and statistically significant relationship with energy consumption per capita in most regions, particularly in BKK&VIC, CE, EA, WE, NE, NO, and SO.

From the analysis of short-run and long-run impacts, the findings for the whole region suggest that the effect of economic growth on energy consumption per capita is greater in the short term than in the long term. This implies that as the economy continues to grow, energy consumption will initially increase more sharply but may gradually level off or decline over time. Moreover, population density is expected to have a more favorable impact on energy consumption in the long run, indicating improved efficiency over time. In contrast, no significant short- or long-term impacts of economic growth on energy consumption per capita were observed in the BKK&VIC, CE, EA, WE, NE, NO, and SO regions.

From the GAM analysis, it can be concluded that provinces with high GPP per capita, land surface temperature at night, rainfall, wind speed, and road density also tend to have high energy consumption per capita. These indicators reflect levels of urbanization and human habitation, as well as the presence of business districts, shopping malls, markets, and industrial estates [57,59,61,62,82,83], which are often concentrated in only a few provinces.

The inflection point analysis based on the EKC theory reveals that, in 2022, the average GRP per capita was USD 40,857 for the BKK&VIC, CE, EA, and WE regions, and USD 2724 for the NE region. The analysis indicates that the inflection point for the BKK&VIC, CE, EA, and WE regions is USD 6130, while for the NE region it is USD 2785. As shown in Appendix G, Table A7, these regions are currently experiencing a continuous increase in energy consumption and inefficient energy use. This trend may lead to energy shortages or other environmental problems resulting from the increased demand for electricity generation.

This study has some limitations. First, future research should conduct more detailed analyses at the district or sub-district levels to better understand localized issues related to energy consumption. Second, additional variables such as innovation and technology may influence energy consumption and should be considered in future studies to improve the model's accuracy. Third, this study does not include forecasts of future energy consumption trends. However, forecasting is crucial for planning and should be explored further in future studies.

5. Conclusions

This study investigates the relationship between economic growth and energy consumption, utilizing the EKC theory. It utilizes data from government agencies and satellite sources to identify additional factors influencing energy consumption. Additionally, we employed hot spot analysis to identify clusters of energy consumption per capita across various regions. We also used the Generalized Additive Model (GAM) to evaluate the combined effects of influential factors. This analytical approach enhances the interpretability of the SDPD IV results, providing deeper insights. These techniques offer greater value for policy implications than relying solely on the SDPD IV model.

The study results indicate that energy consumption in Thailand is spatially clustered, meaning it tends to spill over into nearby provinces and is concentrated in specific regions. Factors that positively influence energy consumption include GPP per capita, population density, and road density. The EKC analysis also reveals a U-shape relationship between GPP per capita and energy consumption in the BKK&VIC, CE, EA, WE, and NE regions. In the early stages of economic growth, energy consumption tends to decrease as GPP per capita increases. However, once the economy surpasses a certain threshold, further increases in GPP per capita lead to higher energy consumption. As shown in Appendix G, Table A7, these regions are experiencing continuous increases in energy consumption, reflecting inefficient energy use and contributing to various environmental problems due to the reliance on non-renewable sources for that generation. For reference, the average GRP per capita in 2022 was USD 40,857 for the BKK&VIC, CE, EA, and WE regions, and USD 2724 for the NE region. Moreover, areas with favorable climate conditions, well-developed infrastructure, and high levels of economic activity tend to exhibit higher energy consumption per capita.

For policy implications, Thailand's current environmental policies are largely informal and decentralized. Therefore, the country should develop rigorous, long-term plans for efficient and sustainable energy use, similar to those implemented in developed countries [43], particularly in regions with high-energy-consumption clusters and in provinces characterized by high levels of urbanization, economic activity, and industrial production (such as BKK&VIC, CE, EA, WE, and NE) that align with the results of this study. Moreover, policies should promote the increased use of renewable energy sources, such as solar, wind, hydro, geothermal, and biomass, to reduce dependence on electricity generated by conventional government-operated systems. Since electricity in Thailand is primarily generated from fossil fuels and natural gas, a significant amount of electricity is produced to meet population demand and support economic development. As a result, this leads to the release of large quantities of pollutants that harm the environment. The use of renewable energy can help alleviate the challenges associated with high energy consumption and mitigate environmental problems related to electricity production [84–87]. Additionally, the transfer of advanced technology from developed countries through investment by multinational companies in Thailand will enable the country to acquire more modern, energy-efficient, and environmentally friendly technologies.

Future studies should explore alternative methods to identify strategies for efficient energy consumption. Forecasting future energy consumption trends is crucial and should be included to better understand how energy demand may evolve over time. Moreover, the application of machine learning algorithms to examine the relationship between economic growth and energy consumption is expected to become increasingly important in the future. This approach enables a comparison of the accuracy between econometric analysis methods and machine learning techniques, ultimately leading to more concrete findings. These insights can also serve as valuable policy recommendations for Thailand. Additionally, cost–benefit analysis remains essential for evaluating the cost-effectiveness of energy-related projects, helping to identify the most efficient methods, those with the lowest costs, those yielding the highest returns, and those offering the best value for policy planning.

Author Contributions: T.S.: writing—review, visualization, investigation, sample collection, methodology, conceptualization. K.A.: investigation, supervision, writing—review and editing, methodology, funding acquisition, conceptualization. All authors have read and agreed to the published version of the manuscript.

Funding: This research received no external funding.

Data Availability Statement: The datasets used and analyzed during the current study are available from the corresponding author upon reasonable request.

Acknowledgments: We appreciate The Government of Thailand for providing the data used.

Conflicts of Interest: The authors declare no competing interests.

Appendix A

Table A1 shows the results of the panel unit root test. The Levin, Lin, and Chu (LLC) test and the Im, Pesaran, and Shin (IPS) test were applied. The results indicate that most variables are stationary at the level, i.e., I(0). However, some variables are stationary at the first difference, for example, *Pop_dens* in the northeastern (NE) region and *Landtemp_night* in the southern (SO) region, i.e., I(1).

Table A1. Panel unit root test.

	Whole Region			
	Level		First Difference	
	LLC	IPS	LLC	IPS
Energycons_pc	−20.843 ***	−5.099 ***	−32.089 ***	−8.568 ***
GPPpc	−41.192 ***	−4.535 ***	−36.113 ***	−4.585 ***
GPPpc^2	−46.642 ***	−4.546 ***	−36.768 ***	−4.578 ***
Pop_dens	-1.1×10^2 ***	4.317	-3.3×10^3 ***	−7.257 ***
Landtemp_night	−74.822 ***	−4.932 ***	-1.2×10^2 ***	−6.653 ***
Rainfall	−24.257 ***	−6.548 ***	−14.518 ***	−6.081 ***
Windspeed	−31.093 ***	−4.568 ***	−67.697 ***	−8.924 ***
Roaddens	−13.813 ***	−3.645 ***	−45.701 ***	−7.309 ***

Table A1. *Cont.*

Bangkok and Vicinity, Central, Eastern, Western Region (BKK&VIC, CE, EA, WE)				
	Level		First difference	
	LLC	IPS	LLC	IPS
Energycons_pc	−13.999 ***	−2.068	−30.020 ***	−4.375 ***
GPPpc	−34.520 ***	−2.682 ***	−95.010 ***	−2.676 ***
GPPpc^2	−40.884 ***	−2.691 ***	−99.332 ***	−2.672 ***
Pop_dens	−75.664 ***	1.532	-8.6×10^3 ***	−4.316 ***
Landtemp_night	−13.322 ***	−3.644 ***	−0.961	−4.528 ***
Rainfall	−14.845 ***	−2.770 ***	−0.203	−1.784
Windspeed	−19.331 ***	−3.221 ***	−42.500 ***	−4.994 ***
Roaddens	−10.677 ***	−2.908 ***	−22.670 ***	−4.249 ***
Northeastern Region (NE)				
	Level		First difference	
	LLC	IPS	LLC	IPS
Energycons_pc	−8.409 ***	−3.063 ***	−7.952 ***	−4.599 ***
GPPpc	−20.121 ***	−2.572 **	−32.806 ***	−2.892 ***
GPPpc^2	−18.695 ***	−2.570 **	−43.748 ***	−2.884 ***
Pop_dens	1.638	3.212	-1.5×10^2 ***	−4.104 ***
Landtemp_night	−64.689 ***	−2.537 *	−94.600 ***	−2.384
Rainfall	−11.857 ***	−4.045 ***	−25.310 ***	−4.292 ***
Windspeed	−15.992 ***	−3.280 ***	−13.360 ***	−5.065 ***
Roaddens	−6.294 ***	−2.565 **	−6.463 ***	−3.687 ***
Northern Region (NO)				
	Level		First difference	
	LLC	IPS	LLC	IPS
Energycons_pc	−13.889 ***	−2.343	−12.131 ***	−4.241 ***
GPPpc	−22.972 ***	−1.821	−14.358 ***	−1.740
GPPpc^2	−23.102 ***	−1.834	−13.084 ***	−1.738
Pop_dens	−25.097 ***	1.390	-2.9×10^2 ***	−3.704 ***
Landtemp_night	−72.246 ***	−2.279	−37.632 ***	−2.454
Rainfall	−15.180 ***	−2.761 ***	11.471	−2.645 **
Windspeed	−12.966 ***	0.047	−47.949 ***	−3.683 ***
Roaddens	−5.660 ***	−0.196	−28.630 ***	−3.445 ***
Southern Region (SO)				
	Level		First difference	
	LLC	IPS	LLC	IPS
Energycons_pc	−7.532 ***	−2.899 ***	−12.988 ***	−3.962 ***
GPPpc	−3.985 ***	−1.903	−50.419 ***	−1.732
GPPpc^2	−3.951 ***	−1.901	−54.422 ***	−1.735
Pop_dens	−6.109 ***	2.666	-4.0×10^2 ***	−2.148
Landtemp_night	0.504	−1.057	3.258	−3.881 ***
Rainfall	−4.052 ***	−3.705 ***	−6.460 ***	−3.785 ***
Windspeed	−13.483 ***	−2.456	−28.477 ***	−4.011 ***
Roaddens	−4.817 ***	−1.303	−26.790 ***	−3.149 ***

Note: * $p < 0.10$, ** $p < 0.05$, *** $p < 0.01$.

Appendix B

From Table A2, it can be seen that most of the Breusch–Godfrey test statistics for all models are statistically significant, leading to the rejection of the null hypothesis (H_0) and indicating the presence of serial correlation. Therefore, using a dynamic model is more appropriate for analyzing these models than relying on the OLS model. Additionally, the mean VIFs values for all models are below 5, suggesting that multicollinearity is not a significant concern in these models.

Table A2. Breusch–Godfrey test and variance inflation factor (VIFs) statistics.

Models	Breusch–Godfrey F-Stat.		VIFs Mean
Whole region	0.255		1.75
BKK&VIC, CE, EA, WE	0.000	***	2.41
NE	0.000	***	1.65
NO	0.000	***	2.27
SO	0.000	***	1.92

*** $p < 0.01$.

Appendix C

From Table A3, it can be observed that the correlation values between *LnEnergycons_pc* and *LnLandtemp_night*, *LnRainfall*, and *LnWindspeed* fall mostly within the moderate-to-weak range. This suggests that these meteorological variables are less correlated with *LnEnergycons_pc* compared to other variables such as *LnGPPpc*, *LnGPPpc*^{2}, *LnPopdens*, and *LnRoaddens*. Although previous research (e.g., [64]) has shown that meteorological factors such as nighttime land surface temperature, rainfall, and wind speed have a direct effect on energy consumption, other studies (e.g., [54–59,61,62]) have indicated that these factors also influence economic growth, labor productivity, and residential settlement patterns, which in turn affect energy consumption. Therefore, this study uses *LnLandtemp_night*, *LnRainfall*, and *LnWindspeed* as instrumental variables, thereby expanding the scope of research by incorporating meteorological factors as instruments in the model. Road density has a negative correlation with population density because densely populated areas often have limited space, making it challenging to construct long-distance roads. In contrast, provinces with larger land areas have more space available for building longer and more extensive road networks.

Table A3. Correlation matrix.

	LnEnergy cons_pc	LnGPPpc	LnGPPpc^2	LnPop_dens	LnRoad dens	LnLandtemp _night	LnRainfall	LnWindspeed
LnEnergycons_pc	1.000							
LnGPPpc	0.921	1.000						
LnGPPpc^2	0.917	0.999	1.000					
LnPop_dens	0.520	0.550	0.552	1.000				
LnRoaddens	0.044	0.084	0.082	−0.056	1.000			
LnLandtemp_night	0.552	0.542	0.538	0.674	−0.236	1.000		
LnRainfall	−0.114	−0.009	−0.008	−0.022	−0.217	−0.088	1.000	
LnWindspeed	0.279	0.213	0.217	0.563	−0.062	0.507	−0.102	1.000

Appendix D

Table A4 presents the descriptive statistics for each variable used in this study, including the minimum and maximum values, mean, and standard deviation (Std.). It is important to note that *Landtemp_night*, *Rainfall*, and *Windspeed* are instrumental variables in the Spatial Dynamic Panel Lag IV Model (SDPD IV).

Table A4. Descriptive statistics.

	Whole Region (n = 616)			
	Min	Max	Mean	Std.
Energycons_pc (kWh/person)	488.41	11,335.86	2151.62	1906.60
GPPpc (baht/person)	49,296.39	1,060,571.10	163,114.05	148,340.36

Table A4. *Cont.*

Pop_dens (unit/km^2)	17.86	5776.47	303.59	751.09
Roaddens (length)	564,563.62	44,665,593.00	7,992,667.82	6,299,861.39
Landtemp_night (Celcius)	14,316.81	14,668.83	14,498.34	70.42
Rainfall (mm/y)	1.97	11.92	4.86	1.87
Windspeed (m/s)	1.00	4.77	2.30	0.72
BKK&VIC, CE, EA, WE (n = 208)				
Energycons_pc (kWh/person)	1144.30	11,335.86	3860.83	2320.39
GPPpc (baht/person)	63,731.97	1,060,571.10	277,110.95	200,107.08
Pop_dens (unit/km^2)	42.12	5776.47	660.68	1199.79
Roaddens (length)	607,745.85	26,271,510.00	6,916,773.18	5,024,772.79
Landtemp_night (Celcius)	14,424.08	14,668.83	14,552.61	55.07
Rainfall (mm/y)	1.97	10.47	4.27	1.40
Windspeed (m/s)	1.60	3.49	2.59	0.48
NE (n = 160)				
Energycons_pc (kWh/person)	528.91	2507.83	981.53	367.97
GPPpc (baht/person)	48,296.39	134,338.26	78,464.42	17,647.38
Pop_dens (unit/km^2)	50.68	162.20	111.68	23.82
Roaddens (length)	564,563.92	44,665,593.00	10,118,021.27	8,307,641.85
Landtemp_night (Celcius)	14,335.12	14,531.52	14,446.47	42.19
Rainfall (mm/y)	2.54	8.98	4.54	1.09
Windspeed (m/s)	1.26	4.13	2.72	0.66
NO (n = 136)				
Energycons_pc (kWh/person)	488.41	3559.13	1303.96	591.44
GPPpc (baht/person)	53,692.75	232,558.92	103,838.57	32,960.50

Table A4. *Cont.*

Pop_dens (unit/km^2)	17.86	123.02	69.84	27.09
Roaddens (length)	1,650,247.80	34,436,118.00	9,113,139.56	5,827,785.85
Landtemp_night (Celcius)	14,316.81	14,593.78	14,454.02	67.63
Rainfall (mm/y)	2.56	5.64	3.78	0.69
Windspeed (m/s)	1.00	2.53	1.50	0.37
SO (n = 112)				
Energycons_pc (kWh/person)	521.67	5089.04	1678.24	947.59
GPPpc (baht/person)	53,802.99	422,668.67	144,310.91	76,855.01
Pop_dens (unit/km^2)	60.58	1099.11	198.42	246.66
Roaddens (length)	887,209.69	16,903,140.00	5,593,965.81	3,959,558.91
Landtemp_night (Celcius)	14,450.81	14,606.54	14,525.49	33.34
Rainfall (mm/y)	4.58	11.92	7.73	1.69
Windspeed (m/s)	1.21	4.77	2.16	0.64

Appendix E

Table A5 shows that most of the Pedroni test statistics indicate rejection of the null hypothesis of no cointegration for the Whole region, BKK&VIC, CE, EA, WE, NE, NO, and SO models. The Kao test also confirms that the variables are cointegrated in these models. Therefore, all models except one strongly satisfy the preconditions for cointegration-based estimation. The only exception is the SO model in the Kao test, where the null hypothesis of no cointegration cannot be rejected. This suggests that the variables in the SO model are likely not cointegrated.

Table A5. Panel cointegration tests.

	Whole Region	**BKK&VIC, CE, EA, WE**	**NE**	**NO**	**SO**
Pedroni Test	**Test Statistics**	**Test Statistics**	**Test Statistics**	**Test Statistics**	**Test Statistics**
Panel PP-Statistic	−18.779 ***	−13.317 ***	−4.094 ***	−12.626 ***	−9.917 ***
Panel ADF-Statistic	24.378 ***	11.672 ***	14.820 ***	12.344 ***	9.950 ***
Kao Test	**Test Statistics**	**Test Statistics**	**Test Statistics**	**Test Statistics**	**Test Statistics**
Panel ADF-Statistic	3.346 ***	1.710 **	3.088 ***	1.697 **	−0.800

Note: ** $p < 0.05$, *** $p < 0.01$.

Appendix F

Table A6 shows that some of the first-stage equation results meet the validity criteria for instrumental variables, with F-statistics greater than 10 and high R-squared (R-sq) values, and some equations indicate the presence of weak instrumental variables. When examining the R-squared values, most variables exhibit high explanatory power in the first-stage regression model. However, for variables that appear to be relatively weak instruments, we rely on supporting theories from previous studies (e.g., [54,56,57,59–63]), which have shown that meteorological factors influence economic growth, labor productivity, and population settlement. These equations are then used in the second-stage regression. This approach aims to broaden the scope of the study by incorporating meteorological factors as instrumental variables, in line with the conceptual framework established in the aforementioned studies.

Table A6. First-stage regression statistical results.

	Whole Region (n = 616)		
First Stage	**R-sq**	**F-Stat**	**Endogeneity**
LnGPPpc	0.999	22.230	0.000
$LnGPPpc^2$	0.999	20.039	0.000
LnPop_dens	0.587	132.811	0.000
LnRoaddens	0.204	46.058	0.000
	BKK&VIC, CE, EA, WE (n = 208)		
First stage	**R-sq**	**F-stat**	**Endogeneity**
LnGPPpc	0.999	0.264	0.000
$LnGPPpc^2$	0.999	0.220	0.000
LnPop_dens	0.722	118.838	0.398
LnRoaddens	0.197	7.459	0.000
	NE (n = 160)		
First stage	**R-sq**	**F-stat**	**Endogeneity**
LnGPPpc	0.999	8.867	0.142
$LnGPPpc^2$	0.999	8.924	0.159
LnPop_dens	0.349	23.208	0.047
LnRoaddens	0.485	8.280	0.000
	NO (n = 136)		
First stage	**R-sq**	**F-stat**	**Endogeneity**
LnGPPpc	0.999	11.080	0.253
$LnGPPpc^2$	0.999	11.346	0.258
LnPop_dens	0.451	11.221	0.083
LnRoaddens	0.413	19.342	0.198
	SO (n = 112)		
First stage	**R-sq**	**F-stat**	**Endogeneity**
LnGPPpc	0.999	4.180	0.009
$LnGPPpc^2$	0.999	3.947	0.010
LnPop_dens	0.564	29.308	0.001
LnRoaddens	0.286	7.812	0.015

Appendix G

As shown in Table A7, the affluence elasticities of impact coefficients (EE_{IA}) were estimated. The results indicate that the EE_{IA} values for income levels ranging from USD 2755 to USD 6125 in the BKK&VIC, CE, EA, and WE regions, and from USD 1990 to USD 2755 in the NE region, are negative. This means that as GPP per capita increases within these ranges, energy consumption per capita continues to decline, although at a diminishing rate. However, once income surpasses the inflection point—USD 6130 for the BKK&VIC,

CE, EA, and WE regions, and USD 2785 for the NE region—energy consumption per capita begins to rise as GPP per capita continues to grow. For reference, the average GRP per capita in 2022 was USD 40,857 for the BKK&VIC, CE, EA, and WE regions, and USD 2724 for the NE region.

Table A7. The affluence elasticity of impact coefficients (EE_{IA}) for various values of affluence from the Spatial Dynamic Panel Lag IV Fixed-Effects Regression model.

	BKK&VIC, CE, EA, WE			
	GPP per Capita (USD)			
	USD 2755	**USD 3675**	**USD 6125**	**USD 6130**
Energycons_pc	−0.510	−0.326	−0.001	0.000
	NE			
	USD 1990	**USD 2145**	**USD 2755**	**USD 2785**
Energycons_pc	−0.316	−0.245	−0.003	0.008

Note: Exchange rate of the Bank of Thailand as of 12 June 2025.

References

1. World Bank. *Electric Power Consumption (kWh Per Capita)*; World Bank Group: Washington, DC, USA, 2022. Available onlin https://data.worldbank.org/indicator/EG.USE.ELEC.KH.PC?end=2023&name_desc=false&start=2023&year=2022 (accessed o 30 March 2025).
2. World Bank. *Electricity Production From Oil Sources (% of Total)*; World Bank Group: Washington, DC, USA, 2022. Available onlin https://data.worldbank.org/indicator/EG.ELC.PETR.ZS?end=2023&name_desc=false&start=2023&year=2022 (accessed o 30 March 2025).
3. World Bank. *Combustible Renewables and Waste (% of Total Energy)*; World Bank Group: Washington, DC, USA, 2022. Available or line: https://data.worldbank.org/indicator/EG.USE.CRNW.ZS?end=2023&name_desc=false&start=2023&year=2022 (accesse on 30 March 2025).
4. Tunpaiboon, N. *Business/Industry Trends 2025–2027: Electricity Generation Business*; Krungsri Research: Bangkok, Thailand, 202 Available online: https://www.krungsri.com/th/research/industry/industry-outlook/energy-utilities/power-generation/io power-generation-2025-2027 (accessed on 29 March 2025).
5. EPPO. *Electricity Consumption by Sectors 2015–2022*; Energy Policy and Planning Office: Bangkok, Thailand, 2022. Availabl online: https://www.eppo.go.th/index.php/en/en-energystatistics/electricity-statistic/Electricity%20Consumption%20for% 20the%20Whole%20Country%20(Classified%20by%20Sector) (accessed on 30 March 2025).
6. Phadkantha, R.; Yamaka, W. The nonlinear impact of electricity consumption on economic growth: Evidence from Thailanc *Energy Rep.* **2022**, *8*, 1315–1321. [CrossRef]
7. EPPO. *Growth Rate of Electricity Demand and Economic Growth 2015-2022*; Energy Policy and Planning Office: Bangkok, Thailanc 2022. Available online: https://www.eppo.go.th/index.php/en/en-energystatistics/electricity-statistic/Power%20Generation% 20Classified%20by%20Fuel%20Type%20(Detail)%20-%20Graph (accessed on 30 March 2025).
8. EPPO. *Power Generation (GWh) by Fuel 2015–2022*; Energy Policy and Planning Office: Bangkok, Thailand, 2022. Availabl online: https://www.eppo.go.th/index.php/en/en-energystatistics/electricity-statistic/Power%20Generation%20Classified% 20by%20Fuel%20Type (accessed on 30 March 2025).
9. Chokchaichamnankit, S. *Energy Projects Under the ASEAN Economic Community*; Thai Climate Justice Organization: Chiang Ma Thailand, 2012. Available online: https://www.thaiclimatejustice.org/sites/default/files/2012-08_%E0%B9%82%E0%B8%84% E0%B8%A3%E0%B8%87%E0%B8%81%E0%B8%B2%E0%B8%A3%E0%B8%94%E0%B9%89%E0%B8%B2%E0%B8%99%E0%B8 %9E%E0%B8%A5%E0%B8%B1%E0%B8%87%E0%B8%87%E0%B8%B2%E0%B8%99%E0%B8%A0%E0%B8%B2%E0%B8%A2 %E0%B9%83%E0%B8%95%E0%B9%89%E0%B8%9B%E0%B8%A3%E0%B8%B0%E0%B8%8A%E0%B8%B2%E0%B8%84%E0 %B8%A1%E0%B9%80%E0%B8%A8%E0%B8%A3%E0%B8%A9%E0%B8%90%E0%B8%81%E0%B8%B4%E0%B8%88%E0%B8 %AD%E0%B8%B2%E0%B9%80%E0%B8%8B%E0%B8%B5%E0%B8%A2%E0%B8%99.pdf (accessed on 3 January 2025).
10. EGAT. *Get to Know ASEAN Electricity*; Electricity Generating Authority of Thailand: Nonthaburi, Thailand, 2016. Available online https://www.thaigov.go.th/uploads/document/66/2017/02/pdf/Coal_power_plant_5.pdf (accessed on 30 March 2025).
11. Fiscal Policy Office. *Regional Fiscal Economic Report for November 2022*; Fiscal Policy Office: Bangkok, Thailand, 2022. Available online https://www.fpo.go.th/main/Economic-report/Monthly-regional-economic-situation/17261.aspx (accessed on 3 January 2025).

NESDC. *Gross Regional and Provincial Product (GPP)*; Office of the National Economic and Social Development Council: Bangkok, Thailand, 2022. Available online: https://www.nesdc.go.th/nesdb_en/more_news.php?cid=156&filename=index (accessed on 3 January 2025).

EPPO. *Electricity Consumption Per Capita*; Energy Policy and Planning Office: Bangkok, Thailand, 2022. Available online: https://www.eppo.go.th/index.php/en/en-energystatistics/electricity-statistic/Electricity%20Consumption%20in%20PEA%20Area%20(Classified%20by%20Sector) (accessed on 30 March 2025).

Arouri, M.; Shahbaz, M.; Onchang, R.; Islam, F.; Teulon, F. Environmental Kuznets curve in Thailand: Cointegration and causality analysis. *J. Energy Dev.* **2013**, *39*, 149–170.

Azam, M.; Khan, A.Q.; Zaman, K.; Ahmad, M. Factors determining energy consumption: Evidence from Indonesia, Malaysia and Thailand. *Renew. Sustain. Energy Rev.* **2015**, *42*, 1123–1131. [CrossRef]

Kyophilavong, P.; Shahbaz, M.; Anwar, S.; Masood, S. The energy-growth nexus in Thailand: Does trade openness boost up energy consumption? *Renew. Sustain. Energy Rev.* **2015**, *46*, 265–274. [CrossRef]

Maneejuk, N.; Ratchakom, S.; Maneejuk, P.; Yamaka, W. Does the environmental Kuznets curve exist? An international study. *Sustainability* **2020**, *12*, 9117. [CrossRef]

Phrakhruopatnontakitti, P.; Watthanabut, B.; Jermsittiparsert, K. Energy consumption, economic growth and environmental degradation in 4 Asian Countries: Malaysia, Myanmar, Vietnam and Thailand. *Int. J. Energy Econ. Policy* **2020**, *10*, 529–539. [CrossRef]

Maneejuk, P.; Yamaka, W.; Sriboonchitta, S. Does the Kuznets curve exist in Thailand? A two decades' perspective (1993–2015). *Ann. Oper. Res.* **2021**, *300*, 545–576. [CrossRef]

Kuznets, S. International differences in capital formation and financing. In *Capital Formation and Economic Growth*; Princeton University Press: Princeton, NJ, USA, 1955; pp. 19–111.

Grossman, G.M.; Krueger, A.B. Economic Growth and the Environment. *Q. J. Econ.* **1995**, *110*, 353–377. [CrossRef]

Stern, D.I. The environmental Kuznets curve after 25 years. *J. Bioeconom.* **2017**, *19*, 7–28. [CrossRef]

Panayotou, T. Demystifying the environmental Kuznets curve: Turning a black box into a policy tool. *Environ. Dev. Econ.* **1997**, *2*, 465–484. [CrossRef]

Stern, D.I. The rise and fall of the environmental Kuznets curve. *World Dev.* **2004**, *32*, 1419–1439. [CrossRef]

Suri, V.; Chapman, D. Economic growth, trade and energy: Implications for the environmental Kuznets curve. *Ecol. Econ.* **1998**, *25*, 195–208. [CrossRef]

Lorente, D.; Alvarez-Herranz, A. An approach to the effect of energy innovation on environmental Kuznets curve: An introduction to inflection point. *Bull. Energy Econ.* **2016**, *4*, 224–233.

Dasgupta, S.; Laplante, B.; Wang, H.; Wheeler, D. Confronting the environmental Kuznets curve. *J. Econ. Perspect.* **2002**, *16*, 147–168. [CrossRef]

Ang, J.B. CO_2 emissions, energy consumption, and output in France. *Energy Policy* **2007**, *35*, 4772–4778. [CrossRef]

Halicioglu, F. An econometric study of CO_2 emissions, energy consumption, income and foreign trade in Turkey. *Energy Policy* **2009**, *37*, 1156–1164. [CrossRef]

Chandran, V.; Tang, C.F. The impacts of transport energy consumption, foreign direct investment and income on CO_2 emissions in ASEAN-5 economies. *Renew. Sustain. Energy Rev.* **2013**, *24*, 445–453. [CrossRef]

Al-Mulali, U.; Saboori, B.; Ozturk, I. Investigating the environmental Kuznets curve hypothesis in Vietnam. *Energy Policy* **2015**, *76*, 123–131. [CrossRef]

Pablo-Romero, M.d.P.; De Jesús, J. Economic growth and energy consumption: The energy-environmental Kuznets curve for Latin America and the Caribbean. *Renew. Sustain. Energy Rev.* **2016**, *60*, 1343–1350. [CrossRef]

Kurniawan, R.; Managi, S. Coal consumption, urbanization, and trade openness linkage in Indonesia. *Energy Policy* **2018**, *121*, 576–583. [CrossRef]

Nguyen, K.H.; Kakinaka, M. Renewable energy consumption, carbon emissions, and development stages: Some evidence from panel cointegration analysis. *Renew. Energy* **2019**, *132*, 1049–1057. [CrossRef]

Hien, P. Excessive electricity intensity of Vietnam: Evidence from a comparative study of Asia-Pacific countries. *Energy Policy* **2019**, *130*, 409–417. [CrossRef]

Kibria, A.; Akhundjanov, S.B.; Oladi, R. Fossil fuel share in the energy mix and economic growth. *Int. Rev. Econ. Financ.* **2019**, *59*, 253–264. [CrossRef]

Shahzad, Q.; Aruga, K. Spatial effect of economic performance on the ecological footprint: Evidence from Asian countries. *Environ. Dev. Sustain.* **2024**. [CrossRef]

Ozcan, B. The nexus between carbon emissions, energy consumption and economic growth in Middle East countries: A panel data analysis. *Energy Policy* **2013**, *62*, 1138–1147. [CrossRef]

Wang, S.S.; Zhou, D.Q.; Zhou, P.; Wang, Q. CO_2 emissions, energy consumption and economic growth in China: A panel data analysis. *Energy Policy* **2011**, *39*, 4870–4875. [CrossRef]

40. Dyrstad, J.M.; Skonhoft, A.; Christensen, M.Q.; Ødegaard, E.T. Does economic growth eat up environmental improvements Electricity production and fossil fuel emission in OECD countries 1980–2014. *Energy Policy* **2019**, *125*, 103–109. [CrossRef]
41. Deichmann, U.; Reuter, A.; Vollmer, S.; Zhang, F. The relationship between energy intensity and economic growth: New evidenc from a multi-country multi-sectorial dataset. *World Dev.* **2019**, *124*, 104664. [CrossRef]
42. Borozan, D. Unveiling the heterogeneous effect of energy taxes and income on residential energy consumption. *Energy Polic* **2019**, *129*, 13–22. [CrossRef]
43. Aruga, K. Investigating the energy-environmental Kuznets Curve hypothesis for the Asia-Pacific region. *Sustainability* **201** *11*, 2395. [CrossRef]
44. Srisaringkarn, T.; Aruga, K. The Spatial Impact of PM2.5 Pollution on Economic Growth from 2012 to 2022: Evidence fro Satellite and Provincial-Level Data in Thailand. *Urban Sci.* **2025**, *9*, 110. [CrossRef]
45. Aslanidis, N.; Xepapadeas, A. *Smooth 'Inverted-V-Shaped' & Smooth 'N-Shaped' Pollution-Income Paths*; University of Cret Rethymno, Greece, 2004.
46. Sinha Babu, S.; Datta, S.K. The relevance of environmental Kuznets curve (EKC) in a framework of broad-based environmenta degradation and modified measure of growth–a pooled data analysis. *Int. J. Sustain. Dev. World Ecol.* **2013**, *20*, 309–316. [CrossRef
47. Mahmood, H.; Alkhateeb, T.T.Y.; Tanveer, M.; Mahmoud, D.H. Testing the energy-environmental Kuznets curve hypothesis in th renewable and nonrenewable energy consumption models in Egypt. *Int. J. Environ. Res. Public Health* **2021**, *18*, 7334. [CrossRef [PubMed]
48. Zhang, B.; Wang, B.; Wang, Z. Role of renewable energy and non-renewable energy consumption on EKC: Evidence from Pakista *J. Clean. Prod.* **2017**, *156*, 855–864. [CrossRef]
49. Marques, A.C.; Fuinhas, J.A.; Leal, P.A. The impact of economic growth on CO_2 emissions in Australia: The environmenta Kuznets curve and the decoupling index. *Environ. Sci. Pollut. Res.* **2018**, *25*, 27283–27296. [CrossRef]
50. Mahmood, H. The effects of natural gas and oil consumption on CO_2 emissions in GCC countries: Asymmetry analysis. *Enviror Sci. Pollut. Res.* **2022**, *29*, 57980–57996. [CrossRef]
51. Anselin, L. *Spatial Econometrics: Methods and Models*; Springer Science & Business Media: Dordrecht, The Netherland, 1988.
52. Wang, Y.; Li, K.; Xu, X.; Zhang, Y. Transport energy consumption and saving in China. *Renew. Sustain. Energy Rev.* **2014**, *2* 641–655. [CrossRef]
53. Neves, S.A.; Marques, A.C.; Fuinhas, J.A. Is energy consumption in the transport sector hampering both economic growth an the reduction of CO_2 emissions? A disaggregated energy consumption analysis. *Transp. Policy* **2017**, *59*, 64–70. [CrossRef]
54. Auffhammer, M.; Mansur, E.T. Measuring climatic impacts on energy consumption: A review of the empirical literature. *Energ Econ.* **2014**, *46*, 522–530. [CrossRef]
55. Brown, C.; Meeks, R.; Ghile, Y.; Hunu, K. *An Empirical Analysis of the Effects of Climate Variables on National Level Economic Growt* World Bank Policy Research Working Paper; World Bank: Washington DC, USA, 2010; Available online: https://ssrn.com abstract=1645727 (accessed on 22 May 2025).
56. Henseler, M.; Schumacher, I. The impact of weather on economic growth and its production factors. *Clim. Change* **2019**, *15* 417–433. [CrossRef]
57. Iamtrakul, P.; Chayphong, S. Factors affecting the development of a healthy city in Suburban areas, Thailand. *J. Urban Mana* **2023**, *12*, 208–220. [CrossRef]
58. Khanthavit, A. Instrumental-Variable Estimation of Bangkok-Weather Effects in the Stock Exchange of Thailand. *Asian Aca Manag. J. Account. Financ.* **2017**, *13*, 83–111. [CrossRef]
59. Nunta, J.; Sahachaisaeree, N. Cultural landscape, urban settlement and dweller's perception: A case study of a vernacular villag in Northern Thailand. *Procedia-Soc. Behav. Sci.* **2012**, *42*, 153–158. [CrossRef]
60. Sarkodie, S.A.; Ahmed, M.Y.; Owusu, P.A. Ambient air pollution and meteorological factors escalate electricity consumption. *Sc Total Environ.* **2021**, *795*, 148841. [CrossRef]
61. Siriwashiraporn, A. *Factors Affecting the Existence of Traditional Settlement of Pak Klong Wat Pradu Community, Samut Songkhram an Ratcha Buri Provinces*; Chulalongkorn Univeristy: Bangkok, Thailand, 2007.
62. Soytong, P.; Janchidfa, K.; Chayhard, S.; Phengphit, N. *Analysis of Climate Change Impacts on Water Resource, Urban, and Huma Settlements in Eastern Thailand: Geo-Informatics Approach for Climate Change Studies*; Faculty of Geoinformatics, Burapha Unive sity: Chon Buri, Thailand, 2019. Available online: http://dspace.lib.buu.ac.th/xmlui/handle/1234567890/3888 (accessed o 22 May 2025).
63. Wan, W.; Wang, J. The Impact of Weather on Economic Growth: County-Level Evidence from China. *Sustainability* **2024**, *16*, 998 [CrossRef]
64. Zhang, M.; Zhang, K.; Hu, W.; Zhu, B.; Wang, P.; Wei, Y.-M. Exploring the climatic impacts on residential electricity consumptio in Jiangsu, China. *Energy Policy* **2020**, *140*, 111398. [CrossRef]
65. Moran, P.A. The interpretation of statistical maps. *J. R. Stat. Society. Ser. B (Methodol.)* **1948**, *10*, 243–251. [CrossRef]
66. Getis, A.; Ord, J.K. The analysis of spatial association by use of distance statistics. *Geogr. Anal.* **1992**, *24*, 189–206. [CrossRef]

- Baltagi, B.H. *A Companion to Theoretical Econometrics(Chapter Fourteen Spatial Econometrics)*; Wiley Online Library: Hoboken, NJ, USA, 2001.
- Elhorst, J.P. Unconditional maximum likelihood estimation of linear and log-linear dynamic models for spatial panels. *Geogr. Anal.* **2005**, *37*, 85–106. [CrossRef]
- Yu, J.; De Jong, R.; Lee, L.-f. Quasi-maximum likelihood estimators for spatial dynamic panel data with fixed effects when both n and T are large. *J. Econom.* **2008**, *146*, 118–134. [CrossRef]
- Elhorst, J.P. *Spatial Econometrics: From Cross-Sectional Data to Spatial Panels*; Springer: Berlin/Heidelberg, Germany, 2014. [CrossRef]
- LeSage, J.; Pace, R.K. *Introduction to Spatial Econometrics*; Chapman and Hall/CRC: Boca Raton, FL, USA, 2009. [CrossRef]
- Bouayad-Agha, S.; Vedrine, L. Estimation strategies for a spatial dynamic panel using GMM. A new approach to the convergence issue of European regions. *Spat. Econ. Anal.* **2010**, *5*, 205–227. [CrossRef]
- Kripfganz, S.; Sarafidis, V. Estimating Spatial Dynamic Panel Data Models with Unobserved Common Factors in Stata. *J. Stat. Softw.* **2025**. [CrossRef]
- Murshed, M. Are Trade Liberalization policies aligned with Renewable Energy Transition in low and middle income countries? An Instrumental Variable approach. *Renew. Energy* **2020**, *151*, 1110–1123. [CrossRef]
- Chen, X.; Chen, M. Energy, environment and industry: Instrumental approaches for environmental regulation on energy efficiency. *Environ. Impact Assess. Rev.* **2024**, *105*, 107439. [CrossRef]
- Siddiquee, M.; Ahamed, R. Exploring water consumption in Dhaka city using instrumental variables regression approaches. *Env. Process* **2020**, *7*, 1255–1275. [CrossRef]
- Karmaker, S.C.; Barai, M.K.; Sen, K.K.; Saha, B.B. Effects of remittances on renewable energy consumption: Evidence from instrumental variable estimation with panel data. *Util. Policy* **2023**, *83*, 101614. [CrossRef]
- Li, L.; Sun, W.; Hu, W.; Sun, Y. Impact of natural and social environmental factors on building energy consumption: Based on bibliometrics. *J. Build. Eng.* **2021**, *37*, 102136. [CrossRef]
- Hastie, T.; Tibshirani, R. Generalized additive models. *Stat. Sci.* **1986**, *1*, 297–310. [CrossRef]
- Bank of Thailand. *Thai Economic Condition Report for 2022*; Bank of Thailand: Bangkok, Thailand, 2022. Available online: https://www.bot.or.th/content/dam/bot/documents/th/thai-economy/state-of-thai-economy/annual-report/annual-econ-report-th-2565.pdf (accessed on 10 January 2025).
- NESDC. *Thai Economic Performance in Q4 and 2022 and Outlook for 2023 (English Edition)*; Office of the National Economic and Social Development Council: Bangkok, Thailand, 2022. Available online: https://www.nesdc.go.th/nesdb_en/article_attach/65Q4%20Press%20Eng%20Q4-2022%20(2102%2015.46).pdf (accessed on 10 January 2025).
- Trakulchavalit, C. *A Comparative Study on Housing and Behavior of Graduated From USA UK and Japan*; Chulalongkorn University: Bangkok, Thailand, 1998.
- Chevaporanapiwat, C. *Factors Affecting Changes in Human Settlements in Bang Lum-Poo District*; Chulalongkorn University: Bangkok, Thailand, 2001.
- Algarni, S.; Tirth, V.; Alqahtani, T.; Alshehery, S.; Kshirsagar, P. Contribution of renewable energy sources to the environmental impacts and economic benefits for sustainable development. *Sustain. Energy Technol. Assess.* **2023**, *56*, 103098. [CrossRef]
- Aruga, K.; Islam, M.M.; Jannat, A. Assessing the CO_2 emissions and energy source consumption nexus in Japan. *Sustainability* **2024**, *16*, 5742. [CrossRef]
- Dincer, I. Renewable energy and sustainable development: A crucial review. *Renew. Sustain. Energy Rev.* **2000**, *4*, 157–175. [CrossRef]
- Mohtasham, J. Renewable energies. *Energy Procedia* **2015**, *74*, 1289–1297. [CrossRef]

MDPI AG
Grosspeteranlage 5
4052 Basel
Switzerland
Tel.: +41 61 683 77 34

Energies Editorial Office
E-mail: energies@mdpi.com
www.mdpi.com/journal/energies

www.ingramcontent.com/pod-product-compliance
Lightning Source LLC
LaVergne TN
LVHW070458061126
851858LV00011B/253
* 9 7 8 3 7 2 5 8 8 1 9 9 4 *